Vehicle Thermal Management Systems Conference (VTMS 13)

London, United Kingdom
17 - 18 May 2017

ISBN: 978-1-5108-7386-5

Printed from e-media with permission by:

Curran Associates, Inc.
57 Morehouse Lane
Red Hook, NY 12571

Some format issues inherent in the e-media version may also appear in this print version.

Printed by Curran Associates, Inc. (2019)

For permission requests, please contact Institution of Mechanical Engineers (IMechE) at the address below.

Institution of Mechanical Engineers (IMechE)
One Birdcage Walk
London, SW1H 9JJ
United Kingdom

library@Imeche.org
www.imeche.org

Additional copies of this publication are available from:

Curran Associates, Inc.
57 Morehouse Lane
Red Hook, NY 12571 USA
Phone: 845-758-0400
Fax: 845-758-2633
Email: curran@proceedings.com
Web: www.proceedings.com

CONTENTS

ENGINES

C6389/151 Influence of sump design on oil warm-up and cool-down behaviour 3
C Zhao, P J Shayler, University of Nottingham, UK; R Gilchrist, Jaguar Land Rover, UK

C6389/154 A benchmark analysis of subcooled heat transfer in a novel half-unit-cylinder-head for verifying in-vehicle engine evaporative cooling under dynamic conditions 17
M Langari, J F Dunne, S Jafari, J-P Pirault, C A Long, J Thalackottore Jose, University of Sussex, UK; Z Yang, University of Derby, UK

C6389/152 Influence on warm-up of split head and block circuits for oil and coolant and the effect of main gallery location 29
C Zhao, P J Shayler, L Cheng, University of Nottingham, UK; R Gilchrist, Jaguar Land Rover, UK

ENGINE PERFORMANCE AND TEMPERATURE CONTROL

C6389/142 Fuel consumption assessments of using a hot water storage strategy on a diesel engine at different ambient temperatures 47
H Sara, D Chalet, J-F Hétet, Ecole Centrale de Nantes, LHEEA Lab. (ECN/CNRS), France; M Cormerais, T Marimbordes, MANN+HUMMEL, France

C6389/169 Prediction of transient charge cooling behaviour and the effect on engine performance 57
R Owen, Bentley Motors Ltd, UK; A Price, (Former) Bentley Motors Ltd, UK; E Tate, Z Yang, Exa Corporation, USA; V Staelens, Exa Corporation, UK

HEAT EXCHANGE

C6389/143 Mechanism of oxide film destruction in fluxless brazing of aluminum under an inert atmosphere 79
T Yamayoshi, Y Itoh, A Fukumoto, UACJ Corporation, Japan

C6389/161 Flux residues of aluminium brazing and engine coolants 87
L Orman, H W Swidersky, U Seseke-Koyro, Solvay Fluor GmbH, Germany; D Lauzon, Solvay Fluorides, LLC, USA

WASTE HEAT RECOVERY

C6389/160 Extending the benefits of exhaust gas recirculation (EGR) in gasoline direct injection (GDI) engines with on-board fuel reforming 97
D Fennell, C Hergueta, M Bogarra, A Tsolakis, M Wyszynski, M Herreros, University of Birmingham, UK

C6389/147 Thermo-economic optimization during preliminary design phase of organic Rankine cycle systems for waste heat recovery from exhaust and recirculated gases of heavy duty trucks 109
L Guillaume, A Legros, R Dickes, V Lemort, University of Liège, Belgium

C6389/157 Waste Heat Recovery Technologies: A technical comparative assessment of EHRS, TEG, ORC and TAC 127
S Uppuluri, H Khalane, A Naiknaware, Computational Sciences Experts Group, USA; D Sahoo, A Kotrba, Tenneco Inc, USA

C6389/158 Thermal and power analysis of a modularised thermoelectric power generator for passenger vehicles 141
K Huang, B Li, University of Nottingham, UK; Y Yan, University of Nottingham, UK and University of Nottingham Ningbo, China; J Li, China FAW Group Corporation R&D Centre, China

SIMULATION AND CONTROL

C6389/150 Development of an HIL rig for vehicle thermal management systems design and optimisation 155
M V Morganti, S Longo, Cranfield University, UK; R M Shah, Charge Automotive, UK

C6389/162 Model predictive control for Organic Rankine Cycle applied to hybrid vehicles 171
A Agurto-Goya, Jaguar Land Rover Ltd, UK; J Kitte, I Trivic, T Tietze, IAV GmbH, Germany

C6389/167 Thermoelectric devices for automotive applications to improve efficiency and reduce CO_2 emissions 185
J Siviter, A Montecucco, P Mullen, M Garland, A Knox, Thermoelectric Conversion Systems Limited and University of Glasgow, UK

C6389/168 Matlab/Simulink based in-vehicle predictive thermal environment model 195
B Stegmann, N Dutta, I Hughes, W Jansen, Jaguar Land Rover, UK

C6389/164 Combined engine-ORC thermodynamic analysis based on a second law and thermo-economic approach 205
S Lion, M G Momesso, Ricardo Deutschland GmbH, Germany and University of Trieste, Italy; I Vlaskos, Ricardo Deutschland GmbH, Germany; C Rouaud, Ricardo UK Ltd, UK; R Taccani, University of Trieste, Italy

ENERGY MANAGEMENT

C6389/153 Oil thermal management during engine transients from a cold state 221
D Di Battista, R Cipollone, University of L'Aquila, Italy

HEATING AND AIR COOLING

C6389/165 Predicting thermal comfort and draught discomfort for an automobile cabin cooldown 233
M Hepokoski, S Peck, C Gingras, A Curran, ThermoAnalytics, USA; S Gullman, D Jacobsson, Volvo Car Corporation, Sweden

C6389/155 Active charge air cooling 243
C D Whelan, Air Cycle Technology Ltd, UK; J M Allport, University of Huddersfield, UK

AUTHOR INDEX

VTMS 13 ORGANISING COMMITTEE

Darren Butler (Chair)	Nissan
Jugurtha Benouali	Valeo
David Bridge	HORIBA MIRA
Jon Caine	Ford Motor Company
Salvio Chacko	Tata Motors
Richard Davies	Calsonic Kansei Technical Centre
Bernd Gruenenwald	MAHLE Behr
Morgan Heikal	University of Brighton
Steve Jones	Bentley Motors
Cedric Rouaud	Ricardo
Paul Shayler	University of Nottingham
Ben Wicksteed	Jaguar Land Rover

ENGINES

Influence of sump design on oil warm-up and cool-down behaviour

C Zhao, P J Shayler
University of Nottingham, UK
R Gilchrist
Jaguar Land Rover, UK

ABSTRACT

The effect of sump design and oil pathway to the main gallery on gallery oil temperature during engine warm-up and cool-down has been investigated using five variants of sump design for a 2litre gasoline turbocharged direct injection engine. Oil warm-up behaviour was recorded for drive cycle conditions with initial temperatures close to 20°C. The variants covered differences in sump shape, material, baffling and external insulation. Under the same test conditions and at the same time after start-up, oil temperatures at inlet to the oil pick-up differed by up to 10 degrees Celsius. However, these temperature differences are effectively eliminated along the oil pathway to the main gallery. Nearly all the temperature moderation occurs in the engine oil pump due to heat exchange between the oil and the pump, which is a gear pump design with integral pressure relief valve. The potentially high impact of pump heat transfer has not been recognised previously. Sump design features had a stronger influence during engine cool-down when the heat transfer between bulk oil in the sump and the ambient is predominant. The sump with lower wetted surface area to volume ratio gives the slower oil cool-down, which is beneficial for the initial conditions of engine restart. Where spatial variations in oil temperature and local heat transfer exist in the sump, heat transfer along the sump wall can act to redistribute the internal energy of the sump oil. This can be advantageous and disadvantageous to slowing the temperature fall at pump inlet, depending on the nature of the temperature stratification.

1 INTRODUCTION

There is a significant vehicle fuel consumption penalty associated with running an engine in a cold state. Over the New European Drive Cycle (NEDC), if the engine is cold-started at the beginning of the cycle, total fuel consumption is typically several per-cent higher than that if the engine is started fully-warm [1, 2]. This is largely due to the difference in engine friction between cold and fully-warm engine states, and in turn this reflects the dependence of engine friction on the viscosity of the lubricating oil. Viscosity is strongly dependent upon temperature and is higher at low temperatures. The development of multi-grade engine oils with reduced sensitivity to temperature has reduced but not eliminated the fuel consumption penalty. In real world driving, vehicles are often used for short journeys and do not reach fully warm condition [3, 4]. Hence, there is a continuing need for, and interest in, thermal management developments which promote the faster warm-up of the oil. These include system modifications designed to exploit exhaust gases and coolant as heat sources in EGR-to-oil (or coolant-to-oil) heat exchangers

[2,5,6,7,8], design modifications to reduce heat transfer from oil to the engine structure, for example by raising contact resistance at the back of bearing shells [9], and sumps designed to raise oil temperature at pump inlet either by directing the returning flow of hot oil towards this [10] or by reducing the amount of oil in circulation. The Toyota sump-within-sump system [11] is an example of the latter approach, which gave a reported 0.8% reduction in fuel consumption over the NEDC in the target application. Other investigators [12] have investigated the effectiveness of insulation to reduce heat losses from sump oil, and note that this has less impact on rates of oil warm-up than on cool down after engine shut down.

In the work reported here, the influence of five sump design variations on oil warm-up and, after engine shut-down, oil cool-down has been investigated experimentally. The same 2.0l, 4 cylinder gasoline turbocharged direct injection (GTDI) spark ignition engine was used throughout the study. The temperature in the sump oil at entry to the pick-up is the key indicator of effect on cool-down. For warm-up, the temperature of oil in the main gallery is more important. Generally, the main gallery is the last common section of the oil system before distribution to different sites within the engine, the temperature of oil in this can be used as an indicator of how successful measures upstream to promote oil warm-up have been. The flow path from the sump to the main gallery is through a pick-up to the oil pump, then to a filter and oil-coolant heat exchanger assembly. Importantly, as will be shown, these components can influence the effectiveness of sump design changes in promoting warm-up, but not cool down. In warm-up tests, oil temperatures were measured in the sump at the pump pick-up inlet, at pump outlet, and in the main gallery. The warm-up tests were carried out for the patterns of engine speed and load shown in Figure 1. These represent an approximation to NEDC test conditions (aNEDC) for a compact SUV with an automatic transmission. In all the cases presented, the oil pump was driven externally by an AC motor and controlled to provide a constant pressure of 3.5barA in the main gallery. Data acquisition was continued after engine shut-down to record the oil temperatures during engine cool-down. The same grade of lubricant (5w-30) was used for all the tests.

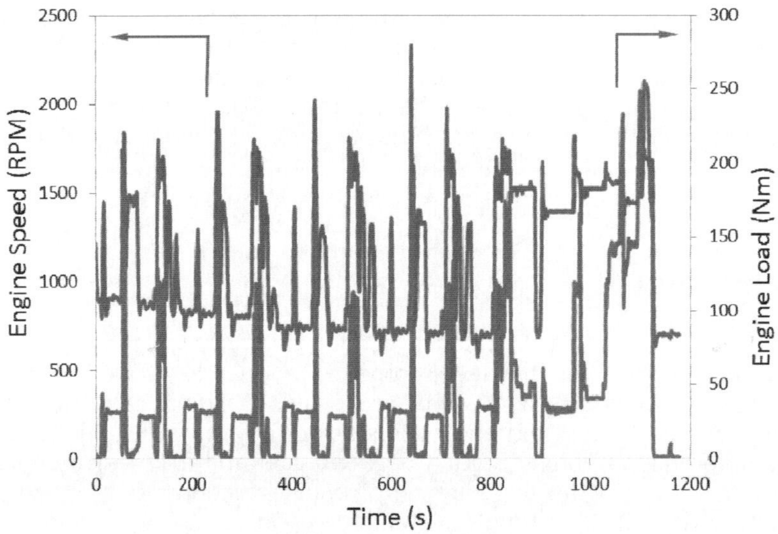

Figure 1: The pattern of engine speed and load over aNEDC

The five sump design variants are shown in Figure 2 and the distinguishing features are summarised in Table 1. Sump#1 and sump#2 have the form of sumps used on the engine in compact SUV applications. Sump#3, #4 and #5 have the form of sumps used in sports saloon applications. The compact SUV sumps have a relatively flat floor. The sports saloon sumps have a sloping floor leading to a deeper well, as can be seen by inspection of Figure 3. The more contoured shape of the sports saloon sumps gives a higher ratio of wetted surface area to volume. The position of the inlet to the pump pick-up is marked with an 'x' in Figure 3; the inlet is lower in the sports saloon sump. Sumps #1, #3 and #4 are aluminium and sump#2 and 5# are polyamide, which is an insulating polymer. Sump#4 has a coating of insulating material (Nylon 11) painted on the inner and outer surfaces and a baffle-and-mesh insert designed to direct returning hot oil to the pump pick-up.

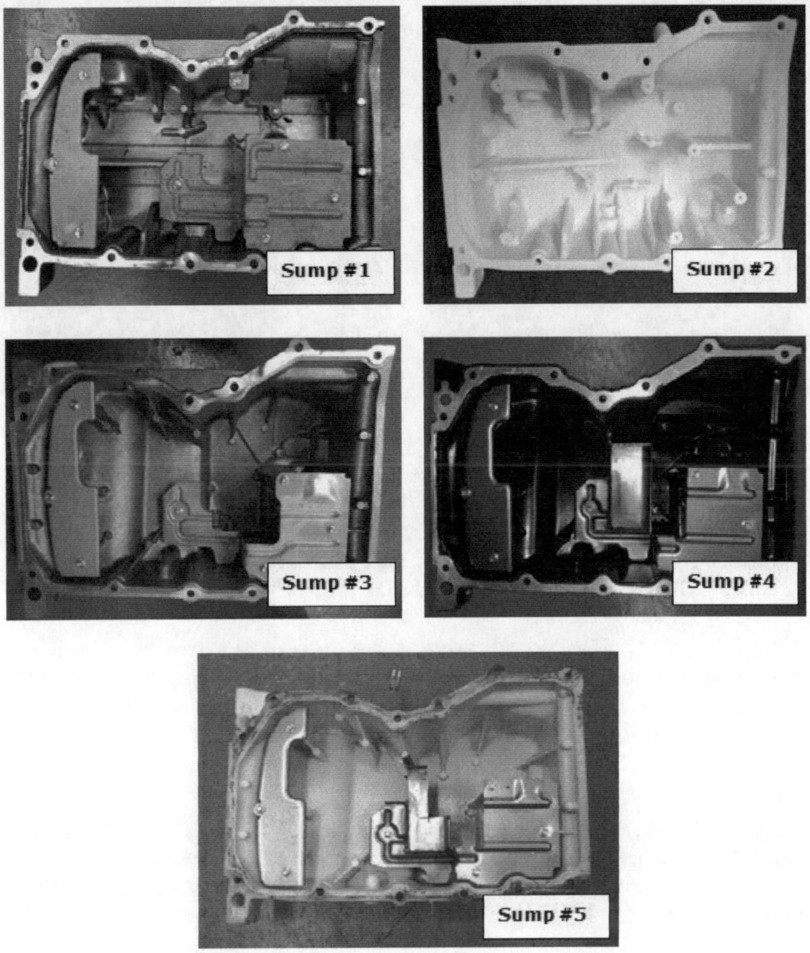

Figure 2: Five sump design variants

Table 1: Summary of the features for five sump design variants

	Sump Material	Baffle Mesh	Insulation Coating	Floor Profile	Pick-up Material	Mass (kg)
Sump#1	Aluminium	-	-	Flat	Plastic	5.80
Sump#2	Polyamide	-	-	Flat	Plastic	2.85
Sump#3	Aluminium	-	-	Step Change	Steel	4.50
Sump#4	Aluminium	✓	✓	Step Change	Steel	5.15
Sump#5	Polyamide	✓	-	Step Change	Steel	2.25

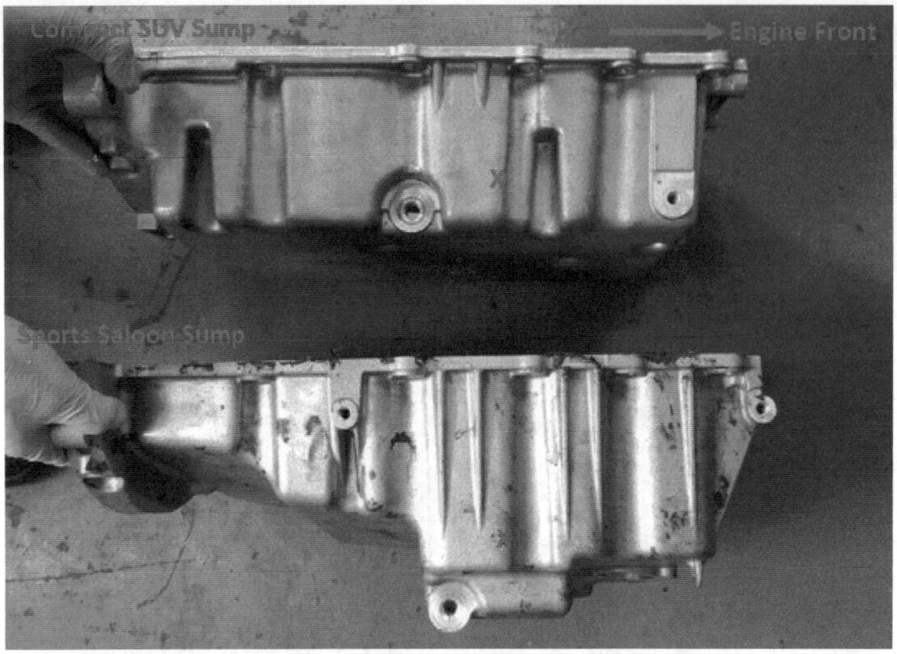

Figure 3: The comparison of the sump floor profile and the pump pick-up position between standard compact SUV sump and sports saloon sump

2 INFLUENCE OF SUMP DESIGN ON OIL TEMPERATURE AT INLET TO PUMP PICK-UP DURING WARM-UP AND COOL-DOWN

If a sump design is to offer some advantage during warm-up, it will produce a higher temperature at the oil pump pick-up. This temperature sets the potential of the sump to influence oil temperatures at the rubbing surfaces of the engine, though downstream influences of work input and heat exchange with the oil flow will moderate the effect. The five sumps available allow the effect of contrasts in shape (#1, compact SUV, versus #3, sports saloon), material (#1, aluminium, versus #2, polyamide), and the effect of the inclusion of baffle-and-mesh in combination with internal and external surface coatings of insulation (#3 versus #4) or with a change from aluminium to polyamide sump material (#3 versus #5). The baffle-and- mesh insert is shown in Figure 4.

Comparisons of the rise of oil temperature at pump pick-up with time for three of the four cases are given in Figure 5(a) –(c). In all cases, test conditions followed the same engine speed and load pattern approximating to the NEDC for a compact SUV. As shown in Figure 5(a), the oil temperature of sump#1 starts to increase earlier than the oil temperature of sump#3. Although #3 has the higher wetted surface area to volume ratio, the convergence of temperatures of the two sumps over time suggests differences in temperature stratification in the sumps is more important than heat transfer losses. As shown in Figure 5(b) however, the change in sump material from aluminium to an insulating material, polyamide, does raise the rate of temperature increase and this can be attributed to reduced heat losses. Finally it can be seen in Figure 5(c), that inserting the baffle-and-mesh into sump#3 to give sump#4 largely eliminates the delay before oil temperature starts to rise. The baffle- and- mesh modify the flow to or circulation of oil near the pump pick–up and by implication the stratification of oil temperature in the sump.

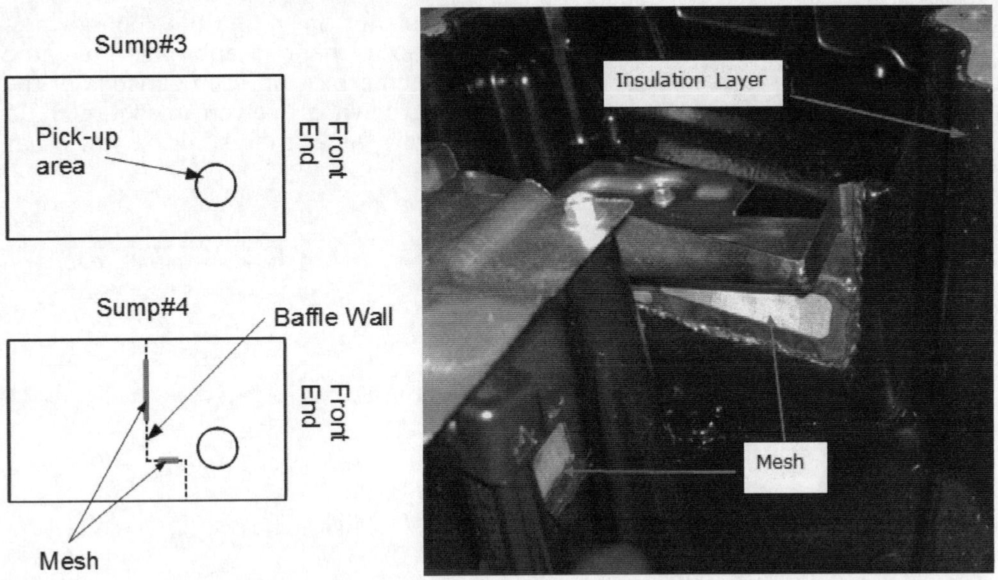

Figure 4: The location of baffle wall (left, top view of the sump) and mesh (right)

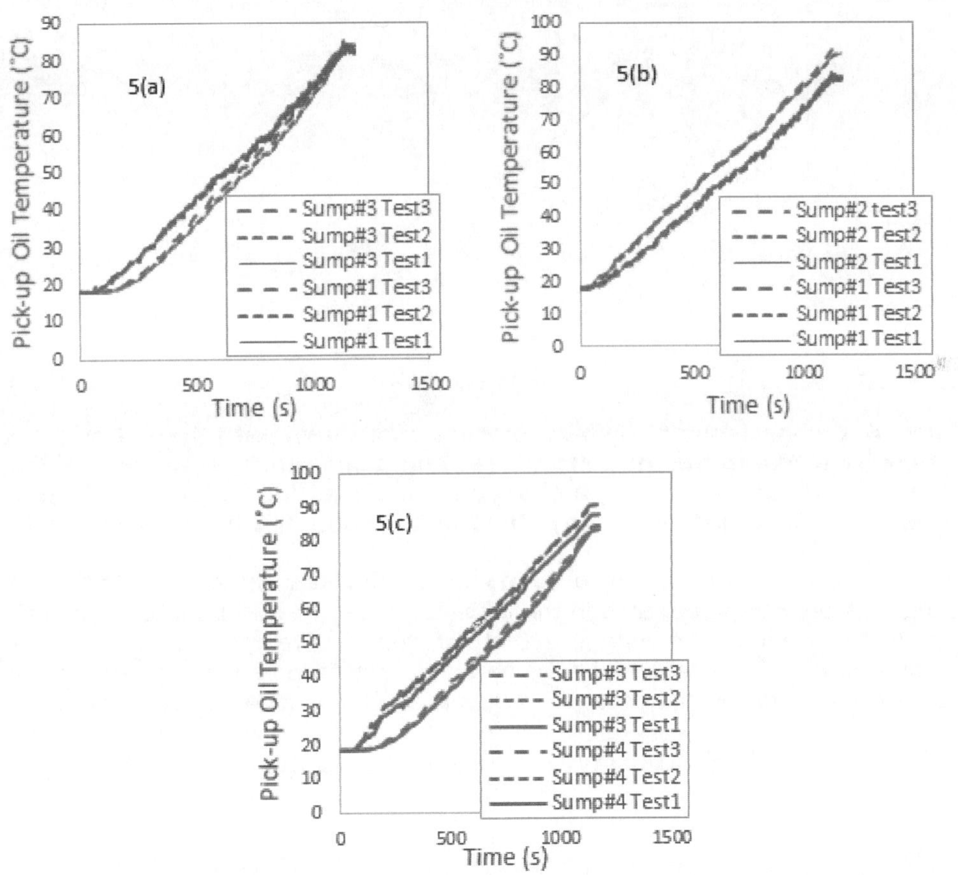

Figure 5: Comparisons of oil temperature rise with time at oil pump pick-up over a cold-started aNEDC showing the influence of (a) sump shape, (b) sump material and (c) internal and external coating of insulation and a baffle-and- mesh insert

At the end of the drive cycle test, the engine was shut down and the cool-down of sump oil recorded. The thermocouple position was unchanged and gave the same measurement of temperature at inlet to the oil pump pick up. Comparisons of the fall in temperature with time since engine shut down are given in Figure 6. In Figure 6(a), sump#3 can be seen to cool marginally faster than sump#1 which has the lower wetted surface area to volume ratio.

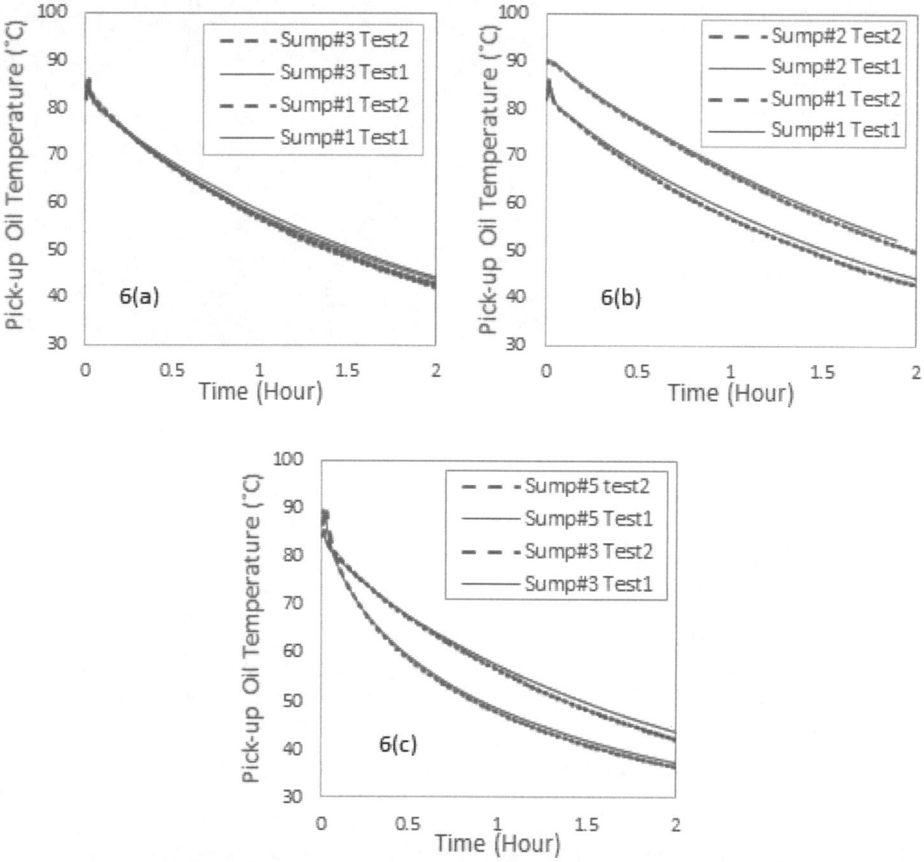

Figure 6: Comparisons of oil temperature cool-down with time at oil pump pick-up showing the influence of (a) sump shape, (b) sump material (compact SUV sump) and (c) a change from aluminium to polyamide sump material (sports saloon sump) plus the insertion of a baffle-and- mesh

In Figure 6(b), results for aluminium (#1) and polyamide (#2) versions of the same sump shape are compared, though the initial temperature offset obscures the effect of the material change. Comparing the rate of change of temperature as a measure of heat loss from the oil at a given temperature difference between oil and ambient (Figure 7) shows this is higher for the aluminium sump throughout cool down.

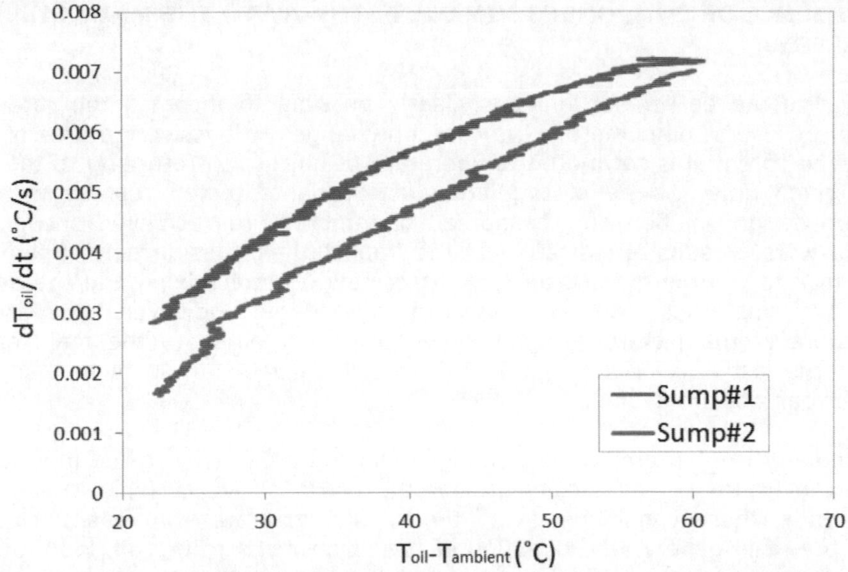

Figure 7: Comparison of rates of temperature change versus oil- to-ambient temperature difference showing higher rates of change for the aluminium (#1) version of the sump

Finally, the apparent effect of changing sump material from aluminium (sump#3) to polyamide and inserting the baffle-and-mesh (sump#5) is shown in Figure 6(c) to accelerate the rate of temperature fall at the oil pick up location. This is unexpected, given the insulting nature of the polyamide, which has a thermal conductivity of 0.25 W/m^2K compared to 210 W/m^2K for aluminium. However, repeating the tests with additional thermocouples in the sump shows the degree of temperature stratification is substantially higher in the polyamide sump (Figure 8). This suggests the higher heat conduction in the aluminium sump wall, particularly in the floor of the sump close to the pick-up, acts to moderate oil temperature stratification in the sump.

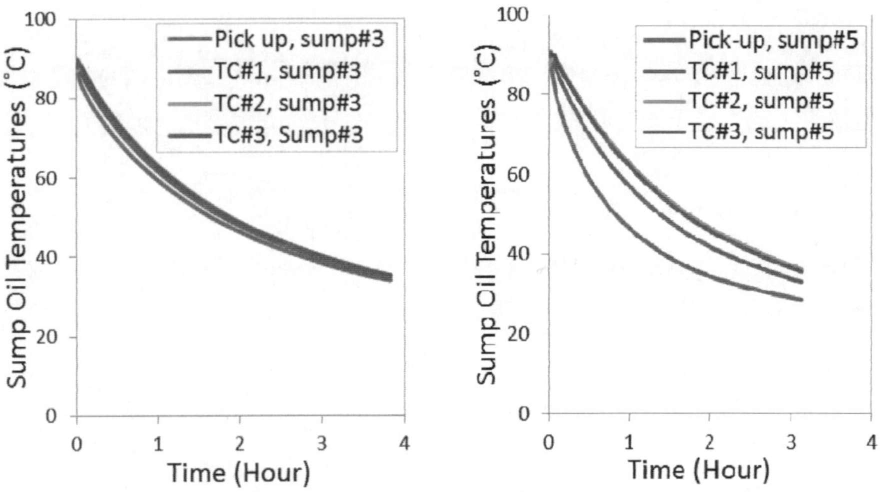

Figure 8: Comparison of oil temperatures between sump#3 (left) and sump#5 (right) at different locations during engine soak down period

3 INFLUENCE OF COMPONENTS IN OIL PATHWAY TO THE MAIN OIL GALLERY

As shown in the earlier section it is clearly possible to increase the rate of oil temperature rise at oil pump pick-up through changes to passive features of sump design. The potential is set by the temperature of hottest oil returning to the sump at any given time. None of the five sump variants tested represents a fully optimised design solution but the spread of temperatures achieved between the best and worse results is indicative of the impact these design details can have. Low surface to volume ratio, inserts to direct flow or limit mixing and the use low conductivity materials can be used to advantage. However, increases in temperature at the pick-up are not necessarily maintained to the main gallery, which is the last part of the system before the oil pathway splits to feed oil to particular parts of the engine.

An example of the oil temperature changes along the pathway to the main gallery is shown in Figure 9. The results are for the aNEDC drive cycle, cold started at ~20°C. Under these conditions, at all times, oil temperature increases along the path to the main gallery. The direction of heat transfer is to the oil, indicating the local engine structure and coolant are at a higher temperature than the oil, and the pump surfaces in contact with the oil are also higher than the local oil temperature. It is not clear, however, to what extent a temperature increase at pump pick up achieved through sump design can be maintained as an increase in oil temperature in the main gallery. The problem is illustrated in Figure 10 through a comparison of results for sumps#1 and #3. Although sump#1 quickly develops an oil temperature at the pick-up which is several degrees higher than the corresponding oil temperature of sump#3, at pump outlet, the oil temperature is effectively the same for both sumps, and thereafter the same downstream in the main gallery. As indicated by the coolant temperature for the two cases the test conditions were very similar. By implication, thermal coupling between oil flow through the pump and the metal of the pump rotor and casing produces sufficiently high rates of heat transfer to eliminate the oil temperature difference at pump inlet.

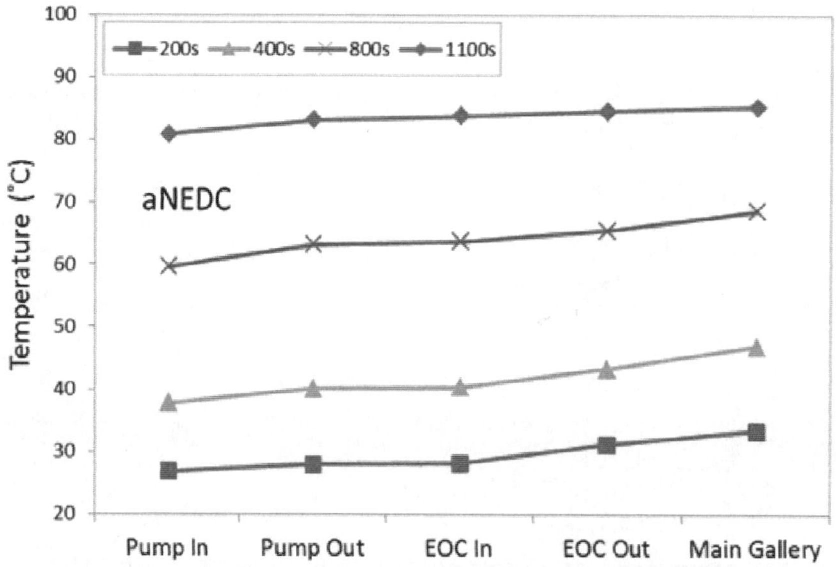

Figure 9: Oil temperature along the pathway to the main gallery at various times during an aNEDC test

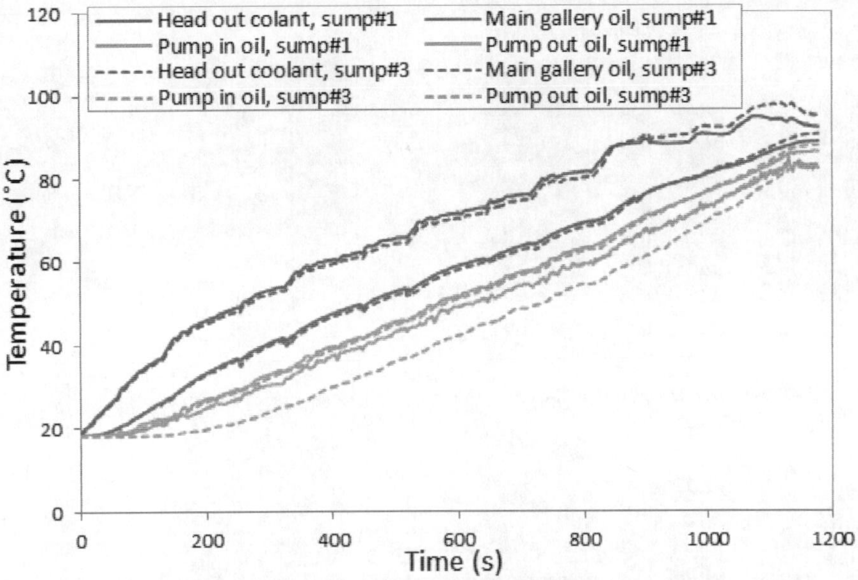

Figure 10: Comparison of temperature variations for sump#1 and sump#3 during engine warm-up over the aNEDC

Across the set of sumps tested, the spread of temperatures achieved at the pick-up point provided an opportunity to compare how well this spread was maintained, amplified or diminished as a set of temperatures in the main gallery. The results are enlightening, showing that for the size and type of gear pump with integral relief valve used, heat transfer between the oil and the pump rotor and body almost eliminates temperature differences of the magnitude produced by sump design changes. In Figure 11, oil temperatures along the path to the main gallery are compared for the five variants of sump, alongside temperature changes across the pump and between the pump and the oil gallery. The comparisons are made at four times since the start of test. The spread of temperatures at pump inlet across the set of sumps is around 10 degrees Celsius after 200s, and this is maintained for the rest of the test period. At pump outlet, the spread of temperature is reduced to 2 or 3 degrees Celsius. Both positive and negative changes in oil temperature occur between inlet and outlet. This points to rates of heat transfer being high enough to adjust oil temperature towards the metal temperature of the pump. Downstream of the pump the oil passes through an oil-coolant heat exchanger in a filter-cooler unit before entering the main gallery. The temperature change of oil in the heat exchanger across the set of sumps is relatively constant in comparison with the changes across the oil pump, and the spread at delivery to the main gallery is approximately 1 degree Celsius. The temperature changes in the heat exchanger are highest in the first half of the test period, when the difference between coolant and oil temperature is greatest. Over most of the test period, after 100s, the effectiveness of the heat exchanger is in the range of 0.3-0.4 as illustrated in Figure 12. The results of tests using the different sump designs virtually overlay.

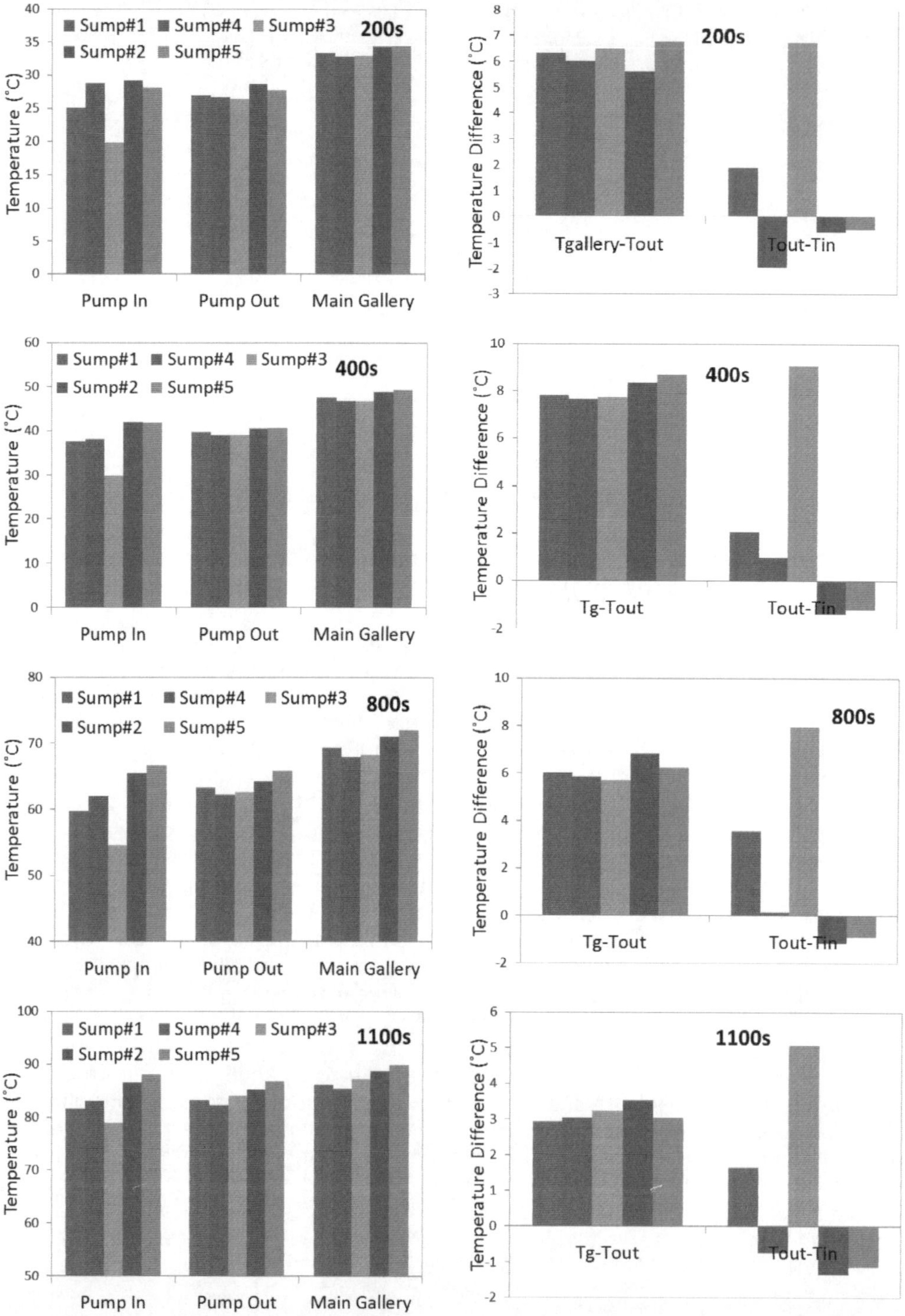

Figure 11. (Left column) Oil temperature at pump-in, pump out and main gallery and (right column) temperature rise between pump-in and pump-out (Tout – Tin), and between pump-out and main gallery (Tg – Tout) for five sump design variants. Test cycle is aNEDC; times refer to times since start-up.

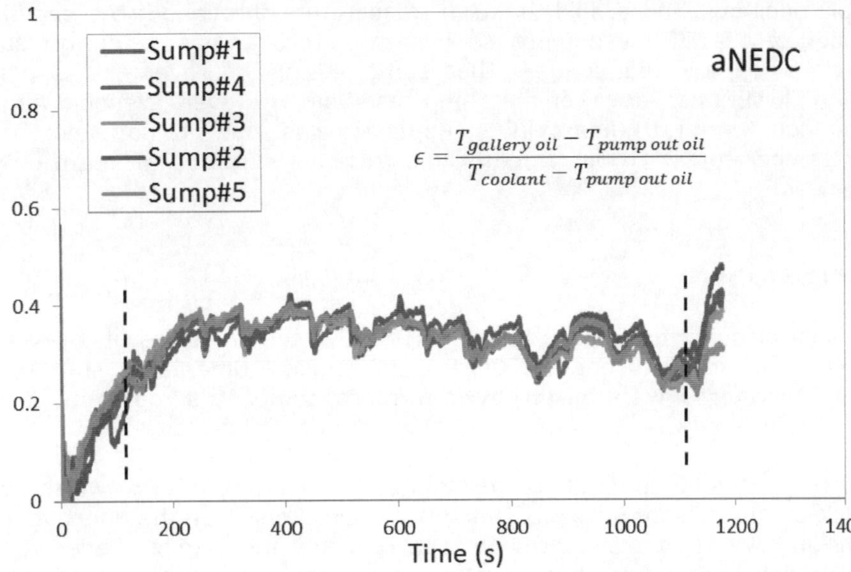

Figure 12: Variation of effectiveness of the oil-coolant heat exchanger

4 DISCUSSION

For a given volume of engine oil fill, the design of the sump influences the temperature of the oil at entry to the pump pick-up both during warm-up and cool down. The five sump design variants examined include two production versions providing a baseline and three others which cover material changes, insulating coatings and baffle-and- mesh inserts intended to promote faster oil warm-up and slower cool-down after engine shut down through the effect on heat transfer and stratification of oil temperature in the sump. The five variants gave a spread of temperatures at oil pick-up of up to 10 degrees Celsius at given times since starting for a drive cycle approximating to the NEDC, diminishing towards the end of the cycle when the engine was fully warm. The spread was largely accounted for by differences in sump shape. The best performing sump was the polyamide version of the compact SUV sump (#2). This was better than the standard aluminium version (#1) but only marginally. The worse performing sump was the standard aluminium sports saloon sump (#3), which has a higher wetted area to volume ratio than the compact SUV sump. The modifications made to the sports saloon sump improved performance without quite matching the compact SUV results. Importantly, these differences in temperature produced by variations in sump design were largely eliminated by heat transfer in the oil pump; the direction of heat transfer could be to or from oil, consistent with small temperature differences between these. The strong thermal coupling undermines the benefit of sump optimisation during warm-up.

The cool-down of sump oil after engine shut down is also effected by variants in sump design, and the benefit is not immediately undermined by heat transfer in the oil pump. After engine shut down, cool-down was slower for the more compact flat bottomed compact SUV sump than the sports saloon sump and slower still for the polyamide version. However, the change to polyamide for the sports saloon had the opposite effect, accelerating the fall in oil temperature at the pick-up location due to changes in stratification and indicating the importance of heat redistribution by conduction along the sump wall. Thus although some effects of design changes on bulk temperature are easily anticipated from simple considerations of heat transfer, effects on stratification are less easily anticipated.

For initial conditions with a mild, uniform temperature throughout the engine, the results suggest it is difficult to improve oil warm-up through passive design features or material changes, and suggest that sump design improvements should be focused on slowing cool-down of the sump reservoir. However, it should be noted that in previous work [10], improving sump design was shown to have more impact at lower temperatures, yielding measurable friction reductions at warm-ups at -10°C or lower.

5 CONCLUSIONS

Five variants of sump design produced a spread of temperatures at entry to the pump pick-up of up to 10 degrees Celsius at the same times after start up. The highest temperatures were given by the more compact sump, with a small advantage associated with a change from aluminium to polyamide.

Heat transfer in the oil pump almost eliminates the differences in oil warm up rate at the pick-up due to sump design. This has a major impact on the effectiveness of sump design revisions on oil warm-up. The temperature changes between pump inlet and outlet could be positive or negative. Downstream of the pump, the oil-coolant heat exchanger promoted oil warm-up without or changing the ranking order of temperatures for the set of sumps.

The cool-down of oil after engine shut down was slowest for the more compact, polyamide sump. Temperature stratification in the oil was higher due to reduced heat redistribution by conduction in the sump wall; depending how the stratification effects the local temperature at the oil pick-up, this can outweigh the benefit of reduced heat transfer to ambient.

ACKNOWLEDGMENT

This work was carried out as part of the Innovate IDP10 project, VIPER2 (Vehicle Integrated Powertrain Energy Recovery) with consortium partners Ford Motor Company, Jaguar Land Rover, and European Thermodynamics Ltd.

REFERENCES

[1] K. Kunze, S. Wolff, I. Lade, J. Tonhauser, A Systematic Analysis of CO_2 Reduction by an Optimised Heat Supply during Vehicle Warm-up, SAE paper 2006-01-1450, 2006

[2] T. Mueller, H. Hans, W. Krebs, S. Smith, A K. Adam, *Thermal Management on Small Gasoline Engines*, SAE paper 2011-01-0314, 2011

[3] L. Avery, *National Transport Survey 2010*, UK Department of Transport 2011, https://www.gov.uk/government/publications/national-travel-survey-2010

[4] M. Andre, *In Actual Use Car Testing: 70,000 Kilometers and 10,000 Trips by 55 French Cars under Real Conditions*, SAE Technical Paper 910039, 1991, doi:10.4271/910039

[5] R. Burke, C. Brace, A. Cox, A. Lewis, J G. Hawley, I. Pegg and R. Stark. *System Approach to Improvement of Engine Warm-up Behaviour*, Proc. IMechE Vol. 225 Part D: J. Automobile Engineering, 2010

[6] F. Will, A. Boretti, *A New Method to Warm Up Lubricating Oil to Improve the Fuel Efficiency during Cold Start*, SAE Paper 2011-01-0318, 2011

[7] J P. Zammit, P J. Shayler, R. Gardiner, I. Pegg,' *Investigating the Potential to Reduce Crankshaft Main Bearing Friction during Engine Warm-up by Raising Oil Feed Temperature*, SAE Paper 2012-02-1216, 2012

[8] G E. Andrew, A M. Ounzain, H. Li, M. Bell, J. Tate, K. Ropkins, *The Use of a Water/Lube Oil Heat Exchanger and Enhanced Cooling Water Heating to Increase Water and Lube Oil Heating Rates in Passenger Cars for Reduced Fuel Consumption and CO_2 Emissions During Cold Start*, SAE paper 2007-01-2067, 2007

[9] P J. Shayler, W S. Baylis, M. *Murphy, Main Bearing Friction and Thermal Interaction during the Early Seconds of Cold Engine Operation*, Journal of Engineering for Gas Turbine and Power (DOI:10.115/1.1804538), 2005

[10] T Law, P J Shayler, I Pegg, *'Investigations of Sump Design to Improve the Thermal Management of Oil Temperature During Engine Warm-up',* IMechE Paper 640/044/2007,Proc. SAE/IMechE Int. Conf VTMS8, Nottingham, May 2007

[11] M. Higashitani, H. Kobayashi, K. Yamada, and K. Yoshijima, *Dual-Chamber Oil Pan for Improved Engine Warm-up Performance*, MTZ, 12/ 2012, Volume 73, 2012

[12] E. Bent, P J. Shayler, A La. Rocca, C. Rouaud, *'The effectiveness of stop-start and thermal management measures to improve fuel economy'* Proc. of IMechE Int Conf C1365, VTMS 11, C1365, paper C027, Coventry, 2013

A benchmark analysis of subcooled heat transfer in a novel half-unit-cylinder-head for verifying in-vehicle engine evaporative cooling under dynamic conditions

M Langari, J F Dunne, S Jafari, J-P Pirault, Z Yang*, C A Long, J Thalackottore Jose
Department of Engineering and Design, University of Sussex, UK
*Department of Engineering, College of Engineering and Technology, University of Derby, UK

ABSTRACT

Evaporative cooling (EC) is a potentially attractive method of thermal management in highly downsized engines. As a benchmark for robust EC system development, a novel half-unit-cylinder-head has been designed to study the heat transfer achievable with evaporative cooling strategies under realistic hot metal conditions with an appropriate coolant flow field. Multi-phase computational fluid dynamics (CFD) and conjugate heat transfer (CHT) analysis are performed under subcooled flow boiling conditions to achieve both the design requirements for the hot-side surfaces and the specification of heating arrangements. Analysis is undertaken to ensure that thermal diffusion and flow conditions are representative of coolant flow in highly-boosted IC engines. The CFD model predictions show good correlation with available experimental data for both heat flux and metal temperatures in a typical downsized pressure-charged spark-ignition engine cylinder head.

1 INTRODUCTION

Progressive trends in downsizing and pressure-charging spark ignition engines have resulted in increasingly high thermal loads. Consequently, high heat rejection rates are needed to avoid operation at dangerously elevated engine metal temperatures. This trend has pushed conventional (single phase) cooling systems to their absolute limit of being able to achieve efficient engine thermal management and temperature control. Reaching this limit has justified further fundamental research on alternative cooling strategies. Evaporative cooling (EC) is a potentially attractive strategy owing to: 1) the high rates of heat transfer available from the coolant phase change during boiling, and 2) from the significant functional advantages that a robustly-controlled EC system can potentially provide. However, despite extensive research to date, no closed-loop evaporation-based IC engine cooling system has been successfully implemented on a production vehicle owing presumably to unsolved technical challenges and insufficient evidence of successful operation at representative engine test conditions.

Predictions of both heat transfer and the associated flow field in the presence of coolant phase-change is of significant importance in the development of downsized internal combustion engines. In a typical downsized internal combustion engine, heat rejection to the cooling system must take place at rates of order 1.2 MW/m^2. With such high heat transfer rates, boiling may be unavoidable in some regions under certain coolant flow conditions. Subcooled boiling has been attractive in engine cooling system applications [1]–[4] where the high heat transfer rates of nucleate boiling may be utilised without use of extra components (e.g. a vapour separator and condenser). For safe operation, and to achieve optimised design of the engine and cooling system, it is crucial to accurately predict the heat transfer rates and vapour fractions within the coolant passages.

Despite extensive research on the phenomenon, nucleate boiling heat transfer mechanism is still not fully understood, and thus a fundamental model without empirical correlations is yet to be developed. Computational Fluid Dynamics, together with mechanistic or empirical boiling models, is a viable tool to provide thermal predictions of nucleate boiling flows. Several experimental and numerical studies have examined the heat transfer regimes leading to the nucleate boiling. Boiling flow however involves various events e.g. bubble nucleation, growth, detachment, coalescence, collapse, and condensation that occur on different space and time scales, making boiling a complex phenomenon. Owing to unknown characteristics of nucleation sites, mechanistic models based on bubble formation and departure phenomena, are not yet able to predict the nucleate boiling heat transfer. Empirical correlations are therefore more often used in practical situations. The boiling correlation of Kandlikar [5] describes subcooled boiling flow of ethylene-glycol-water mixtures in engine-like geometries. It is however mainly developed for heat transfer estimation in fully-developed boiling regime and is known to perform poorly in the nucleate boiling transition region [6]. The correlations of Rohsenow [7] and Chen [8] describe the total heat flux as a combination of a convective component and a nucleate boiling component providing predictions in both fully convective and boiling region. These models have been widely used in automotive applications [6], [9]–[11]. Robinson *et al.* [9] emphasized the significance of accurate prediction of the (single phase) convective component for reliable overall heat transfer estimations in both convective and boiling regimes. Appropriate thermal boundary layer and vapour formation modelling is also known to play a significant role in accurate predictions of heat transfer in the boiling regime [12]–[15].

As part of an extensive research study to address underlying EC system challenges a novel experimental rig has been designed. At the heart of this rig is a representative half-unit-cylinder-head which is being used to study the heat transfer achievable with evaporative cooling strategies under realistic conditions. In particular, the half-unit-cylinder-head can be excited dynamically (with a shaker) to provide measurements of evaporative-cooling system thermal performance in the presence of vibration levels and frequencies typical of a boosted IC engine in a vehicle on the road. Extensive computational fluid dynamics (CFD) simulation is being undertaken to assist the rig design and to understand the performance of both evaporative-cooling strategies, and the conventional cooling systems that they could replace. In this paper, multi-phase CFD and conjugate heat transfer (CHT) analysis of this half-unit-cylinder-head model is examined under subcooled flow boiling conditions. This is to serve as a benchmark for robust EC system development under realistic engine conditions. The study focuses on the thermal conditions for both the coolant and hot-gas sides which is needed for hot-side surface design and for gas heating specification. Simulations are performed by means of the finite-volume CFD solver STAR-CCM+ (version 10) where the VOF (Volume-of-Fluid) algorithm [16] is adopted to capture the liquid-vapour interface. The paper is structured as follows. First, the modelling approach is presented. Then the CFD simulations are described and a comparison made against experimental data.

2 GEOMETRY AND BOUNDARY CONDITIONS

An exhaust-side "half-unit-cylinder-head" model has been designed based on consideration of: i) very high steady-state heat-flux capability, ii) the flow-path representation of the cylinder head, iii) curved and arcuate internal coolant jacket surfaces, iv) the feasibility of optical access, and v) matching the temperature gradients of downsized engines. The current design is intended to represent a highly-boosted downsized engine (with 71.4 mm bore and 82 mm stroke). The use of a half cylinder is intended to focus on the highest heat flux zones of a cylinder head and to reduce the mass of the shaken parts of the test rig to enable higher amplitudes and frequencies to be excited as part of assessing the impact of the vibration and agitation on evaporative cooling. The model comprises a central half-unit-cylinder-head piece with a conventional coolant jacket as shown in Figure 1.

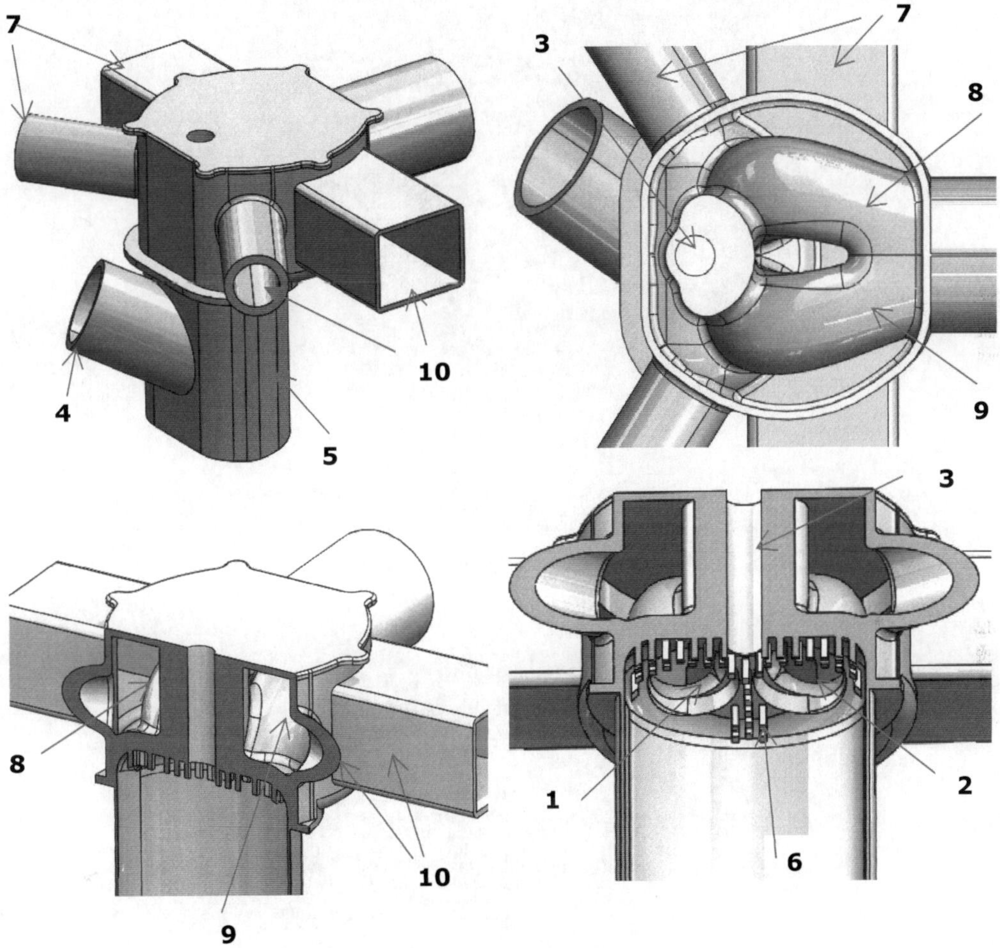

1-2: Exhaust ports openings
3: Spark plug boss
4: Gas by-pass port
5: Gas heater shroud
6: pin fins
7: Coolant inlet
8-9: Exhaust ports
10: Coolant outflow

Figure 1. 'Half-unit-cylinder-head' model showing upper observation plate, conventional cooling jacket, and square heating fins (of dimension 2 mm × 2 mm × 6 mm) on the flame face

There is an uppermost access plate to allow observation and instrumentation, and for eventual use of targeted coolant sprays. The half-unit-cylinder head has three gas-flow paths, i.e. one through the exhaust ports (1 and 2, Figure 1), one through the spark plug boss (3, Figure 1), and a by-pass (4, Figure 1) - all three with throttle controls. Heating on the rig will be achieved using a propane gas burner. The appropriate heat flux levels and three dimensional gradients on the gas face are achieved with 68 distributed pin fins (i.e. No. 6 on Figure 1). Coolant flow arrives via conduit (7, Figure 1) to the exhaust ports (8 and 9, Figure 1) zones. Figure 2 shows the flow-ports and corresponding flow-paths.

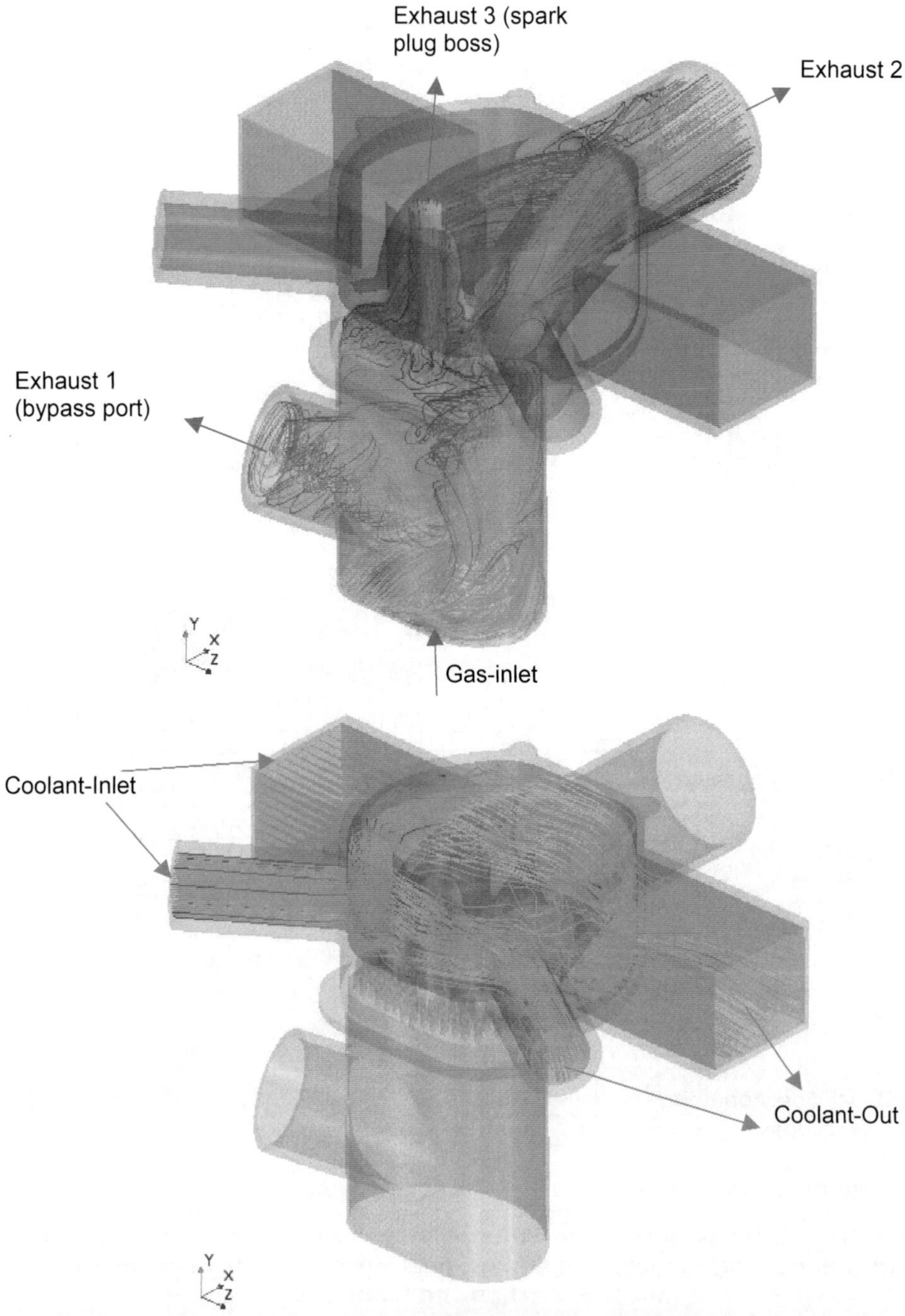

Figure 2. Flow-ports and corresponding flow-paths; a) gas, b) coolant streamlines

The boundary conditions adopted are based on available data relevant to highly-boosted downsized engine operating conditions. The metal properties are set for Aluminium alloy with specific heat 900.0 J/kg-K, and thermal conductivity of 150.0 W/m-K. The fluid domain is a mixture of water and ethylene glycol with 50% volumetric ratios. The coolant inlet is assumed to be 1.225 kg/s at 90°C, based on an assumed value of 1.35 l/s total pump flowrate for a typical 3-cylinder engine. Coolant outlet takes place through outflow boundaries as shown in Figure 2 (Part 10 in Figure 1). These flow outlets will be used selectively to reproduce the flow patterns in the engine cylinder head. Using empirical estimates, 2.4 kW heat transfer through the flame face of one whole unit of a cylinder head at rated power, is envisaged at the rated power of 33.3 kW per cylinder for a fuel energy rate of 103.4 kW. The hot side fin design and hot gas flow conditions were optimised for local heat-flux, representative metal temperature gradients, and integrated heat energy rates. The hot gas inlet flow into the canister (Part 5 in Figure 1) of 0.00214 kg/s at 1970°C is prescribed which corresponds to gas heating with a 6 kW propane heater (given propane enthalpy of combustion 46.1 MJ/kg and stoichiometric air/fuel ratio of 15.5). A swirling effect, to increase heat transfer to the pin fins, is added to the gas inlet boundary by defining the flow direction vector as (*r, theta, z*), for example = (0.1, 0.8, 0.1) in Figure 3. Zero outlet flow to the bypass port, Exhaust 1, is prescribed at the targeted load condition. Atmospheric pressure outlet is imposed on exhaust ports 2-3.

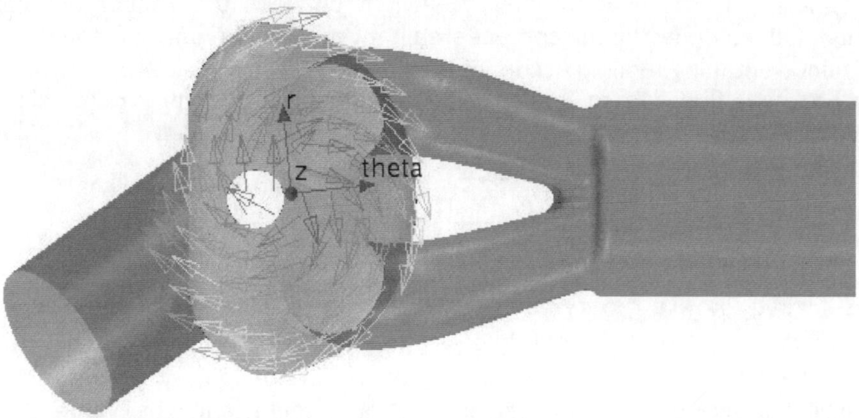

Figure 3. Swirling flow on the gas inlet; flow direction vector (*r, theta, z*): (0.1, 0.8, 0.1)

3 NUMERICAL FRAMEWORK

A steady-state three-dimensional CFD/CHT case with separate fluid, gas and solid regions (interfacing) is used in the analysis presented here. Liquid and vapour flow are simulated using the VOF Multiphase modelling approach within the Eulerian framework [16]. A homogeneous flow model is employed where liquid and vapour bulk properties are defined separately. The governing equations are solved for the mixture flow with shared velocity, pressure, and temperature field assumed to be the same for the two phases. An additional transport equation for the phase volume fraction is solved to obtain the spatial phase concentrations within the domain. Changes in density due to the phase change are therefore taken into account without solving two sets of governing equations for the liquid and vapour phases. The convective part of heat flux is given by:

$$q_c = h_c(T_w - T_c) \tag{1}$$

and

$$h_c = \frac{\rho C_p u_\tau}{T^+|_{y^+(y_c)}} \tag{2}$$

where h_c is the convective heat transfer coefficient, T_w is the wall temperature, T_c is the local near wall cell temperature, ρ and C_p are density and specific heat of the mixture respectively, u_τ is the friction velocity, T^+ is the dimensionless temperature, y_c is the normal distance of the near wall cell, and y^+ is the dimensionless wall distance $u_\tau y_c / v_f$.

The implemented phase change model accounts for the onset of boiling through the use of sub-models. The heat transfer at the wall-fluid boundary is used to calculate the phase change mass transfer rate (evaporation or condensation). The vapour phase temperature is assumed constant at the saturation temperature T_{sat} and the liquid temperature is approximated to the mixture temperature T. The total heat interchange between the liquid and vapour phases is then used to work out the mass transfer between phases:

$$\dot{m}_{ec} = \frac{C_{HTC \times Area}(T - T_{sat})}{h_{fg}} \tag{3}$$

where $C_{HTC \times Area}$ is the heat transfer coefficient between the vapour bubbles and adjacent liquid multiplied by the contact area between the two phases, and h_{fg} is the phase change enthalpy. Boiling occurs at the liquid-solid interface where $T_w > T_{sat}$ and the surface heat flux due to boiling is calculated by the empirical correlation of Rohsenow [7]:

$$q_{bw} = \mu_f h_{fg} \sqrt{\frac{g(\rho_f - \rho_g)}{\sigma}} \left(\frac{c_{p,f}(T_w - T_{sat})}{C_{qw} h_{fg} Pr_f^{1.7}} \right)^{3.03} \tag{4}$$

where μ_f, $c_{p,f}$, Pr_f are the dynamic viscosity, the specific heat, and the Prandtl number of the liquid phase respectively, g is gravitational acceleration, ρ_f is the density of the vapor phase, σ is the surface tension at the liquid-vapour interface, T_w is the wall temperature and C_{qw} is an empirical coefficient, the value of which depends on the liquid-surface (and surface finish) combination. A value of 0.0028 was used in the current study as being representative of engine coolant and aluminium alloy. The vapour mass flow rate per unit area, through boiling on the wall, then follows from:

$$\dot{m}_{ew} = \frac{C_{ew} q_{bw}}{h_{fg}} \tag{5}$$

where C_{ew} is the model constant indicating how much of the boiling heat flux is used for vapour generation.

Liquid and vapour phases are treated as a mixture where one set of turbulent equations are solved for the mixture phase. Steady-state RANS computations were undertaken using the *k-ε* turbulence model with "All y^+" wall treatment approach of STAR-CCM+ for near-wall turbulence quantities - a hybrid wall treatment combining high y^+ wall treatment for coarse meshes, and the low y^+ wall treatment for fine meshes, specially formulated for meshes of intermediate resolution where the near-wall cell lies within $1 < y^+ < 30$. For $y^+ <=1$ no wall function is used and the model assumes that the viscous sublayer is fully-resolved by the mesh. For $y^+ > 30$, the classic wall-function approach is adopted where wall shear stress, turbulent production and dissipation are governed by the equilibrium turbulent boundary layer theory. The "All y+" method blends any turbulence quantity such as dissipation, production, stress tensor, etc. calculated by the "high-y+" approach or by the low-y+ approach using an exponential weighing function where the value for the turbulence quantity Φ is calculated as:

$$\emptyset = w\emptyset_{low} + (1 - w)\emptyset_{high} \tag{6}$$

where w is given by:

$$w = exp\,(-\frac{Re_y}{11}) \tag{7}$$

and where

$$Re_y = \frac{\sqrt{k}y}{v}) \tag{8}$$

is a wall-distance-based Reynolds number, y is the normal distance from the wall to the wall-cell centroid, v is the kinematic viscosity, and k is the turbulent kinetic energy. Boundary layers have been resolved with appropriate near-wall prism layers according to the adopted near-wall treatment model. The near wall resolution in the current study ranges between $12 < y^+ < 25$ (i.e. a near wall cell distance between 0.1 - 0.4 mm). The computational domain is divided into three blocks and polyhedral conformal grids have been generated for the three blocks: hot gas region, aluminium half-unit-cylinder-head region, and coolant flow field region. Figure 4 shows grid cells on the hot gas and coolant blocks. A grid sensitivity study was performed using three different grid resolutions with 600000, 800000, and 910000 cells to ensure the grid resolution does not influence the solution.

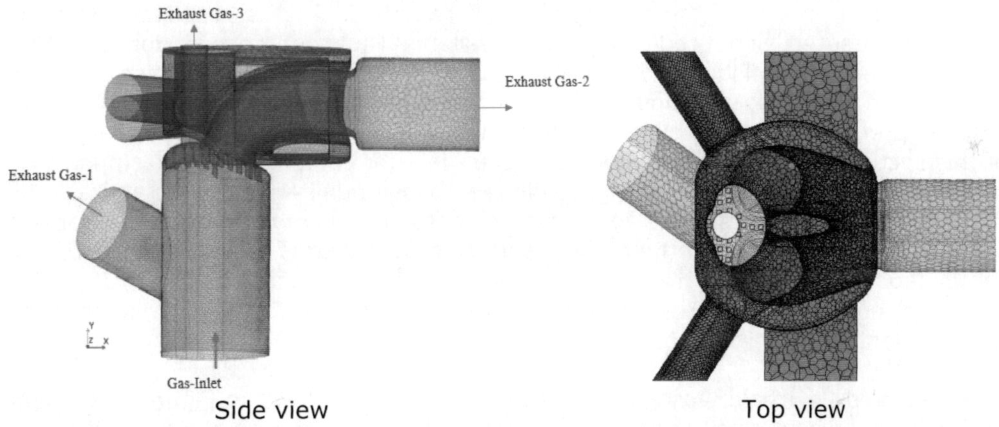

Figure 4. Computational grids on hot gas (grey) and coolant (blue) domains

4 RESULTS AND DISCUSSION

Figure 5 shows the predicted values of surface temperature on both the gas side and the coolant sides of the half-unit-cylinder-head. In these (and all subsequent results in this section) the computations were carried out using a grid with 793,615 cells.

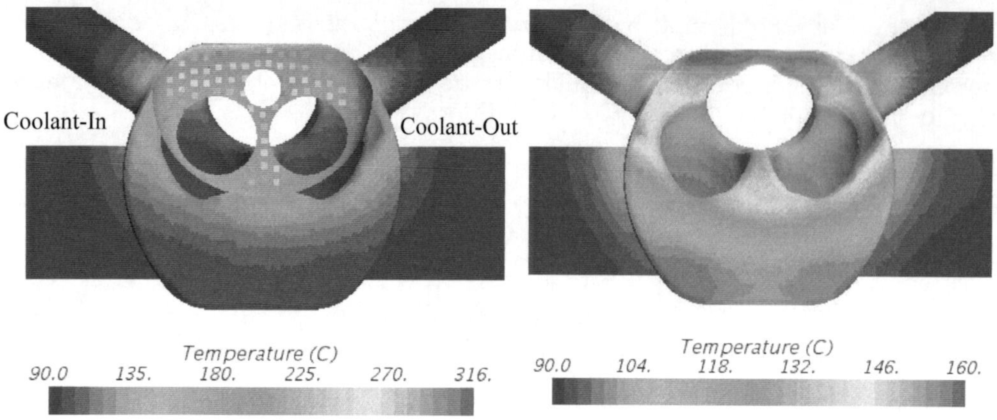

a) Metal temperatures on gas side face **b) Metal temperatures on coolant side face**

Figure 5. Predicted surface temperatures on: a) the gas side, and b) the coolant side

In the interests of clarity, it is considered worthwhile to explain some of the geometric features in figures 5a and 5b as they will also be shown elsewhere in the paper. In Figure 5a, where the view shown looks onto the hot gas surface, the two large circular holes near the centre are the exhaust gas outlets (labelled as Exhaust Gas 2, in Figure 2). The smaller single hole above these is also an exhaust gas outlet (labelled as 3, in Figure 1). Towards the left of the figure are the coolant inlet pipes, and towards the right, the two coolant outlet pipes. Of interest in Figure 5a is the gas side face itself, which comprises the central part of the figure (mostly coloured in green). The 2mm x 2mm heating fins with a typical height of 6 mm can be seen as an array of squares on this hot gas side surface. The maximum temperature of the fins is around 300°C - but most are around 250°C. However, the temperature of the majority of the substrate at the base of the fins is in the region of 180°C to 200°C. This is consistent with the limited (measured and predicted) data available (but not shown) for a similar downsized engine. In Figure 5b, the white blanked-out section (that is also shown in Figure 1) is the spark plug boss with a conduit through it for gas heating. The two exhaust ports are indicated by the voids immediately below, and to the left, and right of this section. Although there is no measured data for comparison, the temperature distributions shown on the coolant-side do follow qualitative expectations. The maximum coolant-side surface temperatures occur in the region above the hot gas side. And within this region the highest surface temperatures occur in the vicinity of the exhaust ports (around 160°C for a bulk coolant temperature of 96°C). It is also worthy of note that there is asymmetry in the temperature field: the temperatures near Coolant-In are higher than those near Coolant-Out.

The predicted flow field, corresponding to Figure 5b, is shown in Figure 6. Coolant enters from the inlet pipe to the cylinder head via two ducts next to the exhaust ports, and then flows to the right and around the exhaust port towards the spark plug boss. The coolant flow is mainly longitudinal, from the inlet pipes on the left, to the exit pipes on the right. The maximum velocities are around 2.0 m/s but over most of the cylinder head they are typically in the region of 0.8-1.4 m/s. The reason for the asymmetry in the temperature field (noted earlier) can now be understood. This is because coolant velocities are lower towards the upper coolant inlet pipe than outflow pipe. The flow pattern and velocities depicted here are qualitatively in broad agreement with the flow pattern in a highly-boosted downsized engine where there is a longitudinal flow in the cylinder head from one cylinder to another.

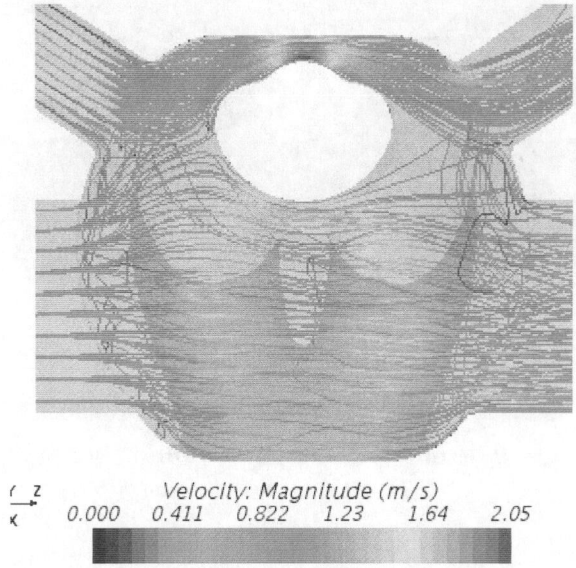

Figure 6. Predicted flow streamlines on the coolant side

Figure 7 shows a contour map of the predicted values of heat flux distribution on the coolant side of the half-unit-cylinder-head. In the coolant inlet and exit pipes the heat fluxes are less than 10^4 W/m^2, which is lower than associated with nucleate boiling. This is to be expected since the surface temperatures here are generally below the saturation temperature and consequently the heat transfer is mostly due to single phase convection.

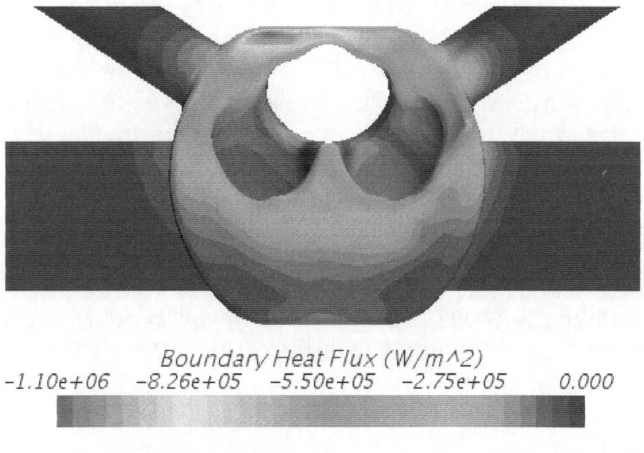

Figure 7. Predicted values of coolant-side heat flux

Elsewhere the heat flux is seen to vary across the cylinder head by several orders of magnitude from 10^4 W/m^2 to 10^6 W/m^2. These values are entirely consistent with those occurring during nucleate boiling. It is also encouraging to note that the predicted maximum heat flux of around 1.1×10^6 W/m^2 which occurs towards the top of the diagram of the half-unit-cylinder-head, is consistent with empirical estimates (of 1.23×10^6 W/m^2) at this location. For the main part of the surface, the heat flux is in the region of 0.5×10^6 W/m^2, which is also consistent with empirical estimates for downsized engines.

Figure 8 shows the predicted values of vapour quality (i.e. fraction of vapour by mass). There are localised pockets of 0.14% vapour fraction which correspond to regions with high heat flux and low coolant velocity. There are also regions of 0.05% vapour fraction in the valve bridge area.

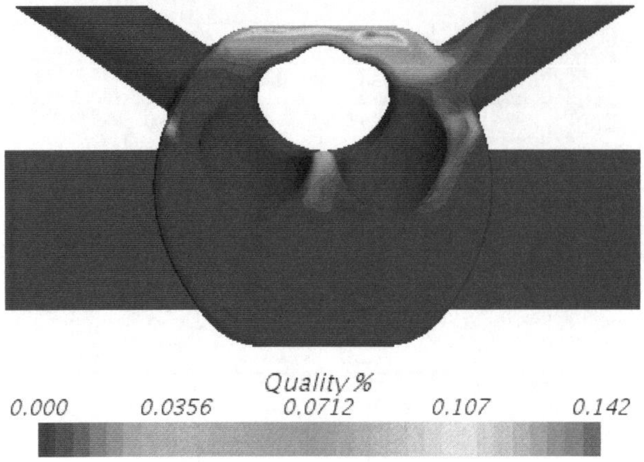

Figure 8. Predicted values of vapour quality

5 CONCLUSIONS

A simulation capability has been developed to obtain coolant flow and associated heat transfer to assist in the design of an experimental half-unit-cylinder-head test piece. This capability is to target conditions that an evaporatively-cooled version of the half-unit-cylinder-head must endure in terms of external heating and cooling, so that peak metal temperature conditions and heat flux levels emulate those found at full-load on a highly-boosted downsized engine cylinder head. Two-phase CFD/CHT simulations, allowing predictions of the temperature field for subcooled boiling conditions on the coolant side, have been used to determine the required gas heating power, and pin-type cooling fin design on the flame-face. Predicted wall temperatures, heat flux, and vapour volume fraction have been presented to confirm that the targeted heat flux on the half-unit-cylinder-head test piece can be achieved. The predicted temperatures and heat flux levels compare well with measured engine data (not shown) at a small number of available discrete points.

ACKNOWLEDGMENTS

The authors wish to acknowledge funding support for this project from the EPSRC under Contract Number: EP/M005755/1. The technical support is also acknowledged of colleagues at Ford Dunton UK and Dearborn USA, Denso Italy, and the Ricardo Technical Centre Shoreham, UK.

REFERENCES

[1] A. J. Torregrosa, A. Broatch, P. Olmeda, and O. Cornejo, "Experiments on subcooled flow boiling in I.C. engine-like conditions at low flow velocities," *Exp. Therm. Fluid Sci.*, vol. 52, pp. 347–354, Jan. 2014.

[2] K. Robinson, N. a F. Campbell, J. G. Hawley, and D. G. Tilley, "A Review of Precision Engine Cooling," *SAE Tech. Pap.*, no. 1999-01–0578, 1999.

[3] N. Ap and M. Tarquis, "Innovative Engine Cooling Systems Comparison," *SAE Tech. Pap.*, no. 2005-01–1378, 2005.

[4] N. A. . Campbell, D. G. Tilley, S. A. MacGregor, and L. Wong, "Incorporating Nucleate Boiling in a Precision Cooling Strategy for Combustion Engines," *SAE Tech. Pap.*, no. 971791, May 1997.

[5] S. G. Kandlikar and M. Bulut, "An Experimental Investigation on Flow Boiling of Ethylene-Glycol/Water Mixtures," *J. Heat Transfer*, vol. 125, no. 2, p. 317, 2003.

[6] H. Steiner, G. Brenn, F. Ramstorfer, and B. Breitschadel, "Increased Cooling Power with Nucleate Boiling Flow in Automotive Engine Applications," in *New Trends and Developments in Automotive System Engineering*, InTech, 2011.

[7] W. M. Rohsenow, "A Method of Correlating Heat Transfer Data for Surface Boiling of Liquids," *Trans. ASME,* vol. 74, no. 1952, pp. 969–976, 1952.

[8] J. C. Chen, "Correlation for boiling heat transfer to saturated fluids in convective flow," *Ind. Eng. Chem. Process Des. Dev.*, vol. 5, no. 3, pp. 322–329, 1966.

[9] K. Robinson, J. G. Hawley, and N. A. F. Campbell, "Experimental and modelling aspects of flow boiling heat transfer for application to internal combustion engines," *Proc. Inst. Mech. Eng. Part D J. Automob. Eng.*, vol. 217, no. 10, pp. 877–889, Jan. 2003.

[10] M. Cardone, A. Senatore, D. Buono, M. Polcino, G. De Angelis, and P. Gaudino, "A Model for Application of Chen's Boiling Correlation to a Standard Engine Cooling System," *SAE Tech. Pap.*, no. 2008-01-1817, Jun. 2008.

[11] S. Fontanesi and M. Giacopini, "Multiphase CFD-CHT optimization of the cooling jacket and FEM analysis of the engine head of a V6 diesel engine," *Appl. Therm. Eng.*, vol. 52, no. 2, pp. 293–303, 2013.

[12] D. Carpentiero, S. Fontanesi, V. Gagliardi, S. Malaguti, S. Margini, M. Giacopini, A. Strozzi, L. Arnone, M. Bonanni, and D. Franceschini, "Thermo-mechanical analysis of an engine head by means of integrated CFD and FEM," *SAE Tech. Pap.*, no. 2007-24-0067, Sep. 2007.

[13] S. Fontanesi, D. Carpentiero, S. Malaguti, M. Giacopini, S. Margini, and L. Arnone, "A New Decoupled CFD and FEM Methodology for the Fatigue Strength Assessment of an Engine Head," *SAE Tech. Pap.*, no. 2008-01-0972, Apr. 2008.

[14] M. Shala, "Simulation of nucleate boiling flow using a multiphase mixture modelling approach," *IMA J. Appl. Math.*, vol. 77, no. 1, pp. 47–58, Feb. 2012.

[15] J. G. Hawley, M. Wilson, N. A. F. Campbell, G. P. Hammond, and M. J. Leathard, "Predicting boiling heat transfer using computational ftuid dynamics," *Proc. Inst. Mech. Eng. Part D J. Automob. Eng.*, vol. 218, no. 5, pp. 509–520, Jan. 2004.

[16] C. W. Hirt and B. D. Nichols, "Volume of fluid (VOF) method for the dynamics of free boundaries," *J. Comput. Phys.*, vol. 39, no. 1, pp. 201–225, 1981.

Influence on warm-up of split head and block circuits for oil and coolant and the effect of main gallery location

C Zhao, P J Shayler, L Cheng
University of Nottingham, UK

R Gilchrist
Jaguar Land Rover, UK

ABSTRACT

A bespoke variant of a 2.0l gasoline turbocharged directly injected engine produced to explore the influence of design features on thermal behaviour has been used to investigate the use of split coolant and split oil circuits. In both cases, the circuits separate recirculating flows to the head and block of the engine. Full factorial experimental tests were conducted to examine the degree of interaction between circuits and explore the effect on heat transfer and thermal state of changing coolant and oil flow rates between high and low flow settings. The flow rate of the fluid in each engine circuit (head cooling, block cooling, head lubricating and block lubricating) has the strongest influence on heat transfer and warm-up behaviour of the fluid. Thermal interaction between circuits was weak. The main oil gallery was immersed in block coolant jacket around the cylinder liners; heat transfer to oil was increased by raising the gallery and enhancing thermal coupling of the gallery to the upper region of the cylinder liner. Comparing results for SAE 5w-30 and SAE 0w-12 grade oils, showed the latter yielded a reduction of engine friction in the early stages of engine warm-up, and marginally higher friction under fully warm engine conditions. However the effect of oil formulation on coolant and oil warm up was negligible. Strategies for coolant and oil pump control promoting the temperature rise of the block oil are discussed.

NOMENCLATURE

$\dot{E}$	Rate of internal energy change
EWP	Electric water pump
$\dot{H}$	Rate of enthalpy change
$\dot{m}$	Coolant mass flow rate
$M_{C,engine}$	Coolant mass located in the engine
$M_{C,total}$	Total coolant mass in the cooling circuit
MGH	Main gallery at high position
MGL	Main gallery at low position
$\dot{V}_{C,B}$	Coolant volumetric flow rate in the block
$\dot{V}_{C,H}$	Coolant volumetric flow rate in the head
$\dot{V}_{O,B}$	Oil volumetric flow rate in the block
$\dot{V}_{O,H}$	Oil volumetric flow rate in the head
HHHH	'High' level setting for $\dot{V}_{C,H}$, $\dot{V}_{C,B}$ $\dot{V}_{O,H}$ and $\dot{V}_{O,B}$
LLLL	'Low' level setting for $\dot{V}_{C,H}$, $\dot{V}_{C,B}$, $\dot{V}_{O,H}$ and $\dot{V}_{O,B}$

1 INTRODUCTION

Engine fuel economy is relatively poor during cold running. Over the New European Drive Cycle (NEDC), there is typically a 5-10% increase in fuel consumed by a private or light commercial vehicle if the engine is cold-started compared to fully warm at the start of the drive cycle [1]. This is mainly due to an engine friction penalty associated with high lubricant viscosity at low temperatures. Generally, engine friction follows a power law relationship with the viscosity of engine lubricant except the early seconds of engine warm-up [2]. Given that vehicles are often used for short journeys during which the engine does not reach fully warm conditions [3-4], measures which accelerate engine warm-up and reduce the cold running penalty are of great importance. For this reason, the potential of passive and active ways of manipulating heat flows in the engine is attracting attention. One approach of interest is using separate cooling and oil circuits for the engine head and block, four in total, allowing flow rates in each to be managed independently and in principle providing a degree of control over heat transfer to and between oil, coolant and engine structure. This has been explored in the experimental study reported here. The aim has been to determine the extent of interaction between the circuits of a bespoke engine design and to examine if a benefit can be achieved, most directly as a friction reduction resulting from an acceleration of oil warm-up. As a benchmark for comparison, similar tests were carried out with two oils, one a standard, SAE 5w-30, and a second a lower viscosity, SAE 0w-12 grade oil.

The study was enabled by the design and manufacture of a bespoke engine prototype, shown in Figure 1, which had the cylinder head of a 2.0l gasoline turbocharged direct injection engine modified to flow coolant along the length of the head from front to back any fluid exchange with coolant circulating in the block. The block was a wet-liner design. The configuration of the core components and internal structure of the engine are shown in Figure 2. The block splits into an upper part (black) containing the cylinder bank (red) made of grey cast iron and cooling jacket, and a lower part (turquois) where the crankshaft (pink) presents. The volume of the cooling jacket is defined by the space left between the cylinder bank and the water jacket inserts (green). The material of the inserts is a silicone elastomer, ACC QSil553, which has a relatively low thermal conductivity (0.68 W/mK) compared to aluminium (~200W/mK) or grey cast iron (40-50W/mK).

Figure 1: The rig setup of VIPER research engine

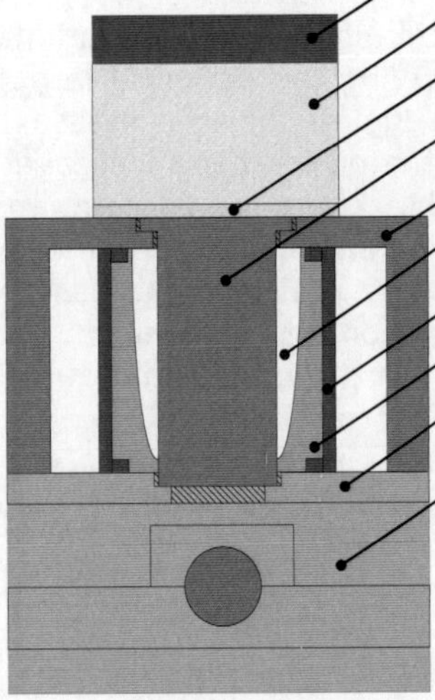

Figure 2: The configuration of engine core components and engine internal structure

Two arrangements of oil flow in the block were possible, as illustrated in Figure 3. In both cases, the distribution of oil to the piston cooling jets and crankshaft bearings was from a principal oil gallery running the length of the engine block. This gallery is low in the engine block in a position typical of main gallery position in a conventional engine design. When the oil feed into the block was connected to the mid-gallery position by a short connecting pipe, this is referred to as the low gallery position. In the second case, a second gallery positioned high in the coolant jacket and fully immersed in the coolant feeds oil to the ends of the primary gallery. The shape of the gallery follows the contour of the cylinder liner bank with a gap up to 1mm. The oil flows into the primary gallery from both ends and is distributed to the piston cooling jets and crankshaft bearings as before. This is referred to as the high gallery position. The second gallery increases the oil path length in the block and increases the gallery wall area exposed to the coolant.

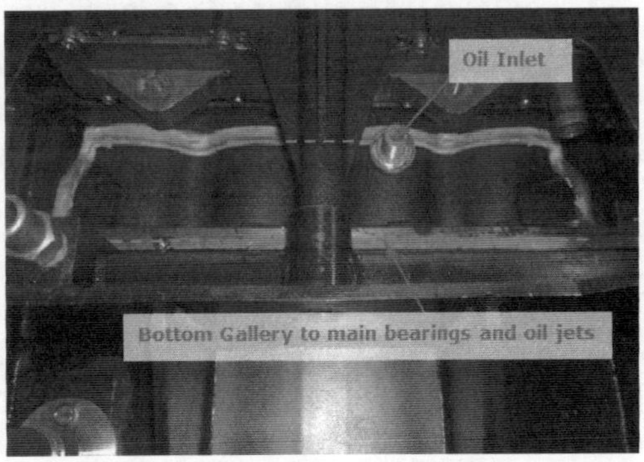

Figure 3: The configurations of main oil gallery at high (right) and low (left) positions

Schematics of the split cooling and split lubricating circuits are presented in Figure 4. The four circuits are completely separate and no mixing takes place. The pump in each circuit was driven electrically. The coolant flows were circulated using 12 volt coolant pumps; the oil pumps were driven using AC motors. Pump speeds were varied to maintain either high or low target flow rates in each coolant circuit, and a high or low gallery pressures in the oil circuits. Targets are given in Table 1. The oil pumps were mounted off the engine; this required the inclusion of additional pipework in both head and block oil circuits and the inclusion of an oil return tank for the head circuit. The unconventional design of the block adopted to allow access to the liner cooling jacket had a larger thermal capacity than a production engine, hence larger engine thermal mass. The combination of increased thermal mass connected to the oil path and increased system volume of oil slow the warm-up rate of oil in the block circuit. To compensate, two heaters providing a fixed input of 2.5kW were installed in the engine sump to give a similar warm up rate as the production engine on which the design was based. A single 1.25 kW oil heater was installed in the oil return tank of the head circuit for the same purpose. These heaters were run throughout the duration of a test.

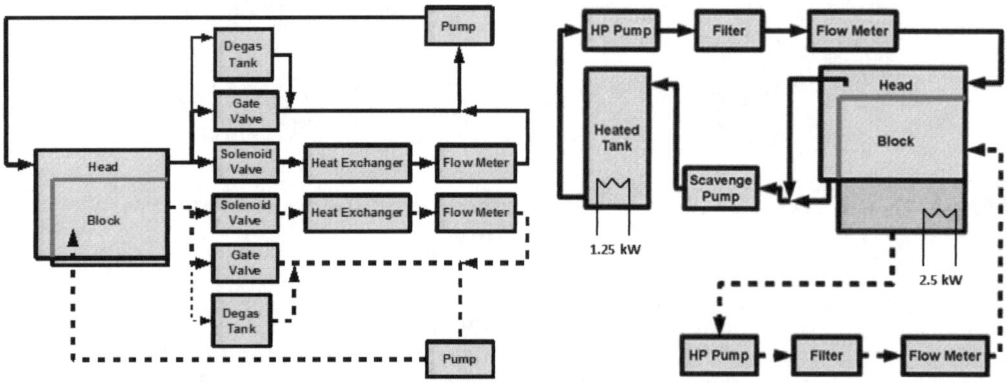

Figure 4: The schematics of the split cooling (left) and split lubricating (right) circuits of the VIPER research engine

Table 1: The high/ low level for each control factor

		High Flow	Low Flow
Cooling Circuit	Cylinder Head	100% EWP 0.57 (l/s)	10% EWP 0.15 (l/s)
	Engine Block	100% EWP 0.43 (l/s)	10% EWP 0.11 (l/s)
Lubricating Circuit	Cylinder Head	MGP 3.5 barA	MGP 2.25 barA
	Engine Block	MGP 3.5 barA	MGP 2.25 barA

2 TEST RESULTS

Tests covering all combinations of high and low settings of coolant flows and oil pressure settings were carried out at 1500 rpm and a BMEP of 2.5bar from a cold-start at laboratory ambient temperatures. The full factorial test plan is given in Table 2. At least 3 tests were carried out for each test number. The high and low settings for oil refer to gallery pressure values. The oil circulation rates change as the oil and engine state change, but for convenience the engine settings are referred to as giving a high or low oil flow rate. In this sequence of tests the block oil gallery was in the low position and the oil used in both head and block oil circuits was SAE 5w-30 grade oil. Tests were run for between 300 and 400seconds, by which time the coolant and oil temperatures in the block had risen by 20-30 degrees Celsius and the head coolant by 60-70 degrees Celsius.

Table 2: Test plan of the full factorial test

Test Number	Head Coolant	Block Coolant	Head Oil	Block Oil
1	H	H	H	H
2	L	H	H	H
3	H	L	H	H
4	L	L	H	H
5	H	H	L	H
6	L	H	L	H
7	H	L	L	H
8	L	L	L	H
9	H	H	H	L
10	L	H	H	L
11	H	L	H	L
12	L	L	H	L
13	H	H	L	L
14	L	H	L	L
15	H	L	L	L
16	L	L	L	L

2.1 Influence on coolant temperatures and heat transfer to coolant

Coolant temperatures at inlet to and outlet of the cylinder head and the block are presented in Figures 5-6. These are plotted as temperature changes from initial values to eliminate noise associated with small test to test variations in laboratory ambient values. The results cover all the combinations given in Table 2 but it is clear that the coolant flow setting for the circuit is the dominant influence on the rate of temperature rise in that circuit. Over the 350s duration of a test, for the head coolant (Figure 5) the highest temperature increases were recorded when the coolant circulation rate was low. For the block coolant (Figure 6) the highest temperature increases were recorded when the coolant circulation rate was high. Note however that the analysis of heat transfer to coolant presented later shows a similar trend for both the head and block coolant in which low coolant circulation rates give lower initial rates of heat transfer to coolant and higher rates later in the test. In both cases, the effects of coolant flow rate in the other circuit, and of settings for the oil circuit were small and no systematic trends were identified.

Since the coolant mass is constant, the rate of increase of coolant temperature in either head or block circuit is proportional to the increase in internal energy and should be in agreement with the net rate of heat input around the circuit. In addition, the rate of heat transfer in the block or head alone can be calculated from the flow rate, temperature rise across the engine and change of internal energy of coolant within that part of the circuit. This provides a means of checking consistency of the data and establishing what proportion of the total heat transfer occurs within the engine itself, which is important given there is external pipe work and the coolant pumps are positioned outside the engine.

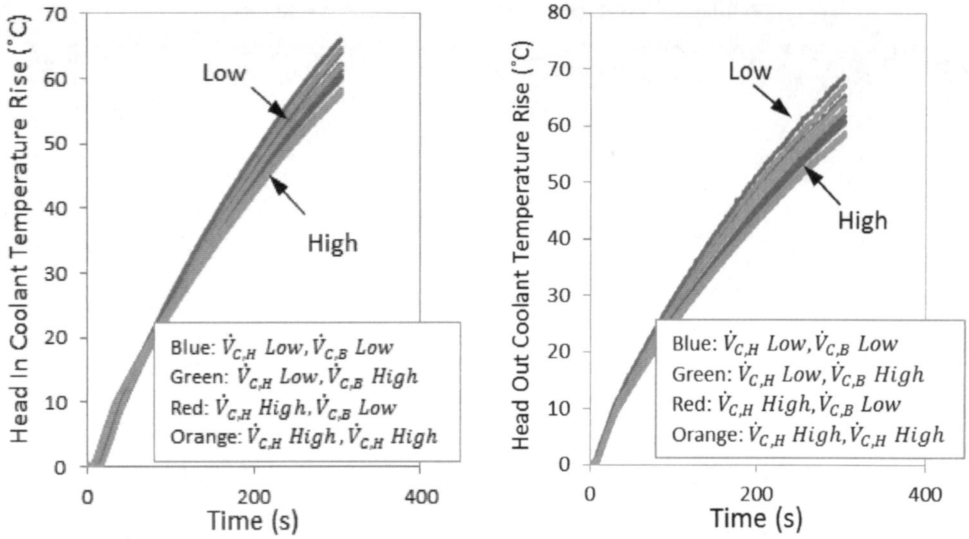

Figure 5: Coolant temperatures at inlet and outlet of cylinder head. Key influence is low or high setting of coolant circulation rate in the head ($\dot{V}_{C,H}$)

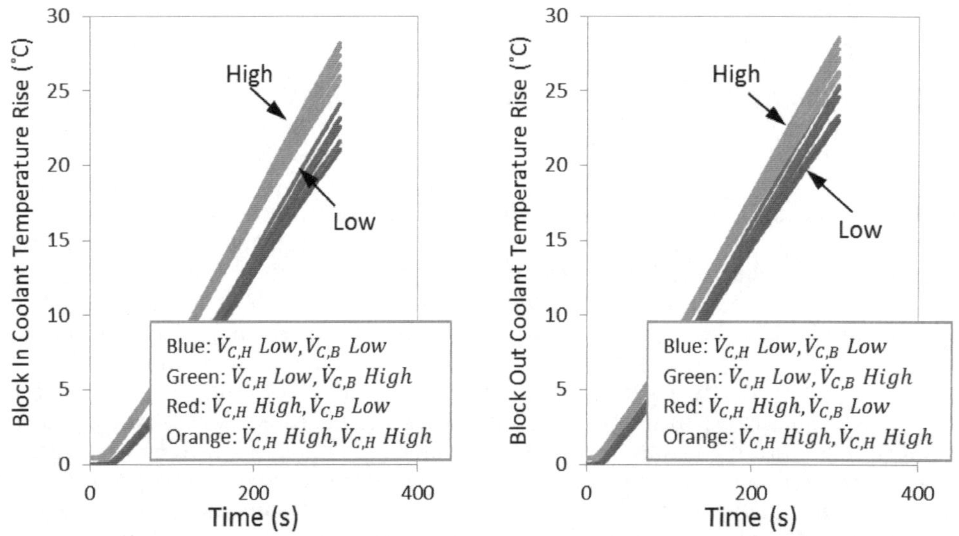

Figure 6: Coolant temperatures at inlet and outlet of engine block. Key influence is low or high setting of coolant circulation rate in the block ($\dot{V}_{C,B}$)

Firstly, the net heat transfer rate to head coolant or block coolant within the engine was estimated. The net heat transfer is the sum of changes in enthalpy and internal energy given by Equations 1 and 2.

$$\dot{Q}_{net,engine} = \Delta\dot{H} + \Delta\dot{E} \tag{1}$$

$$\dot{Q}_{net,engine} = \dot{m}C_p(T_{out} - T_{in}) + M_{c,engine}C_v\frac{d(T_{out}+T_{in})}{2dt} \tag{2}$$

Using measured temperature changes between inlet and outlet, and the flow rate of coolant, the calculated net heat transfer variations are as presented in Figure 7. The net heat transfer to coolant flowing through the head rises quickly during the early seconds of engine operation, to a peak of between 3kW and 4kW as gas-side heat transfer from the combustion system and exhaust ports warms the head and develops a metal-to-coolant temperature difference. The heat transfer rate falls as the coolant temperature rises and the rate of increase of metal temperature decreases. In the block, heat transfer to coolant rises at a similarly quick rate but only to around 1kW. After this, unlike in the head, the rise in heat transfer rate continues although more slowly. The early high rate of rise of heat transfer is driven principally by heat transfer from the upper parts of the cylinder liner. The lower parts of the engine structure warm more slowly and act initially as a heat sink holding down metal temperatures around the lower part of the block coolant jacket.

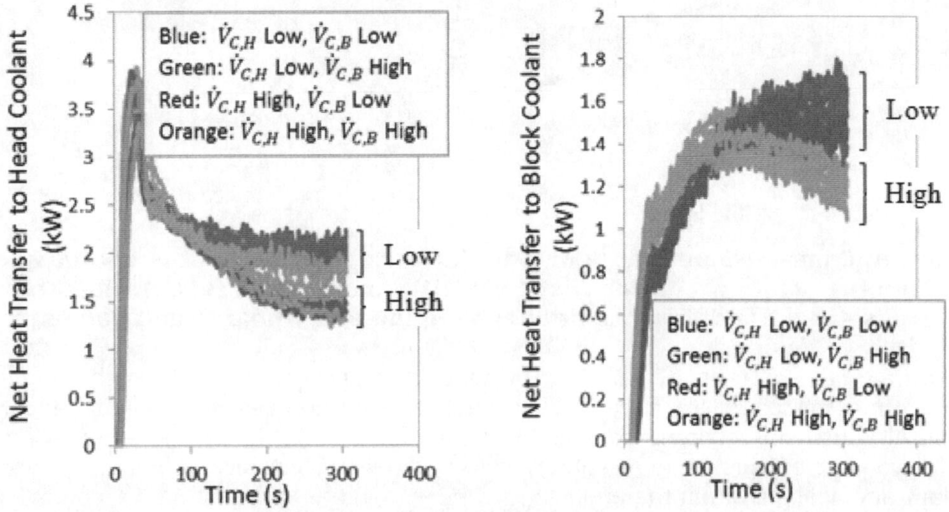

Figure 7: The net heat transfer rate to head coolant (left) for low and high settings of head coolant flow, and block coolant (right) for low and high settings of block coolant flow

The calculated heat transfer levels and patterns in Figure 7 are consistent with the temperature variations given in Figures 5 and 6 and the sensitivity to coolant flow rates. Again, the coolant flow rate in each cooling system is the factor that has the most significant influence and the effect of other factors is marginal. In the head, after the first minute of engine running, the low coolant circulation rate produced the highest net heat transfer rate and a higher coolant temperature. As heat transfer to coolant occurs by forced convection, with a heat transfer coefficient which will be lower when the coolant flow rate is low, the higher rate of heat transfer is associated with an increase in metal temperatures greater than the increase in coolant temperature. In the block coolant circuit, the net heat transfer variation is more complicated. During the first ~100-120s of running, the high coolant flow setting produced the higher net heat transfer rate, after which the low coolant flow rate produced the higher values. Changes in heat transfer coefficient do not account for the transition; instead this should be associated with changes in metal temperatures along the coolant flow path in the block.

A further check on consistency is possible by determining the net heat transfer to coolant around the complete cooling circuit, whether head or block, from the change in internal energy of the total circuit fill. This is given by Equation 3, and includes the influence of heat exchange in the engine and the external cooling circuit. An illustration of the results, and comparison with the net heat transfer occurring just within the engine, is presented in Figure 8. As can be seen, the contribution of heat transfer within the engine dominates the total heat transfer around the circuit during warm-up in both cooling circuits.

$$\dot{Q}_{c,total} = M_{c,total} C_v \frac{dT}{dt} \tag{3}$$

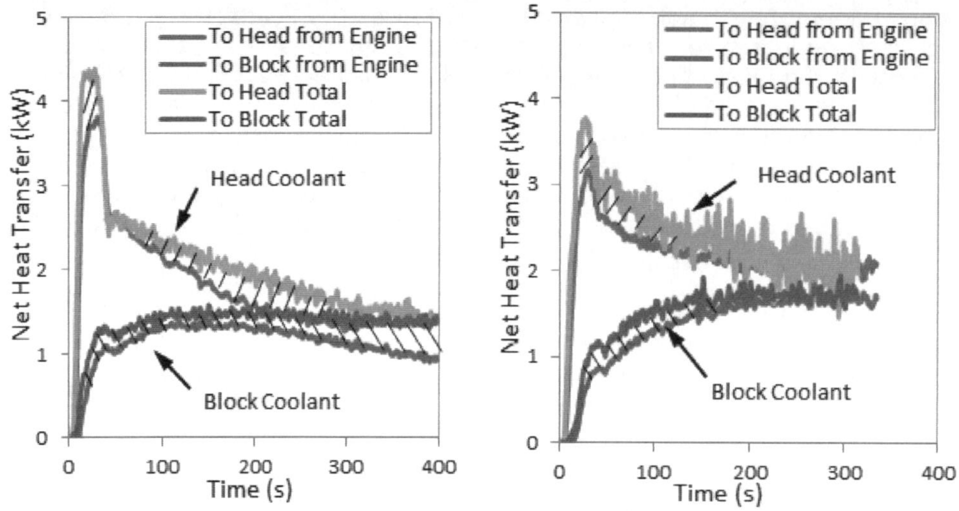

Figure 8: Comparison of net heat transfer to coolant within the engine and within the whole circuit for Test 1 (HHHH) and Test 16 (LLLL). Hatched areas show the difference between engine and whole circuit values

In Figure 8, the patterns of variation of heat transfer to coolant within the engine and around the complete circuit are similar and differences shown hatched are small. Although suggesting a positive heat transfer contribution in the external part of the coolant circuits, it may equally reflect small uncertainty in values of total coolant and coolant within the engine.

2.2 Influence on the lubricating system

The largest contributions to engine rubbing friction are made by friction at the piston-liner interface and friction in crankshaft assembly [5]. Both sources are located in the engine block. The rubbing surfaces operate largely in the hydrodynamic lubrication regime. Frictional losses resulting from shear in the oil film separating surfaces depend upon the viscosity of the oil, which in turn is strongly dependent upon oil temperature. Hence any positive influence on the rate of warm up of oil in the block circuit is of particular interest.

A comparison of oil temperatures in the engine block during warm up is presented for permutations of oil and coolant flow rates in the block in Figure 9. The results separate into two sets, determined by the oil flow rate setting in the block. The most obvious difference between the two sets of results is the longer delay, around 50s, before the block oil temperature starts to rise for the lower setting of oil circulation rate. This difference is associated with the longer time taken for oil to return to the sump, mix and be recirculated when the oil flow rate is lower. At later times in the warm-up, the higher oil flow rate and shorter delay time gave a near 5°C temperature advantage over the low oil flow rate setting. Clearly this is significant, and points to the benefit of minimising oil volume in the circuit, but in the bespoke engine it is grossly exaggerated by the larger than normal volume of oil circulating in the engine block oil circuit. Post-delay, the effect more specifically attributable to the oil circulation rate is a marginally higher rate of temperature rise when the oil circulation rate is low.

The data presented in Figure 9 covers high and low settings of coolant flow in the block. The effect of the coolant flow rate on the rate of temperature rise of oil is small but discernible: The lower rate of coolant circulation, producing lower coolant temperatures in the block, gives a marginally higher rate of oil temperature rise when the oil flow is high. Conversely, coolant and oil settings in the head had no measureable influence of block oil temperature rise.

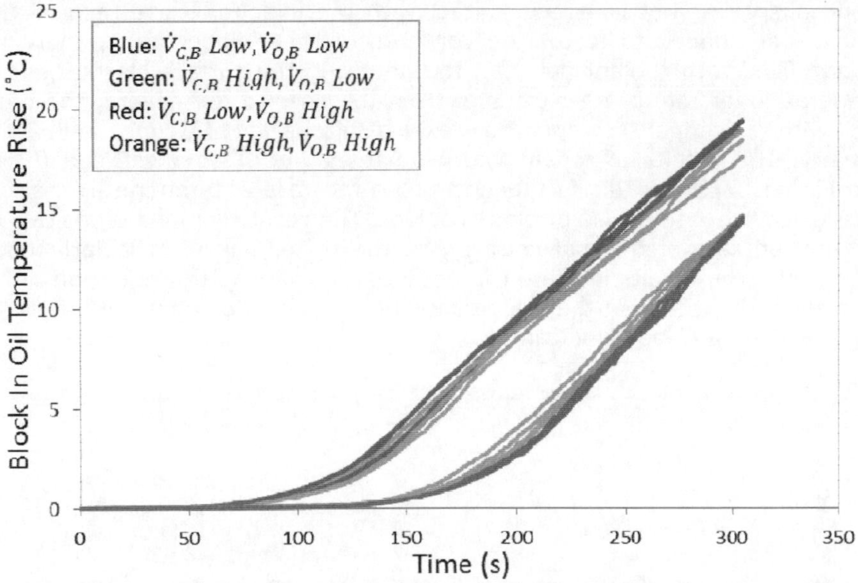

Figure 9: The comparison of oil temperatures in engine block

3 INFLUENCE OF RAISING OIL GALLERY TO HIGH POSITION

Rebuilding the test engine with the block oil circuit in the 'high' configuration increased the path length of the oil feed to the primary (main) gallery through the secondary gallery high in the coolant jacket. The anticipated effect of the increased length and exposure to higher temperatures higher in the engine block was an increase in the block oil warm up rate. Five of the test cases in the original matrix were repeated to examine this, namely with test settings 1,4,11, 13 and 16 in Table 2.

The variations of net heat transfer to coolant in the head and block circuits are shown in Figures 10. When the coolant flow setting is low, the change in oil gallery configuration has no measurable effect. In contrast, when the coolant flow is high, changing to the high gallery setting accelerates the fall in heat transfer to coolant in the head and slows the fall in heat transfer to coolant in the block in the later (>200s) stage of warm-up. This apparent switch of heat transfer from head to block indicates the head metal is slightly cooler and the block metal is slightly warmer, most likely due to restriction of the block coolant flow in the upper block posed by the high secondary oil gallery. The change in heat transfer rate is small and the effect on coolant temperature rise is negligible, as shown in Figure 11, for both low and high coolant flow rates. Raising the gallery does have a positive effect on the block oil temperature when the oil circulation rate is low, as also shown in Figure 11 by the corresponding measurements of oil temperature in the block circuit. When the oil circulation rate in the block is low, giving a longer residence time in the oil gallery, the high gallery configuration gives a higher rate of oil temperature rise.

In the round, changing the oil gallery configuration to the 'high' arrangement did have a positive impact on the block oil temperature rise for the low oil circulation rate without a discernible change in the rates of rise of coolant temperature. Although small changes in heat transfer to coolant were recorded when the coolant flow rates were high, the main influence on the oil temperature is the increase in gallery area and proximity to the higher liner temperatures in the upper block.

The high gallery was not in direct contact with the liner for these tests. Although the gallery was shaped to follow the contours of the cylinder bank, there was a ~1mm gap filled with coolant between the liner and the surface to the gallery. To encourage a higher rate of heat transfer from the liner to the gallery, the gap was plugged with a copper-impregnated resin (Belzona 1111) with high thermal conductivity. The results of repeat warm-up tests with Test 1 settings ('HHHH'), given in Figure 12, show this further increased the rate of temperature rise of the oil, and slowed the rate of rise of block coolant. The results show the positive effect on oil warm-up rate of thermal coupling to the upper liner, particularly between 100s and 200s. The results indicate the potential of coupling to the region of higher liner temperatures, but also the importance of increasing contact area if increases in warm up rate are to be significant.

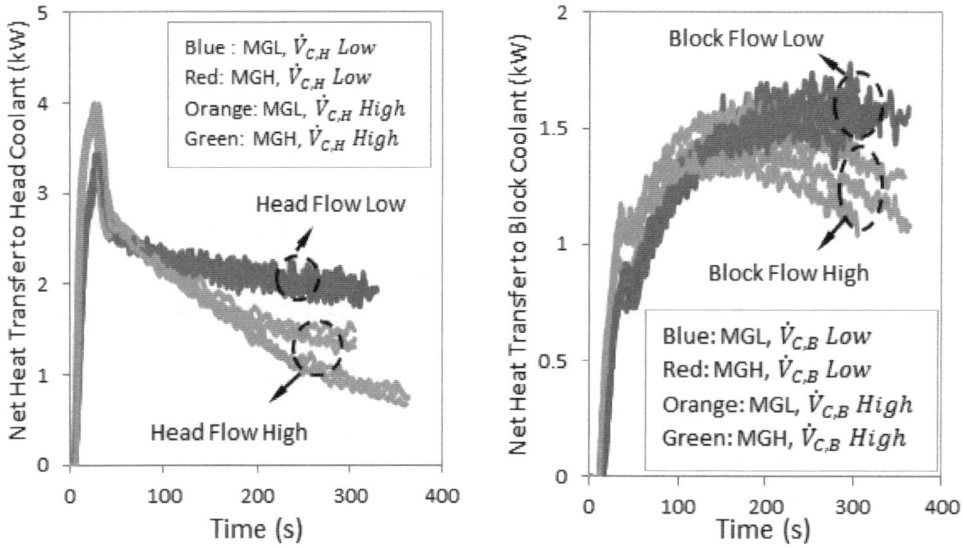

Figure 10: Comparisons of net heat transfer to head coolant (left) and block coolant (right) for Main Gallery low (MGL) and high (MGH) positions

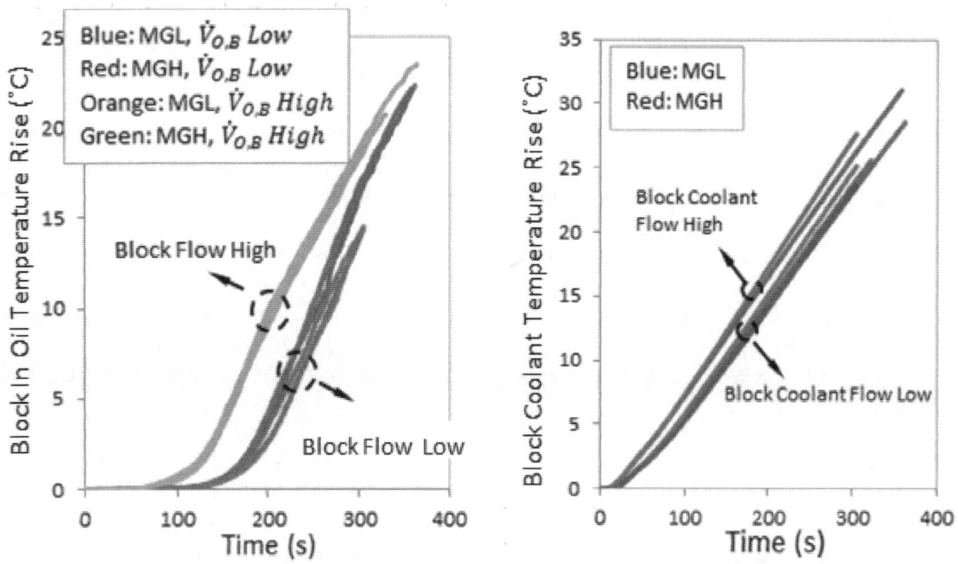

Figure 11: The influence of changing the main oil gallery from low (MGL) to high (MGH) setting on block oil temperature (left) and block coolant temperature (right)

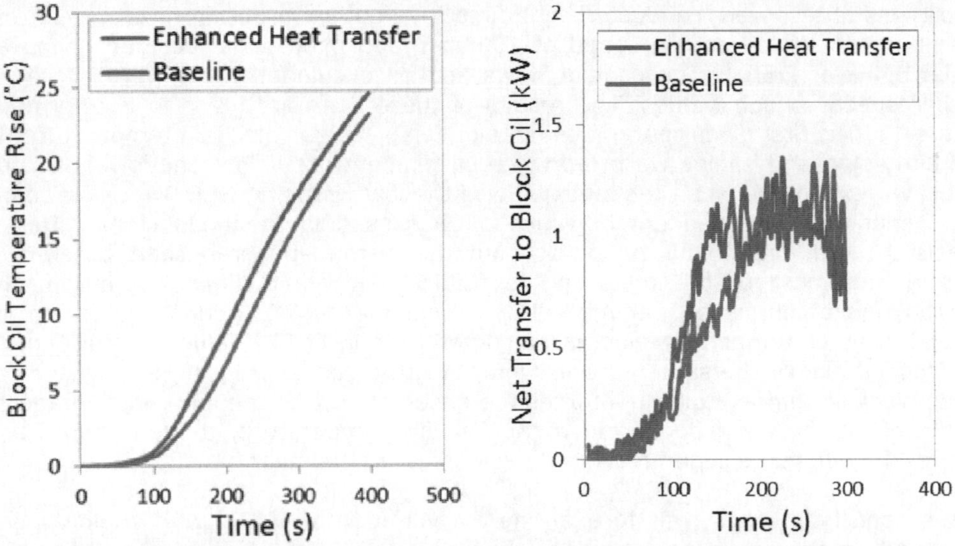

Figure 12: The influence of enhanced thermal coupling of the high oil gallery to cylinder liner on block oil temperature and net heat transfer to oil

4 LUBRICANT TESTS

A further small number of tests were undertaken to examine the influence of lubricant grade used in the head and block oil circuits. In these, an SAE 5w-30 grade oil and an SAE 0w-12 grade oil were used in combination, one in the head circuit and one in the block. The tests covered three of the four possible permutations of block/head oils: 5w-30/5w-30, 0w-12/5w-30, and 0w-12/0w-12. When changing the oil fill in the circuits, a flushing procedure was followed to minimise the risk of residual effects. The procedure included a periods of running fully-warm under mixed speed and load conditions to remove surface films laid down by the previous oil.

The principal aim of these tests was to determine the effect on engine friction. This was evaluated from differences between net indicated and brake mean effective pressures. IMEPn was computed from cylinder pressure measurements. Pressure transducers (Kistler 6052) were fitted to each of the four cylinders. BMEP was calculated from brake torque measurements. The difference yields the sum of rubbing friction and ancillaries load on the engine:

$$FMEP^{\#} = FMEP + AMEP = IMEP_n - BMEP \quad (4)$$

Where

$$IMEP_n = \oint PdV \quad (5)$$

Because the coolant and oil pumps were driven electrically by external supplies and there was no alternator fitted, the only ancillary driven by the engine was the high pressure fuel pump. The fuel pump work could reasonably be expected to be the same for similar tests, and so any change in $FMEP^{\#}$ can be associated with a change in rubbing friction only.

Two types of test were conducted. The first covered warm-ups from a cold started condition, at a fixed engine speed of 1500rpm and fixed pedal position giving an initial BMEP of 2.5bar. The coolant flows and oil circulation rates were all set to their respective 'high' values. The results of these tests are given in two forms in Figure 13. The first gives continuous plots of FMEP# versus block oil temperature. In the early stages of engine warm-up when oil temperature is low, the lowest friction levels were recorded when the block oil was the low viscosity, SAE 0w-12 grade oil. The benefit was independent of which oil was used in the head circuit. This is consistent with friction in the block from the piston and crankshaft assemblies contributing most to the total engine friction, and that these operate in the hydrodynamic lubrication regime with a dependence of friction losses on oil viscosity. As oil temperature increased, differences in FMEP# values diminished. In the second plot of the same data in Figure 13, the comparison of FMEP# values at three block oil temperatures reinforces the observation that the low winter grade oil in the block offers a reduction in friction at low temperature, though at 60C and above, the difference in summer grade values gave no similar benefit.

The second type of test was to examine the variation of FMEP# with engine speed under fully warm operating conditions. In these tests, BMEP was fixed at 2.5bar for all speeds, and the coolant and oil circulation rates were fixed at their high settings. The results are presented in Figure 14. The variation of friction with engine speed is similar for each oil combination, passing through a minimum at around 1500 rpm, and thereafter increasing with increasing engine speed. With regard to the influence of the oil grades, again only the oil in the block has a discernible effect. In this case, using the thinner grade oil (SAE 0w-12) in the block gave higher levels of friction. The likely reason for this is that the low summer grade of the oil gave insufficient lubrication to maintain operation in the hydrodynamic regime under fully warm conditions.

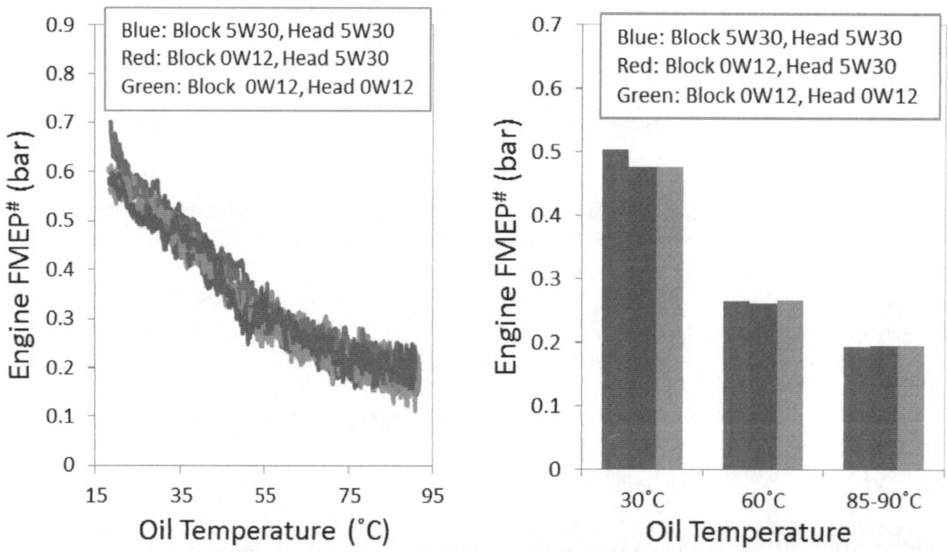

Figure 13: The comparison of engine friction FMEP# under different types of lubricant over fixed pedal warm-up

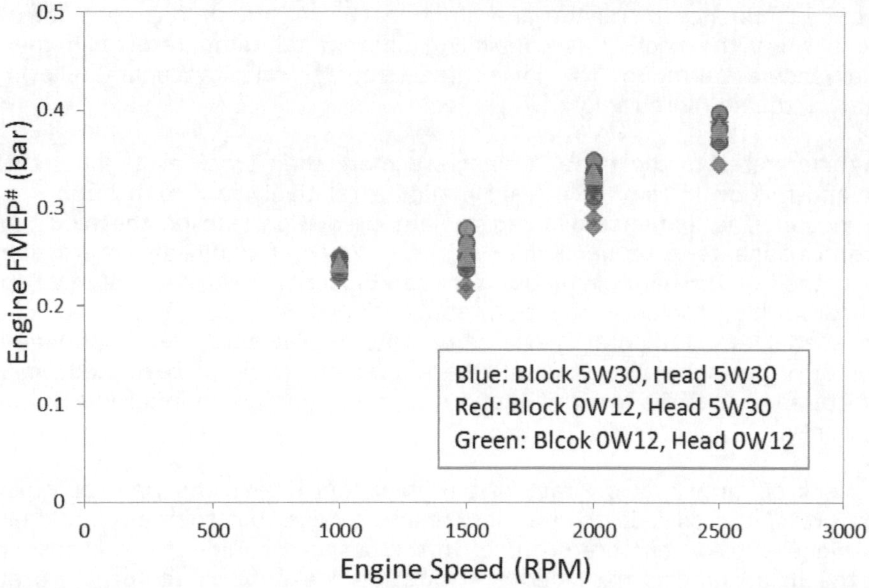

Figure 14: The comparison of FMEP# under fully warm conditions

5 DISCUSSION

The design and manufacture of the engine with its novel architecture and features posed challenges in commissioning and operation. The high mass and thermal inertia of the engine, the increased volumes of oil and coolant in their respective circuits, and the added surface area and thermal inertia of external components differentiate the thermal behaviour of this engine from the typical production engine, and the engine on which the bespoke design was built on. Heaters delivering a fixed heat input to oil in the head (1.25kW) and block (2.5kW) circuits were used to raise warm-up rates of oil to more normal levels, but this is an imperfect form of compensation for the additional thermal inertia of the engine. The effects are clearly evident in Figure 9, in the extension of the time delay before the block oil temperature begins to rise. This delay, associated with the recirculation of oil and mixing in the sump, is more than double observed for production versions of the engine under similar operating conditions. Nevertheless, the engine has provided valuable guidance and insight to factors influencing thermal behaviour which is new.

The use of separate circuits for head and block coolant has shown the dominant factor controlling heat transfer and rates of temperature rise to be the flow rate of coolant in the circuit. The influence of flow rate in the other coolant circuit, or oil flow in the corresponding head and block circuit was negligible. In the early phase of engine warm-up, in both head and block circuits, high coolant circulation rates give higher heat transfer to coolant and a higher coolant temperature. Later, the heat transfer rate drops below that for the low coolant circulation rate and the temperature advantage falls. The complex behaviour is attributable to time-dependent changes in the temperature differences between coolant and engine structure. If the principal aim of setting coolant flow to promote oil warm up, particularly in the early minute or two of engine running when oil viscosity falls most quickly with increasing oil temperature, then low net rates of heat transfer to coolant are desirable. This retains heat in the engine structure and promotes heat transfer to oil. It suggests a low coolant flow circulation rate is particularly desirable in the block during the early period of engine running. This might appear contrary to work showing that coolant temperature, rather than bulk oil temperature, has the stronger influence on piston-liner friction [6]. Although the case for steady conditions when the coolant is controlling liner metal temperature, in the early phase of engine warm up, the lower heat transfer rate to coolant allows liner temperature to rise more rapidly.

Heat transfer rates to the head coolant are more than twice as high as to block coolant and friction in the valve train makes a relatively low contribution to total engine friction. The influence of head coolant circulation rate on the block coolant temperature appears to be negligible. In the absence of significant interaction with the block, the head coolant circulation rate can be set to meet requirements specific to the head. A high coolant circulation rate promotes heat transfer to head coolant, limiting spatial and temporal metal temperature gradients and improving early heater matrix performance, although the effects on levels of regulated emissions might dictate the use of a low circulation rate if for example, minimising HC and CO was a priority.

In the block oil circuit, a low rate of oil circulation delays the time at which the temperature of the oil feed to the block starts to rise, but thereafter this gives a slightly higher rate of temperature rise. In the bespoke engine, for reasons already noted, the delay time is extended and dictates the bulk oil temperature at any given time after start-up. For more typical, shorter delay times, the temperature will reflect both influences suggesting pump delivery should be initially high to minimise the delay and subsequently low to maximise the rate of temperature rise. Generally, oil flow delivery is set to meet requirements for oil feed pressure, either or both for actuator performance or lubrication. In the case of the latter, there is evidence to show that reducing pressure can produce reductions in engine friction by other, none thermal mechanisms which add to any benefit from higher oil temperature [7].

The intent of rebuilding the engine with the block oil gallery in the 'high' position was to examine the possibility of increasing the oil warm up rate through closer thermal coupling to the upper liner. In the first implementation, the gallery was fully immersed in the coolant jacket with a small gap between the gallery and the liner. In warm-up tests, the change produced a small increase in the rate of oil temperature rise when the oil circulation rate was low. This effect was amplified by filling the gap with material with a high thermal conductivity. The results indicate the more direct coupling of the gallery to the liner raised heat transfer to the oil by up to ½kW early in the warm-up. Though the increase in heat transfer rate is relatively small, the results provide an indication of potential. The gallery design and coupling to liner was far from optimised. The benefit shown by the increased gallery length and improved thermal coupling between oil and cylinder liner suggests this is worth further investigation.

The approaches discussed above focus on achieving higher block oil temperature in order to reduce friction. The comparison of results of similar tests with 5w-30 and 0w-12 oils show the significance of block oil and its viscosity in reducing friction during engine warm up. The study reported shows progress on this can be made through thermal manipulation to raise oil temperature, through a combination of coolant and oil flow control and design revisions to increase thermal coupling between oil gallery and the upper liner area.

6 CONCLUSIONS

- The flow rate of coolant or oil in either a head or block circuit is the dominant influence on the rate of heat transfer to the fluid in the circuit. The influence of flow rate in the other three circuits is negligible.

- In the early phase of engine warm up, in both head and block coolant circuits, low coolant circulation rates give the lowest rates of net heat transfer. As warm up progresses, this relationship is inverted.
- A low net rate of heat transfer to block coolant during the early phase of warm-up is desirable and achieved with a low coolant flow rate.
- The block oil temperature at a given time reflects the influence of the delay before any temperature change is observed and then the subsequent rate of rise of oil temperature. The delay is shorter for high rates of oil circulation and the subsequent rate of temperature rise is higher for low rates of oil circulation.
- Positioning the oil gallery high in the block to take advantage of higher liner temperatures increases the rate of oil temperature rise early in the engine warm-up, most clearly when the oil circulation rate is low. Increasing the strength of the thermal coupling of the oil gallery to the liner increases the effect.
- To best promote the rise of block oil temperature, consistent with minimising friction during warm up, the coolant circulation rate in the block should be low. The block oil circulation rate should initially be high to minimise the delay before oil temperature starts to rise and then low to maximise the rate of temperature increase.

ACKNOWLEDGMENT

This work was carried out as part of the TSB IDP4 VIPER and Innovate IDP10 VIPER2 (Vehicle Integrated Powertrain Energy Recovery) projects with consortium partners Ford Motor Company, Jaguar Land Rover, and European Thermodynamics Ltd.

REFERENCES

[1] F. Will, A. Boretti, *A New Method to Warm Up Lubricating Oil to Improve the Fuel Efficiency during Cold Start*, SAE Paper 2011-01-0318, 2011

[2] P. J. Shayler, W. S. Baylis, M. *Murphy, Main Bearing Friction and Thermal Interaction during the Early Seconds of Cold Engine Operation*, Journal of Engineering for Gas Turbine and Power (DOI:10.115/1.1804538), 2005

[3] L. Avery, *National Transport Survey 2010*, UK Department of Transport 2011, https://www.gov.uk/government/publications/national-travel-survey-2010

[4] M. Andre, *In Actual Use Car Testing: 70,000 Kilometers and 10,000 Trips by 55 French Cars under Real Conditions*, SAE Technical Paper 910039, 1991, doi:10.4271/910039

[5] J. B. Heywood, *Internal Combustion Engine Fundamentals*, McGraw-Hill Book Company, 1988

[6] R Gardiner, C Zhao, J Addison, P J Shayler, *Investigations of Thermal State Changes during the Warm up of a Spark Ignition Engine,* Proc. of IMechE Int Conf C1365, VTMS 11, paper C028, Coventry, 2013

[7] P J Shayler, L Cheng, Q Li, E Wahab, *A modified oil lubrication system with flow control to reduce crankshaft bearing friction in a 1.5 litre 4 cylinder diesel engine*, SAE Paper 2016-01-1045, SAE Congress, Detroit, April 2016

ENGINE PERFORMANCE AND TEMPERATURE CONTROL

Fuel consumption assessments of using a hot water storage strategy on a diesel engine at different ambient temperatures

H Sara [1], D Chalet [1], M Cormerais [2], T Marimbordes [2], J-F Hétet [1]
[1] Ecole Centrale de Nantes, LHEEA Lab. (ECN/CNRS), France
[2] MANN+HUMMEL, France

ABSTRACT

Storing hot water from a previous engine run to the next start-up is one of the many thermal management strategies that are based on waste heat recovery. In this paper, numerical simulations have been performed on a four cylinder turbocharged Diesel engine GT-suite model using a 3 liters storage volume of water downstream the thermostat. Numerical simulations were run over a driving cycle. Different ambient temperatures were taken into consideration allowing a full review and evaluation of the hot water strategy. Coolant temperature profiles are presented for the different cases and fuel consumption reduction varied between 0.2% and 1.17% depending on the different conditions.

Keywords: Thermal management, Hot water storage, Coolant circuit, Internal combustion engine, Fuel consumption.

1. INTRODUCTION

Thermal management is one of the fields that proved its importance in minimizing fuel consumption and emissions. Thermal management's target is to let the engine operates at its optimal temperature as quickly as possible but with reliability respect. Engines retain their best performance while operating at their optimal temperature, however functioning in transient phase leads to an overconsumption.

As the statistics in Europe proved that the average length of travel in Europe is for 5-8 km (1) and the engine has a cold start-up. In addition, CO_2 emission are getting severe with the time as the target is set to 95 g CO_2/km in 2020/21 (2); thermal management strategies showed good results in the recent studies.

Different technologies are presented is studies as the electronically controlled smart thermostat (3), the modern multi-way valve (4). The latter is capable to apply a zero-flow strategy, a split cooling strategy and pushes the engine to operate at a higher optimal temperature. Another solution is to replace the mechanical pump by an electrical one to control the coolant flow (5) (6). Exhaust heat recovery showed an interest in improving the coolant and lubricant warm-up temperature (7), as well as the heat recovery from the EGR (8).

Energy storage is an efficient thermal management strategy. It is based on storing heat once the engine reaches its optimal temperature in order to use it on the following cold start-up. As a recent study of energy balance showed that during the permanent phase of the engine during the New European Driving Cycle (NEDC), around 95% of the heat dissipated to the coolant circuit is transferred to the ambient air at the radiator (9), highlighting the best moment to store the heat. Some studies focused on using Phase Change Materials (PCM) to store the heat (10)(11). Hot coolant storage was studied on a hybrid vehicle over different driving cycles without mentioning the ambient temperature (12), or at 5°C but without mentioning the driving conditions (13). It was tested also on a B class vehicle to pre-warm or warm the engine on start-up, leading to fuel consumption on a specific engine operating point due to improvement in combustion efficiency and a reduction in friction. However, no significant impact on the oil temperature rise was obtained (6).

In this study, a 3L hot coolant volume will be studied to assess the gain in fuel consumption. NEDC is chosen as the driving conditions. Ambient temperature will be varied between -7°C and 20°C. And other architecture of the strategy will be proposed at the end to study.

2. GT-SUITE MODEL

The work was carried on a turbocharged 4-cylinder Diesel engine which main characteristics are as follows: it's a compression ignition type with a compression ratio of 15.3, 4 valves per cylinder (4 cylinder in line), and the bore measures 76 mm while the stroke is 80.5 mm. The engine has a common rail fuel injection system, a high EGR loop, a turbocharger with a VGT and a charge air cooler.

Engine tests were held at Ecole Centrale de Nantes in order to calibrate the thermodynamic engine model at first then to calibrate the thermal model (coolant and lubricant circuit). Cold start-up test were performed every morning and registered to regulate the transient phase in the model. Pressures, temperatures and flow rates of the air, coolant and lubricant were registered all along the intake line, the exhaust line, the coolant circuit and the lubricant circuit. Tests were carried on different engine operating points. The engine torque varied with a 20 Nm step and the engine speed with 100 rpm step.

The engine thermodynamic high frequency model was done on GT-Power software. The combustion burn rate model, presented in GT-Power, covers the full operation mode. In order to better represent the process in the engine test bench, and as in NEDC, the engine functions mostly in part load, the combustion model of the engine was rectified based on different part load engine tests. The engine HF model covered all the air circuit. Starting from the compressor, where the air is pressurized and then sent to the charge air cooler to be cooled down, and before entering the cylinder, it is mixed with the EGR-HP. The intake line is modeled in a way to have the best filling efficiency, taking into consideration the unsteady characteristics of the pressure pulsations. In addition, the exhaust line and the EGR line were also modeled. In the engine model, wall heat losses are computed and then sent to the thermal model.

The thermal model consists of the coolant and lubricant circuits. Different parts of the coolant and lubricant circuits were modeled as the coolant pump, the thermostat, the oil pump, the oil tank, the expansion tank and different heat exchangers as the radiator, the EGR cooler and the oil cooler. An engine block of five masses was used from the GT-Suite library to calculate the different heat transfer between the engine masses and the fluid in the system. As the HF engine model provides the heat fluxes to warm-up the coolant and the lubricant, the thermal model computes the fluid and masses temperatures and send them as a feedback to the HF engine model in order to better compute the different losses

(*Figure 1*). At the same time, the lubricant temperature is used to correct the mean effective friction power, as seen in Shayler *et al.* (14) where the T_{ref} is fixed at 90°C and the exponent α is set to 0.35.

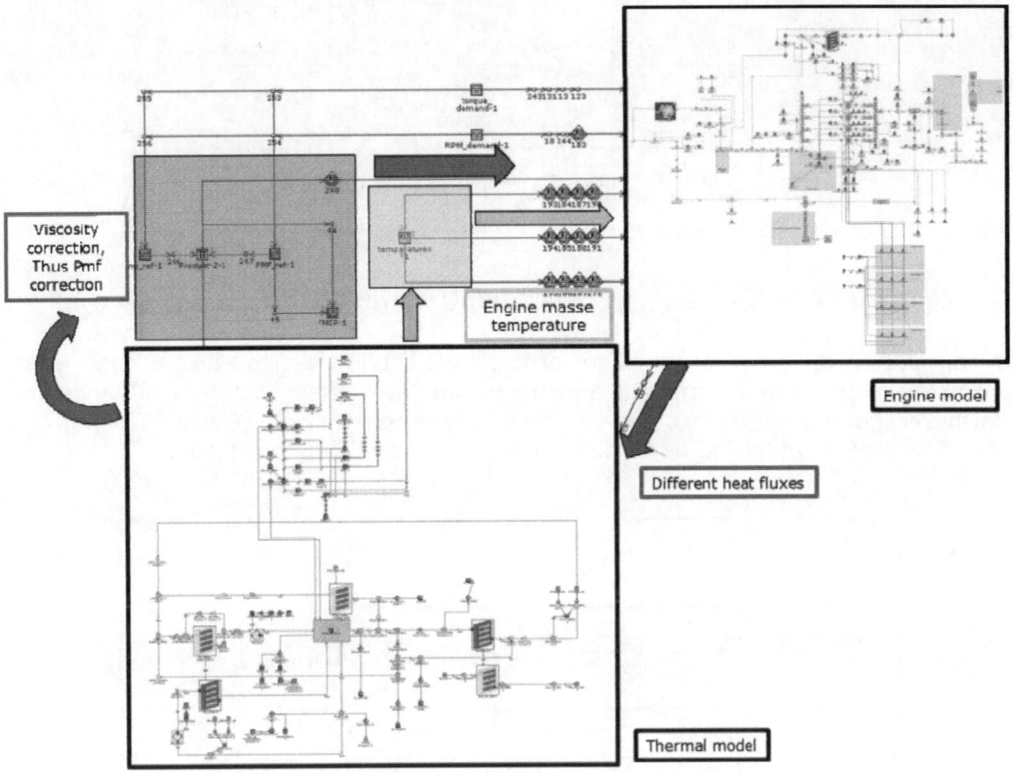

Figure 1 - GT-Suite Model

Engine tests were used to calibrate the thermal model, first every block was regulated, and then the hydraulic circuit was readjusted to have the good pressure losses and different mass flow rates in the branches. After that, the thermal model was first calibrated at static operating points then during transient phases. As shown in *Figure 2* , there is a good agreement between the simulation code and the engine tests for 1500 rpm and 50 Nm, with a maximum temperature difference of 4.8% during the transient phase. The model is considered validated and will be used to assess the hot coolant strategy. The vehicle used during the simulation is a medium passenger vehicle with a mass of 1430 kg.

3. HOT COOLANT STRATEGY

Studies showed that almost one third of the fuel power injected in the engine is transferred to the coolant circuit. This energy is used during the transient phase to warm-up the coolant, thus the engine. Once the engine's optimal temperature is reached, the thermostat opens the radiator branch, and the coolant evacuates the heat transferred from the engine to the ambient air at the radiator. To improve the engine thermal efficiency and the fuel consumption; energy is stored once the optimal temperature is reached. At the next start-up, the stored heat is released to heat-up the engine, leading to a faster warm-up, a minimized temperature difference between the combustion gases and the combustion chamber, thus minimized heat loss to the coolant. In addition, a better warmed-up engine leads to lower friction losses, because the lubricant temperature is increasing with engine's temperature.

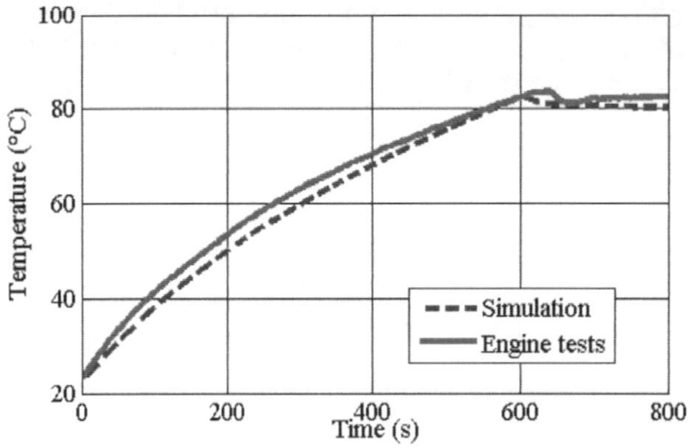

Figure 2 - Coolant Temperature Evolution for 1500 Rpm and 50 Nm

In this paper, the heat is stored in form of hot coolant. A storage tank is placed downstream the wax thermostat and upstream the radiator, to not influence the warm-up phase when the storage tank is at the ambient temperature. The storage tank flow rate is controlled by three simple valves as shown in the *Figure 3*.

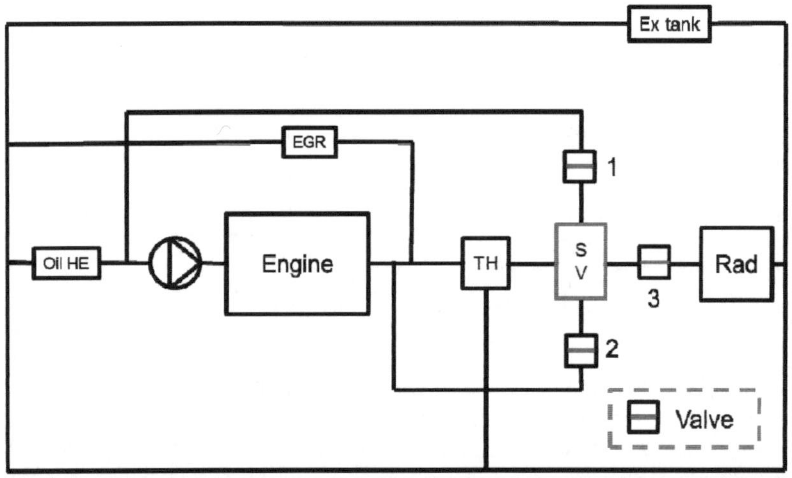

Figure 3 - Hot Coolant Storage Strategy (SV: Storage Volume; TH: Thermostat; Oil HE: Oil Heat Exchanger; Ex Tank: Expansion Tank; Rad: Radiator)

During the first cold start-up, through the transient phase, the thermostat blocks the radiator branch and the coolant circuit is considered as a traditional one, where the hot storage tank does not badly influence the warm-up phase and the coolant flows back to the engine through the by-pass branch or the EGR branch. Once the optimal temperature is reached, the thermostat opens the radiator branch, and the hot coolant flows to the storage tank, and the storage phase begins. The valve number 3 (*Figure 3*) is similar to the thermostat. It opens with when coolant's temperature is higher than the regulation set point and closes when the temperature is lower.

With the next cold start-up, the hot coolant stored will be used to warm-up the engine. At this point, the thermostat will be closed, thus the valve number 3. Valve 1 and 2 function as follows: **If** the temperature of the hot storage volume is greater than engine's temperature **and if** the thermostat is closed, Valve 1 and 2 opens and the coolant pump sucks hot coolant from the hot storage tank and let it circulates through the coolant circuit. Thus, valve 1 and 2 allow the hot coolant to be mixed with the coolant at the ambient temperature already in the system. If one of the two conditions is not met, the two valves (1 and 2) close and the engine acts as there is no storage tank in the system, and goes back to the traditional coolant circuit.

4. SIMULATIONS RESULTS AND DISCUSSIONS

The engine, the model and the thermal management strategy has been described, the results of applying the hot coolant strategy on the (NEDC) will be presented in this part of the paper. Simulation takes into consideration different ambient temperature as 20°C, 10°C, 0°C and -7°C. Results for the two extreme cases will be shown only. Fuel consumption gain will be assessed on these two cases.

At first the influence of the ambient temperature was tested on the traditional coolant circuit (without a storage volume). As seen in the *Figure 4*, the ambient temperature of 20°C representing a normal warm European day is plotted against 10°C, 0°C and -7°C, the latter presenting an extreme cold winter day. The ambient temperature showed a big influence on the coolant temperature, thus the engine temperature. For an ambient temperature of 20°C, the temperature of 60°C of the coolant is reached at 607s, for the ambient temperature of 10°C at 741s, for 0°C at 853s and for -7°C at 960s of the NEDC. An advance of 353s is obtained with a 20°C compared to -7°C as ambient temperature. As for the optimal temperature, the ambient temperature of 20°C showed an advance of 84 seconds in comparison with the coldest temperature of -7°C.

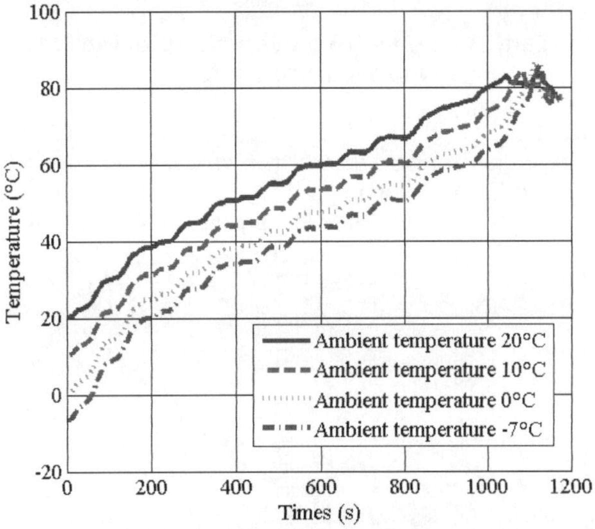

Figure 4 - Coolant Temperature Evolution during NEDC for Different Ambient Temperature

At the coldest temperature, the engine generates more friction power than at 20°C, because of the lowest temperature of the lubricant thus higher viscosity. With lesser ambient temperature, there is a greater temperature difference between the combustion gases and the chamber walls, so greater heat losses to the coolant circuit. Hence, an overconsumption of 19.2% is obtained with the ambient temperature of -7°C compared to 20°C. From now on, studies will be shown only for the two extreme cases of the ambient temperature: 20°C and -7°C.

3L volume of the storage tank was chosen after a study between different storage tank's volumes of: 1L, 2L, 3L, 4L and 5L. The lower volumes didn't show any interest in term of engine warm-up and fuel consumption gain, and the greater volumes were considered too much because it is almost equal to the engine's coolant volume or greater. Furthermore, a study of the storage tank temperature during the soak period of 15 hours, with a light weighted plastic material with a 3mm thickness, showed for a 3L storage tank, a drop of 13°C. The ambient temperature was set to 20°C, the external convection coefficient to 20 W/m².K and the initial temperature of the storage tank to 13°C, the engine's optimal temperature.

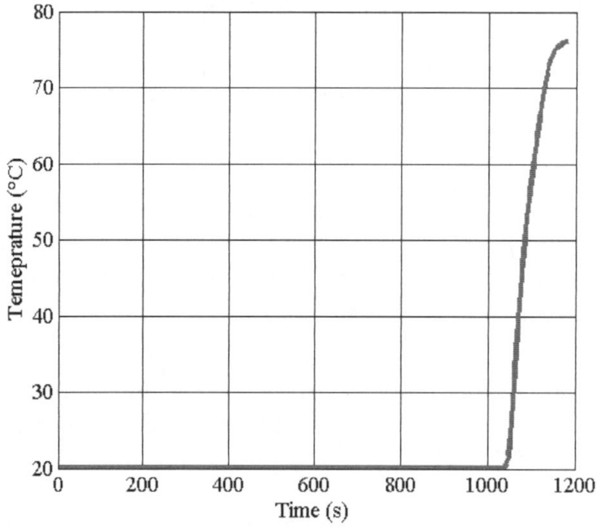

Figure 5 - Storage Tank Temperature Evolution during NEDC at an Ambient Temperature of 20°C

Figure 5 shows that the storage volume of 3L after the NEDC cycle reaches 76°C. To decide the optimal initial temperature of the storage tank for further simulations, a comparison between different starting storage tank temperatures is made and the evolution of the engine coolant temperature is plotted in *Figure 6*.

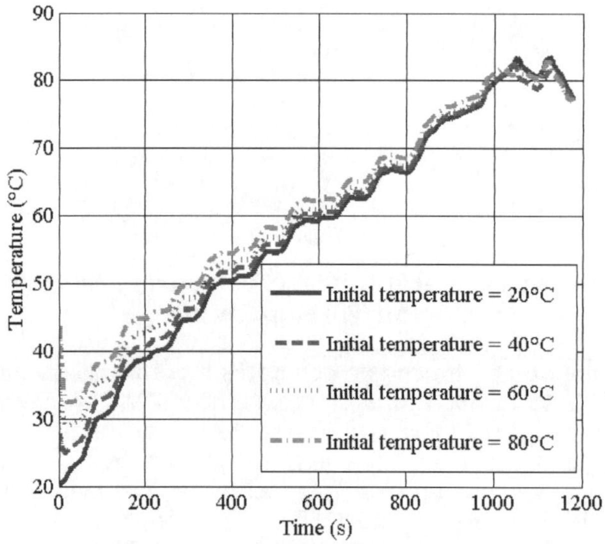

Figure 6 - Evolution of Engine's Coolant Temperature for a Different Initial Storage Tank's Temperatures

In fact, the tank's initial temperature depends on the previous driving cycle and the possibility to reach the engine's regulation temperature or not. To simplify the present work, it is assumed that the energy stored during the regulation of the engine is equal to that within the regulation temperature (83°C for the studied engine). It is shown that the greatest influence is within the greatest initial temperature of 80°C. However, this temperature is considered as a theoretical temperature as it seems difficult to maintain 80°C of storage tank temperature if the optimal temperature, thus the maximum temperature of the storage tank is set to be 83°C. However, on one hand, an initial temperature of 40°C is considered too low as it indicates a bad isolated tank after a soak period of 10-15 hours. On the other hand, simulations carried on with 40°C as an initial storage tank temperature didn't show much improvement in comparison to 20°C (*Figure 6* - the initial temperature of 20°C presents the reference line where the storage tank is at the ambient temperature and not heated-). Hence, 60°C as initial storage tank temperature is considered as the optimal starting temperature where it takes into a consideration some heat loss during the soaking period and as seen in the plot, it promises a good improvement in the coolant temperature evolution during the early period of the transient phase.

The study will be carried now on a 3L of hot coolant storage tank at an initial temperature of 60°C for two different ambient temperatures of 20°C and -7°C.

Starting with an ambient temperature of 20°C, the results of the evolution of coolant and lubricant temperatures are plotted in Figure 7 against the vehicle velocity of the NEDC. The "ref" line in Figure 7 refers to the traditional coolant circuit without the storage tank taken into consideration.

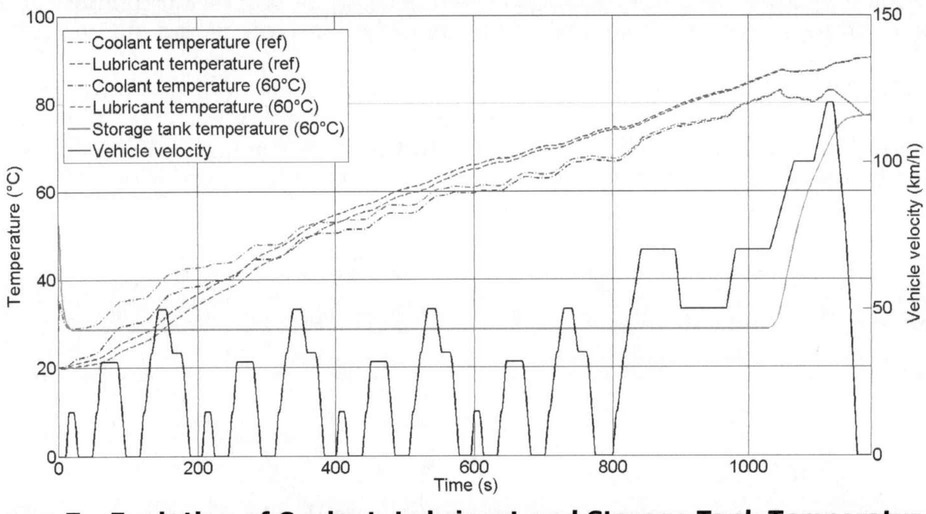

Figure 7 - Evolution of Coolant, Lubricant and Storage Tank Temperatures over NEDC for an Initial Temperature of Storage Tank at 60°C at an Ambient Temperature of 20°C

With the thermal management strategy applied on the engine, the coolant temperature profile shows an improvement in the early stages of the driving cycle. It starts with an advance of 150s at 30°C, reached the temperature of 60°C with an advance of 64s, to attend the optimal temperature at almost the same time. This is due to the fact that with the thermal management strategy of hot coolant storage applied on the engine, a higher temperature difference between the coolant and the lubricant is observed leading to a greater heat transfer between the two fluids. This is seen in the lubricant temperature profile where there is no contact between the hot coolant made in circulation and the lubricant except at the oil heat exchanger where the lubricant at the early stages is heated by the coolant. In addition to that with a higher coolant temperature than at the reference case, the temperature difference between the coolant and the chamber walls becomes lesser leading to a lesser heat transfer to the coolant, thus lesser heat source to warm-up the engine. Hence, the two temperature profiles for the reference case and with applying the hot coolant storage thermal management strategy should not stay parallel through NEDC. Looking into the lubricant temperature profile, reaching 60°C with the hot coolant storage strategy is achieved in advance of 34s; where in this hot coolant storage configuration, the hot coolant tank output is downstream the oil heat exchanger. To have a better influence on the oil temperature, the configuration of the hot coolant storage should be modified in a way to have the output upstream the oil heat exchanger to have a direct influence on the oil. The hot coolant storage strategy at 20°C showed a 0.26% improvement in fuel consumption compared to the tradition coolant architecture where the hot coolant storage tank does not exist.

The storage tank temperature profile over NEDC shows that the discharge of the hot coolant through the first stages of the NEDC is fast where the valve 1 and 2 of the *Figure 3* are opened, and the hot coolant flows through the coolant circuit. The thermostat is still shutting down the radiator branch, thus valve 3 is closed too. Then as seen in the plots, the valves close after 30s where the storage tank temperature becomes lower than the engine's. Once the optimal temperature is reached at the end of the cycle around (1029s), the thermostat opens the radiator branch, the valve 3 opens simultaneously (having valve 1 and 2 always closed) and the hot coolant flows through the storage tank, and the latter's temperature starts to rise. At the end of the NEDC the storage tank is recharged again with hot coolant and rises to 76°C, ready to be used in the upcoming cold start-up.

The evolution of the coolant and lubricant temperatures for applying the hot coolant storage strategy with an ambient temperature of -7°C are presented in Figure 8.

The NEDC cycle is not shown entirely in the plots of Figure 8 in order to highlight the influence of the applied thermal management strategy on the temperature profile. For the coolant it is shown from the start of the cycle until 800s, and for the lubricant a segment between 200s and 800s is presented to show the difference between the two plots.

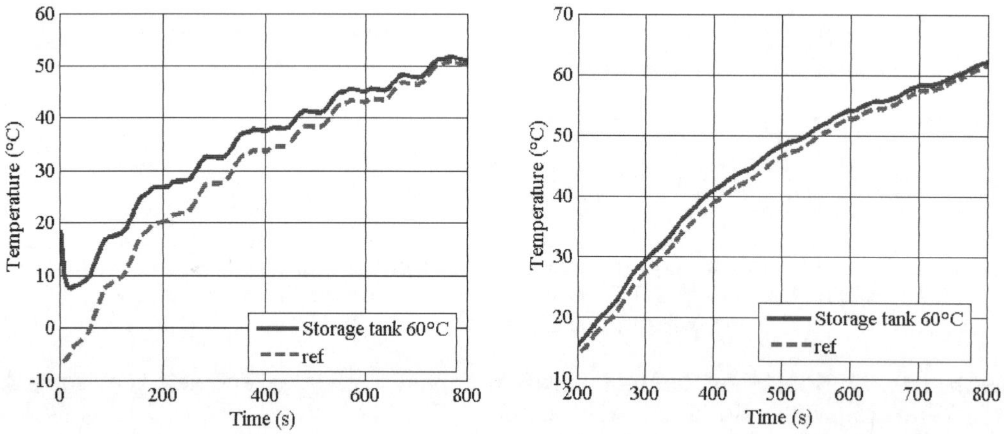

(a) (b)

Figure 8 - Evolution of the Coolant Temperature (A) and Lubricant Temperature (B) through NEDC while Applying the Hot Coolant Strategy with an Ambient Temperature of -7°C

The coolant temperature, with the strategy applied, starts with a temperature greater than 0°C, once the hot coolant is pumped through the system. The 20°C of the coolant is reached with an advance of 60s compared to the temperature of the reference, without having the storage tank existing in the system. Similar to the coolant temperature profile for 20°C, the two plots don't stay parallel. The zoomed plot of the lubricant highlights the temperature difference between the two cases during the transient phase, where the 40°C is reached faster by 27s. The lubricant temperature profile showed a greater temperature profile until around 800s where the lines become closer. A higher lubricant temperature leads to lower mean effective friction power, thus higher engine efficiency. At an ambient temperature of -7°C, the hot coolant strategy reduced the fuel consumption by 1.17%.

At first, the mass of the additional components of the strategy was estimated to 4kg to analyze their impact on the engine, considering that the three valves of the system can be replaced by a multi-way valve. The additional mass has no influence on the engine from a thermal point of view. The evolution of the temperature profile did not show any significant drawback. However the fuel consumption gain was reduced to 0.15%.

The potential of the hot coolant strategy was proved upon the NEDC cycle with an initial temperature of 60°C, taking into consideration 16°C of temperature loss during the soaking period. The further studies should be focused on studying the structure of the hot coolant storage tank to have the best light weighted material, with a low thermal conductivity coefficient, the best insulation method in order to have the lesser heat transfer to the ambient and a higher starting temperature, hence improving the fuel consumption gain.

5. CONCLUSION

A new thermal management product consisting of 3L of hot water tank starting at an ambient temperature of 60°C showed its potential in reducing the engine's fuel consumption.

At first in this study, a HF GT-Suite model of a 4-cylinder turbocharged Diesel engine coupled to the coolant and lubricant circuits is presented and validated over engine tests. A hot coolant storage strategy was proposed in a particular configuration to assess the fuel consumption gain. This thermal management strategy proved its capability to reduce the engine's fuel consumption by 0.27% at an ambient temperature of 20°C and 1.17% at an ambient temperature of -7°C.

The hot coolant strategy in the architecture presented in this work has a direct influence on the coolant and indirect one on the lubricant where the latter temperature evolution did not show a big difference to the reference case compared to that shown in the coolant temperature profile. In order to study the direct influence of the hot coolant storage strategy on the lubricant, other architecture is proposed, where the outcome of the hot coolant tank is upstream the oil heat exchanger, having the hot coolant flowing first through the exchanger before entering the engine. This architecture will be analyzed in further studies. In addition to that, to improve the fuel consumption gain, a study of the heat loss of the hot coolant tank to the ambient air should be done to have the better insulation of the tank, thus lesser temperature transfer to the ambient, thus higher initial storage temperature.

Compared to the engine encapsulation where the whole engine structure is used to store thermal energy, the hot coolant storage strategy does not show the drawback of the necessity of upsizing the radiator to protect the engine from thermal stresses during the steady state phase. However, the availability of space under-hood puts a constraint towards the proposed thermal management strategy in this work.

ACKNOWLEDGEMENTS

The work in this article is done in a joined International Teaching and Research Chair entitled "Innovative Intake and Thermo-management Systems" between MANN+HUMMEL and Ecole Centrale de Nantes.

REFERENCE LIST

(1) C. Rouaud, D. Gerard, D. Issartel, B. Pfeffer and S. Grolet. Thermal management solutions for Diesel engine. *SIA Publications*, 2008.

(2) European Parliament, Council of the European Union. Regulation (EU) No 333/2014 of the European Parliament and of the Council of 11 March 2014 amending Regulation (EC) No 443/2009 to define the modalities for reaching the 2020 target to reduce CO2 emissions from new passenger cars. *OJ L,* vol. 103, pp. 15-21, 5.4.2014. http://data.europa.eu/eli/reg/2014/333/oj

(3) R. Cipollone and C. Villante. A fully transient model for advanced engine thermal management. SAE Technical Paper 2005-01-2059, 2005, doi:10.4271/2005-01-2059.

(4) M. Cormerais, T. Marimbordes, S. Warnery, D. Chalet, H. Mezher and L. Roussel. ACT valve: active cooling thermomanagement valve. SAE Technical Paper 2014-01-0632, 2014, doi:10.4271/2014-01-0632.

(5) D. Jung, J. Yong, H. Choi, H. Song and K. Min. Analysis of engine temperature and energy flow in diesel engine using engine thermal management. J Mech Sci Technol (2013) 27: 583. doi:10.1007/s12206-012-1235-4.

(6) E. Bent, P. Shayler, A. La Rocca, and C. Rouaud, "The effectiveness of stop-start and thermal management measures to improve fuel economy," *Vehicle Thermal Management Systems Conference and Exhibition (VTMS11)*. pp. 27–39, 2013.

(7) E. Barrieu, N. Becker and A. Rossi. Engine Warm-up optimization using innovative components. *SIA Publications*, R-2013-06-06, 2013.

(8) R. D. Burke, C. J. Brace, A. Lewis, A. Cox and I. Pegg. Analysis of energy flows in engine coolant, structure and lubrication during warm-up. *Vehicle Thermal Management Systems Conference and Exhibition, VTMS 10*, Warwickshire, 2011, pp. 167-176.

(9) H. Sara, D. Chalet, M. Cormerais et J.-F. Hétet. A Detailed Study Of An Energy Balance Using Numerical Simumation On A Diesel Engine During Transient And Permanent Phase. *Fisita 2016 World Automtoive Congress - F2016-ESYE-002*, Busan, 2016.

(10) J. Vetrovec. Engine Cooling System with a Heat Load Averaging Capability. SAE Technical Paper 2008-01-1168, 2008, doi:10.4271/2008-01-1168.

(11) M. Gumus. Reducing cold-start emission from internal combustion engines by means of thermal energy storage system. *Applied Thermal Engineering,* vol. 29, pp. 652-660, 2009. http://dx.doi.org/10.1016/j.applthermaleng.2008.03.044

(12) P. Revereault, C. Rouaud and A. Marchl. Fuel economy and cabin heating improvements thanks to thermal management solutions installed in a Diesel hybrid electric vehicle. SAE Technical Paper 2010-01-0800, 2010, doi:10.4271/2010-01-0800.

(13) Y. Kuze, H. Kobayashi, H. Ichinose and T. Otsuka. Development of new generation hybrid system (THS II)- Development of Toyota coolant heat storage system. SAE Technical Paper 2004-01-0643, 2004, doi:10.4271/2004-01-0643.

(14) P. Shayler, S. Christian and T. Ma. A model for the Investigation of temperature, heat flow and friction characteristics during engine warm-up. SAE Technical Paper 931153, 1993, doi:10.4271/931153.

Prediction of transient charge cooling behaviour and the effect on engine performance

R Owen [1], A Price [2], E Tate [3], V Staelens [4], Z Yang [3]
[1] Bentley Motors Ltd, Thermal Systems, UK
[2] (Former) Bentley Motors Ltd, Thermal Systems, UK
[3] Exa Corporation, USA
[4] Exa Corporation, UK

ABSTRACT

3D CFD combined with 1D system modelling can accurately predict cooling medium temperatures during simple drive cycles for vehicles with single air paths. During VTMS 12, a method was shown to extend this to complex drive cycles with non-uniform flow at high ambient temperatures and vehicles with multiple air paths.

In this work for VTMS 13, the method is developed further to overcome the challenge of transient prediction of air-to-air intercoolers due to the non-uniform external air flow, rapid change in cooling medium conditions and the thermal inertia of the intercooler itself. This is done by using a full vehicle 3D CFD simulation at multiple stabilised operating points to characterise the non-uniform airflow boundary conditions at the entrance and exit of the charge air coolers, which can then be applied to the 1D system model.

This paper describes the virtual process taken to accurately predict transient charge air cooler performance, evaluates the pros and cons of the method fidelity, operating point selection, use in 1D engine modelling and application to a vehicle development workflow.

GLOSSARY

ATF – Automatic Transmission Fluid
CAC(s) – Charge Air Cooler(s)
CAD – Computer Aided Design
CAE – Computer Aided Engineering
ECU – Engine Control Unit
LHS – Left Hand Side
NVH – Noise, Vibration and Harshness
PowerFLOW – Exa 3D CFD Simulation Software
RHS – Right Hand Side
RMSE – Route Mean Square Error
Vmax – Maximum vehicle velocity
VTMS 12 – Vehicle Thermal Management Systems Conference 2014

1 INTRODUCTION

The under bonnet heat generated from both the powertrain and the HVAC system is rejected via the vehicle cooling pack to the ambient air. This is done by air passing through the cooling pack, which is either driven by a fan or by vehicle movement.

During the development of a vehicle, the air inlet aperture size needs to be considered carefully; too small and the airflow will be insufficient to meet the cooling requirements, too large and it will generate unnecessary drag and influence on the vehicle styling. CFD simulation has been proven as an accurate and reliable tool to simulate the airflow allowing heat reject calculation of a single path cooling pack through to a multipath cooling pack on a high performance vehicle.

By using CFD simulations to understand and define the aperture size to meet an airflow requirement, a vehicle project can be successfully front loaded during the development stage to give a highly accurate status of the cooling package performance, to within an average 1.5°C to physical testing (1), before any prototype vehicles have been built. This offers the company a faster development time, optimized aperture size and a reduced cost; the cost saving comes from a reduction in physical testing and from a series production saving by avoiding fitting parts with a specification higher or lower than required.

For a customer the benefits of a well CFD optimised aperture and cooling package will be provided by a more fuel efficient vehicle due to reduced cooling drag and minimised weight thanks to appropriately sized heat exchangers. Additional customer benefits come from low NVH from a well sized cooling fan and overall a vehicle that can perform in extreme ambient conditions and harsh drive cycles without triggering obtrusive engine or driveline torque reductions.

Developing simulation tools to allow for this front loading and reduced development time brings many benefits, but in order for these tools to be successfully implemented the level of accuracy has to be consistently high and the results trusted by the users. This paper will focus on the method to improve and validate the simulation of a cooling package's performance by addressing the transient simulation challenges associated with air-to-air charge coolers (CACs) due to the highly dynamic operating conditions and non-uniform external and internal flow they are exposed to during a high load driving cycle. This test cycle used here, is the most aggressive transient test for the charge air coolers cooling performance.

1.1 Progression from Previous Works

The procedure for developing a new powertrain includes completing customer representative driving cycles on a track or in a climatic wind tunnel. These cycles are well defined to reproduce known driving conditions and deliver repeatable and reliable thermal data for both development and sign off criteria. However, their use is limited until the mid to late stages of a project as a physical vehicle is required. To overcome this, simulation has been used to reproduce these drive cycles in a virtual environment allowing the data they generate to be received much earlier in the project to inform layout and component selection prior to design freeze. This is not a new approach, but the accuracy of the data and complexity of the cycles has progressed rapidly from simple cycles with single cooling packs (2) to hot lap simulations with multipath cooling packs that can generate cooling pack data to a high accuracy (1) in a short period of time.

Through the use of these high load transient simulations, it is possible to gather data at a component level before a single vehicle has been made, and when a physical vehicle test is conducted its data can be fed in to the simulation model to validate it and improve the accuracy for future simulation. This concept has been proven during the design and launch of the Bentley Continental GT W12 (1) where the cooling package performance was accurately simulated for a range of operating conditions; however some limitations to the simulation were found. The main limitation was the transient performance of the CACs, which while still successfully modelled, showed a notable delta between the simulation and physical testing (1). Upon further investigation, this delta is believed to be down to the highly dynamic air flow behaviour the CACs are exposed to during high load and high speed drive cycles – during which both the internal and external mass flow can change rapidly. A suggestion from VTMS 12 on how to improve this in future work was to look carefully at how the internal airflow is simulated, and the impact of any thermal inertia the parts may have. Although the parts themselves are subject to rapid heating and cooling, it is likely the temperature is dampened somewhat due to the mass of the parts (3) and could contribute to the outlet temperatures.

To overcome these issues, and to improve the accuracy further, the flow characterization and simulation will be taken a level deeper, with each single tube inside the CACs looked at, and the individual tube performance included in the simulation. This will allow for the internal flow distribution to be mapped against the well-defined external flow distribution. To understand the effect of the thermal inertia on the simulation accuracy, physical measurements will be included as a time trace during the validation of the model and compared against the simulated temperatures, which will give an insight in to the effects and accuracy of the component temperature as well as the charge air temperatures output by the simulation.

With accurately simulated charge air outlet temperatures, the engine performance can be accurately simulated early in the development process by using the correlated CAC model as an input source for engine intake air temperatures, rather than waiting till physical data is available from dynamometer testing. This will allow for the powertrain components to be efficiently developed toward the project's performance and power targets. For example, when sizing a turbocharger the performance of the charge air cooler has a direct influence. By having an accurate simulation method, the turbocharger can be sized with confidence in the early stages of the project. Due to the high investment cost in such a part, late changes are unlikely to be feasible, so any errors from the simulation could result in performance compromises. The smaller the error in the simulated data, the closer they will be to the physical values once a test vehicle has been built, which will result in fewer development loops and cost to reach the optimum design.

2 METHOD

The process for completing a cooling package simulation for a drive cycle can be shown in four stages as per the flow chart in Figure 1. The detailed steps for each of the stages are given in Section 3, and the output produced is the prediction of the cooling package temperatures. As the focus of this work is on the CACs, the simulation will focus on the accuracy of the charge air outlet temperature. Where the method in the paper differs (1) is the inclusion of the internal CAC performance data from a tube level characterisation to ensure that the behaviour both outside and inside the CAC is included. This is done by completing 3D CFD simulations of the airflow behaviour prior to 1D engine simulation. The 3D airflow behaviour is used as an input for the 1D model so that the simulated temperature data is as close to the real world as possible.

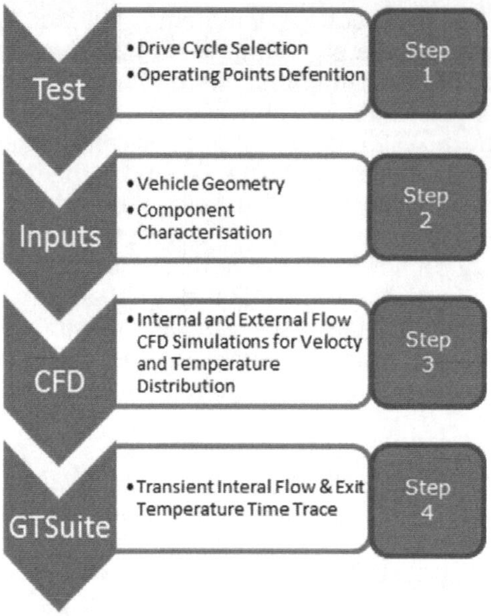

Figure 1 - Process Flow Chart

From physical testing it is known that the external air temperature distribution across the CACs is non-uniform, this has been recorded by fitting a map of five paired thermocouples on the front and rear faces of the Bentayga V8 CACs during a test cycle. The data recorded from these thermocouples, plotted in Figure 2, shows that the temperature delta between the external air entering the front face of the CAC and the air exiting from the rear face is not consistent between the five measurement pairs, it is also worth noting that the thermocouple pairs at the same height show different deltas across the face, for example pair 1 vs. pair 2, which indicated the internal behaviour and heat reject is non-uniform. If this difference in internal behaviour can be reproduced in the simulation, it is thought that the accuracy of the air outlet temperature will improve. Including the external airflow distribution from 3D CFD in 1D engine modelling improved the accuracy in VTMS 12 (1), therefore it is reasonable to predict that capturing the same level of detail on the internal flow, and using it with external flow, will improve the accuracy of the simulation further.

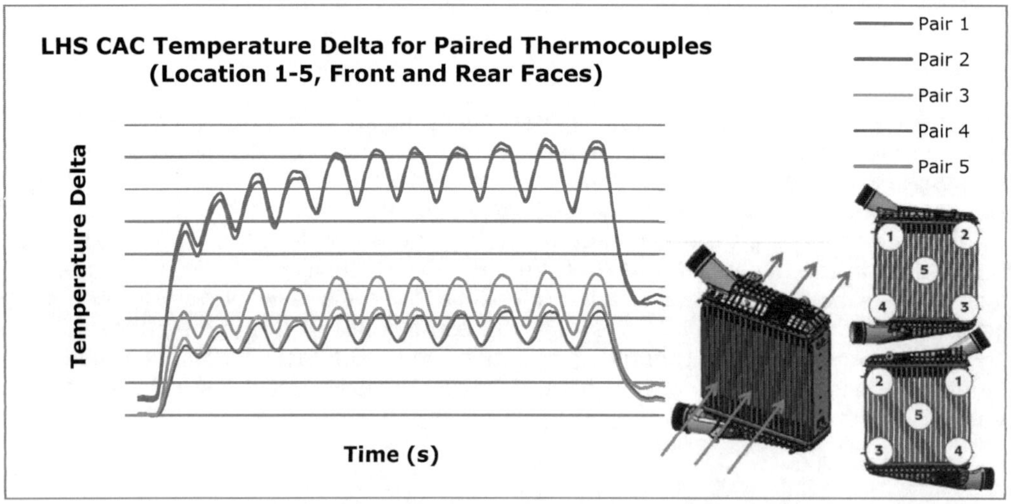

Figure 2 - Thermocouple Pairs Temperature Deltas

Before embarking on the bulk of the simulation, an initial round of CFD simulations were carried out to understand the internal air flow behaviour and where effort should be focused in the pursuit of accuracy. One area that was hypothesised as an impact on the accuracy was the modelling of the CAC core. For this we considered two methods, one with a porous media and the second with individual tubes, both replicating the overall core size and pressure drop, shown in per Figure 3, it can be seen that a large amount of information is lost in the porous media method. The initial condition of the air entering the core of the CAC is similar between both approaches but when looking at the flow in the core the similarity ends, as the flow distribution across the core is lost. Both methods can correctly replicate a CAC or any heat exchange in terms of pressure drop, but it is the tube level method that is predicated to be more accurate for predicting temperatures when combined with the non-uniform planar flow across the face of the core as it will allow for the impact of effective vs. non effective areas of the core to be captured. For example, an area of the porous media core could be showing higher internal flow than experienced in the vehicle, and if this area is coupled to a high external flow from the 3D CFD the simulation could over predict the heat transfer that is occurring.

Once validated, it will be possible to use this method to simulate drive cycle, allowing for quick assessments of component changes, power output increase or airflow changes due to styling, and component selection on future projects.

With accurately simulated charge air outlet temperatures, the engine performance can be accurately simulated early in the development process, rather than waiting till physical data is available from dynamometer testing. This will allow for the powertrain components to be efficiently developed toward the project's performance and power targets. The smaller the error in the simulated data, the closer they will be to the physical values once a test vehicle has been built, which will result in fewer development loops to reach the optimum design.

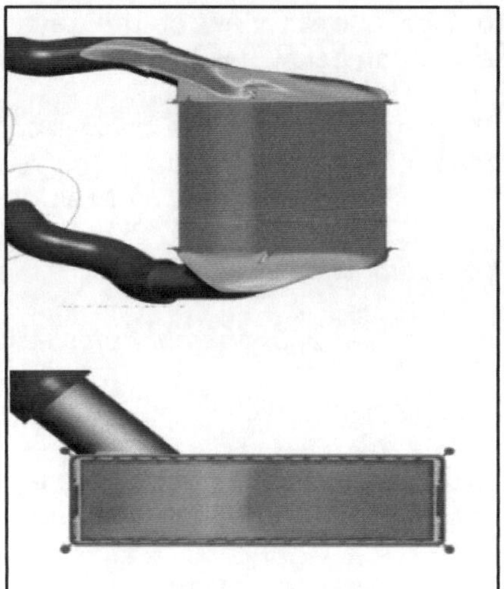

Porous Media Flow

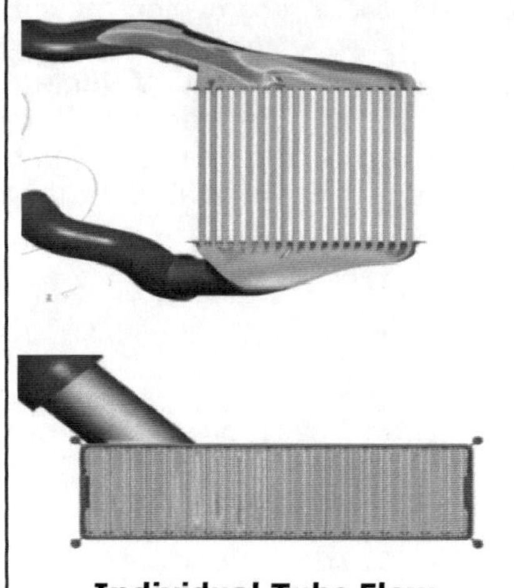

Individual Tube Flow

Figure 3 - Porous Media vs. Individual Tubes

3 INPUT SOURCES

To support the four stages of the simulation method, the inputs shown in the process flow chart, Figure 1, are required. The more accurate this input data, the more accurate the simulation results will be. Therefore the source for each input was carefully considered, explained below.

3.1 Step 1 – Test Profile

The first step in the process is to select the test cycle to simulate and at what point in time, which is used to form the boundary conditions of the simulation. As the focus of this work is the transient performance of the CACs.

A test was chosen where the CACs performance will be pushed to the limits, for this the 'Transient CAC System Performance' drive cycle was selected from the Bentley test specification list. This is used to confirm the CAC performance during a repeated high load cycle where the internal flow is rapidly changing due to a mix of acceleration and braking actions, and the external flow is also changing rapidly due to a repeated changes in vehicle velocity.

From the drive cycle, a series of operating points to reproduce during the simulation needs to be selected. This is done to avoid conducting a CFD simulation for every operating point during the drive cycle which would be both costly and time consuming. From previous studies, (1) & (2), it has been shown that a high level of accuracy can be obtained by selecting operating points that are frequently visited during the test cycle and interpolating between them. This greatly reduces the number of CFD runs required and reduces the overall time and cost of the process. The method for selecting the most appropriate operating points is covered in Section 4.

3.1.1 Transient CAC System Performance Drive Cycle

To give a clear view of the test cycle to be reproduced in the simulation, the description below has been extracted from test specification (4):

"This test examines the charge cooling performance of the vehicle. It provides a standardised drive cycle for the recording of transient vehicle temperatures in hot climates. The target ambient temperature for test is 25ºC.

Preconditioning;

Drive in D at 200kph until fluids have stabilised.

Measurement drive;

With Sport gear selected, accelerate the vehicle with wide open throttle (WOT) from 100kph to 200kph then brake back to 100kph and repeat for 10 cycles."

This test was conducted on a physical car as part of the vehicle development programme; the data collected will allow for a direct comparison of the simulated values to those recorded on the car.

3.2 Step 2 – Inputs: Vehicle Geometry & Component Data

The vehicle selected for this study is the Bentley Bentayga V8, Figure 4, which features a cooling package split in to three separate cooling stacks positioned across the front face of the vehicle, each fed by a separate flow path.

Figure 4 - Bentayga Cooling Package

To avoid compromising the styling with extra vehicle over hand and one large aperture by packaging all of the heat exchangers in a large single air path opening, it is common to separate the package into smaller stacks, positioned across the vehicle's front end, each with its own flow path. It is also typical for there to be more components in a multipath cooling package, which is often the driver behind splitting the pack across multiple air paths in the first place. In a higher performance vehicle this can provide a cooling benefit by providing a feed of ambient temperature air to more coolants, rather than the exit air temperature from one cooler feeding the next. It is not possible to avoid this for all coolers, but splitting the heat exchangers in to multiple cooling stacks reduces the occurrence. The simulation of a multipath cooling package adds complexity as the interaction of the different paths needs to be understood and carefully included in the model, while the boundary conditions for each airflow path can vary. To make sure the airflow interaction between the cooling stacks is captured, all of the cooling pack components were included despite the focus being on the CAC performance, with their geometry exported in to the CFD model from the 3D component models.

3.2.1 Component Layout

Positioning the three different cooling stacks within the Bentayga package requires them to be offset and angled in relation to each other. The largest of the cooling stacks is positioned on the vehicle centerline and consists of the main engine radiator, the HVAC condenser and two electric fans. To the left of the vehicle from the drivers view is the next largest stack which contains a CAC for the LHS engine bank and an auxiliary radiator mounted to the rear of it. On the right of the vehicle is a single CAC for the RHS engine bank.

Each cooling stack is separated, Figure 5, by three sets of ducts that work to target the air flow through each grille aperture on to their respective cooling stacks. For this flow to be captured accurately each of the airflow paths needs to be reproduced accurately in a full vehicle CFD model.

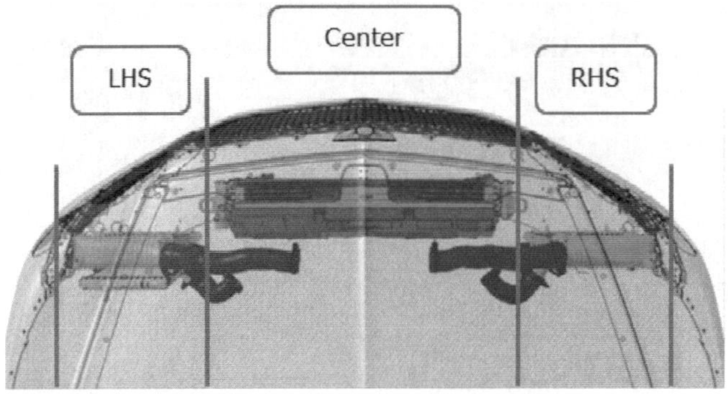

Figure 5 - Cooling Stack Separation

3.2.2 Cooling Package Operation

At the beginning of the drive cycle, the vehicle is brought to a stabilized condition through the preconditioning drive. At this point all of the cooling medium temperatures have reached normal operating temperatures.

Once stabilized, the acceleration and deceleration cycles begin – this is the part of the cycle where the cooling package is subject to high load and high heat reject from the under bonnet systems. This cycle causes the system to produce heat during the high load acceleration portions, but due to the deceleration this heat needs to be dissipated as the airflow decreases, and for engine driven pumps, the internal flow rate will decreases.

By the end of the cycle, the cooling package is reaching its upper operating temperature, which must be compared with the temperature limits defined in the test specification to assess if the vehicle has passed the test.

3.2.3 Charge Air Cooler Characterization

Component characterisation is regularly reviewed during a development cycle and compared against the project specification, with regard to cooling systems; this is typically done with performance data from a flow bench. The internal and external pressure drop is measured over a range of flow rates to understand how the part will perform in the car, and this data can be used when calculating the performance of the part. This is carried out with an airflow applied across the entire face of the heat exchanger so the impact of the vehicle geometry is not captured. The results are given for the performance of the overall part in a series of curves with pressure drop plotted against internal mass flow, and heat reject against external mass flow. This data will be used when creating 3D or 1D simulation model. To support the tube level model, the characterisation data is needed per tube for the part in question. For this study into CAC temperature simulation, a single tube from the multi tube heat exchanger was flow tested so its performance could be quantified and included in the simulation multiple times.

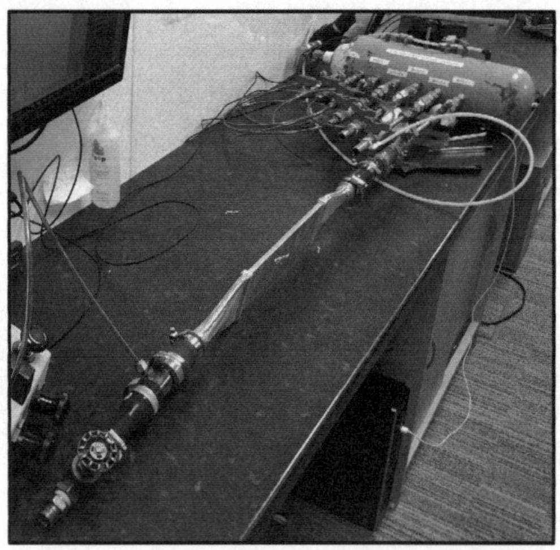

Figure 6 - Tube Characterisation Test Part

This was done by removing a tube from the CAC core assembly and carrying out a flow test to give the internal pressure drop values for one tube instead of the CAC as a whole. To allow for the single tube to be fitted to the flow rig a custom inlet and outlet featuring a smooth taper were fabricated and used to attach the tube to the flow bench as shown in Figure 6. The tube level data is then available as an input for the 3D CFD, which is predicted to improve the accuracy of the internal flow simulation.

3.3 Step 3 – External Charge Air Cooler Flow CFD

Due to the styling requirements for the vehicle front end, the airflow entering each of the three apertures varies as the vehicle geometry produces obstructions in front of the heat exchanges, causing the flow distribution across their faces to be non-uniform. This distribution needs to be included within the simulations in order to produce representative and accurate results. The external flow for the CAC face was found by carrying out a 3D full vehicle simulation in PowerFLOW which shows the airflow distribution through each of the airflow paths while including the impact from blockages from the vehicle aperture size and styling, as per the method demonstrated at VTMS 12 (1).

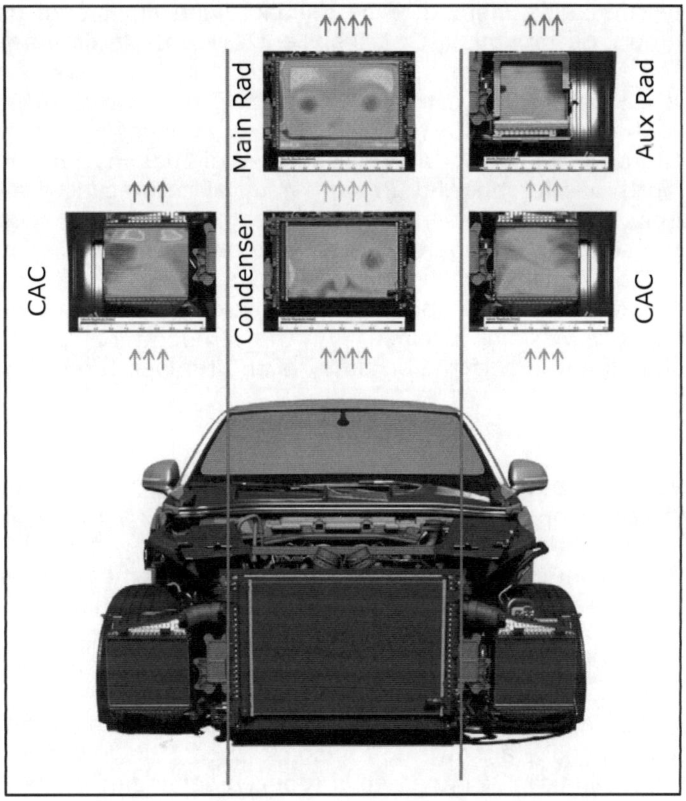

Figure 7 - Cooling Stack Velocity Distribution

The resulting distribution from the CFD is shown as planar velocity distribution maps in Figure 7, with one for each flow path. Looking at the velocity distribution colour maps and comparing them to the vehicle styling shown in Figure 4, it is clear to see the influence of the A-surfaces and the cooling fan motors on the flow. This is best shown on the CACs where the 'moustache' feature of the lower grille apertures can be seen as an area of very low flow velocity, and on the main radiator where the fan motors can be seen as two circular areas of very low flow velocity. It is this detail that must be captured in order to truly represent the cooling package performance during a simulation.

It is also important to make sure the fan blades are correctly included and that the fan rotates at the speed it would on a physical vehicle. In the colour maps shown in Figure 7 the fan can be seen as operating due to the near uniform flow it draws, shown by the two green circles around the motors. If the fan was not modelled as functional the fan blades would be visible as areas of low flow due to blockage, similar to the fan motors. A common solution to overcome this is to remove the fan blades from the 3D model; however the focus of the method is to bring the model closer to the physical car during the drive cycle, so the fan blades need to be retained.

4 STEP 4 - SIMULATION

The results from the CFD simulation are used alongside the rest of the input data listed in Section 3 to create an intercooler model in GT Suite that allows for the engine performance to be combined with the vehicle speed and fan speed based airflow during the chosen test cycle. In this paper the transient part of drive cycle will be simulated in its entirety to give the most accurate results possible.

4.1 Drive Cycle Profile & Operating Points

To allow for the simulation method to be validated the input data used needs to match a real world case, for this a CAC Transient Drive Cycle data set for hot field trial was selected. Any deviation from the specification in the speed or ambient temperature was logged on the car and the physical traces were used as the test profile instead of the exact 100-200kph cycle. The differences are minor, and can be seen in Figure 13, for example the driver decelerated to 103kph rather 100kph before beginning the fourth acceleration, but in the pursuit of accuracy it was deemed best that these differences are included in the simulation to avoid them being a source of error between the simulated temperatures and the physical values. Once validated, the method can be applied against the prescribed cycle when used for future projects. To reduce the length of the simulation to the areas of interest, the preconditioning part of the time trace was not included. As the temperatures at this point were stabilised this should not impact the accuracy of the temperatures simulated at the end of the cycle.

4.1.1 Operating Points

During the drive cycle, several vehicle operating conditions are frequently visited, and due to the repetition of the acceleration cycles this is to be expected. By identifying which conditions are visited most frequently, they can then be assessed to determine which points should be selected for the CFD simulations as mentioned in Section 3.1. These points will be simulated to give the planar airflow conditions for each airflow path, and then interpolated to produce a reference table that can be used by the GT Suite model to look up the operating conditions for all of the values entered in the transient simulation, without the need to simulate each and every condition in 3D CFD. Care needs to be taken to ensure that the airflow due to fan speed is included, as unlike airflow due to vehicle speed this is largely independent of vehicle movement and can be increased or reduced depending on the under bonnet temperature and electric fan motor calibration.

To capture the full spread of operating conditions, the operating points from a physical drive cycle were recorded at a 10Hz frequency to make the data files a manageable size. The fan speed and vehicle speed were then plotted as a convex hull shown in Figure 8 - the opacity of each point plotted was the same. As a condition is visited again and again, the point on the graph is thickened until it becomes a solid colour – this indicated a highly used operating condition during the test cycle and a desirable area to capture in the operating points selected for simulation.

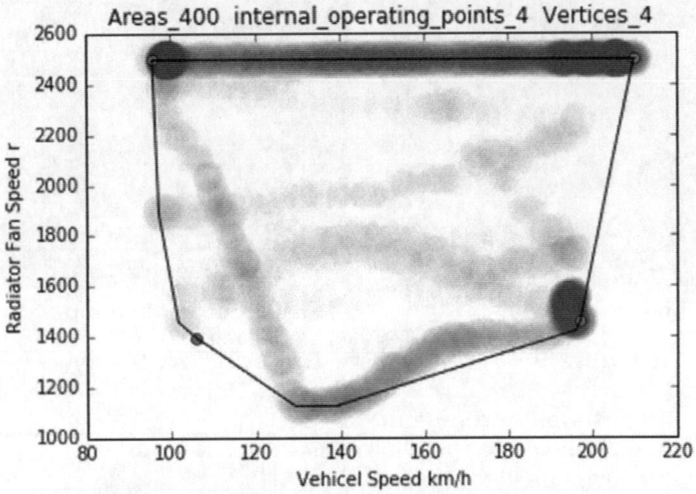

Figure 8 - Operating Points Convex Hull

The vehicle speeds between 100-200kph are shown by the darkest area, which is to be expected with the repeated cycles. With regard to fan speed, it can be seen to sweep between 0 and 2500RPM but spends most of its time operating at 2500RPM.

Table 1 - Selected Operating Points

Operating Point	**Vehicle Speed**	**Fan Speed**	**CAC Mass Flow (LHS)**	**CAC Mass Flow (RHS)**	**Charge Air Temp (LHS)**	**Charge Air Temp (RHS)**
Unit	kph	RPM	kg/hr		degC	
1	197.17	1457.52	210.61	211.41	82.04	79.44
2	106.03	1396.48	97.53	98.26	137.72	131.34
3	95.92	2495.12	123.73	124.17	81.68	75.77
4	210.04	2495.12	243.92	244.81	110.19	107.57

Based on the results shown in the convex hull, four operating points were selected for CFD simulation to produce the planar flow distribution map for each cooling stack. These operating points are shown as the red points on Figure 8 and are listed in Table 1. They were selected to capture both the frequently used points and the extremes of the operating conditions.

4.1.2 Impact of Charge Air Temperature on Flow Distribution

During the physical drive cycle the internal temperature of the CAC flow changes rapidly, so an investigation was carried out to see if this change of internal temperature, changing the temperature and density of the cooling air exiting the rear of the CACs, would have an impact on the blockage and distribution across the CACs.

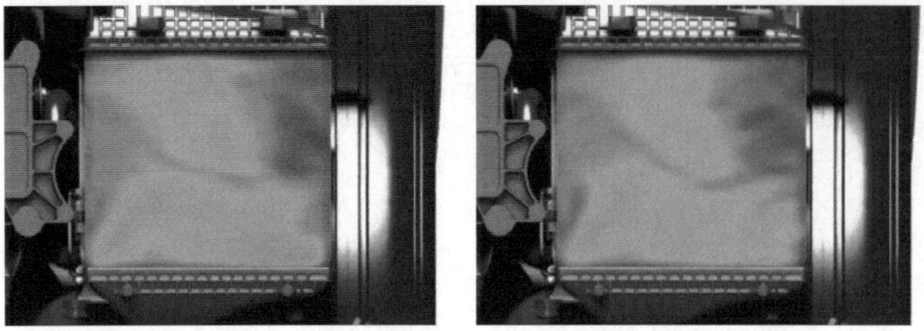

82°C Charge Air Temperature 193°C Charge Air Temperature

Figure 9 - Temperature Impact Flow Study

A pair of simulations was carried out for a portion of the drive cycle where the charge air temperatures were low, one with the boundary conditions as recorded during the physical test, and one with the internal charge temperature at the maximum value recorded on the drive cycle but with the rest of the conditions unchanged. The results of the study are shown in Figure 9, and the outcome was no noticeable difference in blockage with temperature, with less than 2% change in the mass flow over a temperature range of 111degC, so the method of selecting operating points with velocity and fan speed can be used with confidence for the CAC temperatures too.

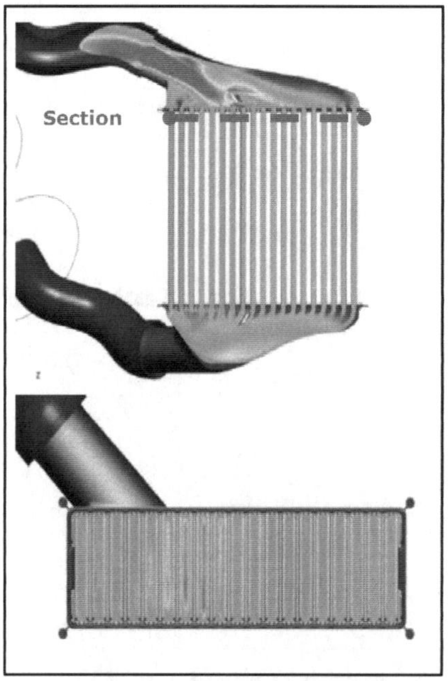

Figure 10 - CAC Internal Flow Distribution

4.2 Conventional Simulation Method

To allow for an accuracy improvement for the method used in this paper to be quantified, a 1D GT suite CAC outlet temperature simulation was completed using the method outlines in VTMS 12 (1) which is referred to as the conventional method. Despite now being the conventional method, it is still an advanced approach to CAC simulation as it uses the 3D CFD to capture and include the external airflow distribution. The conventional part of the name is given based on the use of a default heat exchanger model to represent the CAC on a component level. This default model is built using the physical geometry from the part, but it does not capture the complete internal flow distribution, which is what the new method aims to improve on. The results from the conventional method can be seen in Section 5.

4.3 Internal Charge Air Cooler Flow Distribution Ratio

To overcome the difficulties found in VTMS12 (1), the tube level characterization for the CACs was used in an internal airflow CFD simulation using PowerFLOW. This was carried at the boundary conditions of each of the selected operating points and with the CAC inlet and outlet pipe geometry as per the vehicle installation to reproduce the internal air flow entry and exit, with the results for first operating point shown in Figure 11.

The results from the CFD simulations provided the internal airflow distribution between the CAC tubes as a percentage of the overall mass flow which could be used as an input to the 1D GT Suite model. The results of the CFD showed that the internal distribution between each tube was inconsistent as predicted, with the greatest flow through a tube 68% more than the lowest.

The distribution results for each operating point were noticeably similar, with the percentage difference between the internal distribution nearly consistent for each of the operating points, shown in Figure 11. From this it was decide that a fixed ratio of the average percentage flow for each tube from the four operating points would be applied during the transient 1D simulation to optimise the runtime.

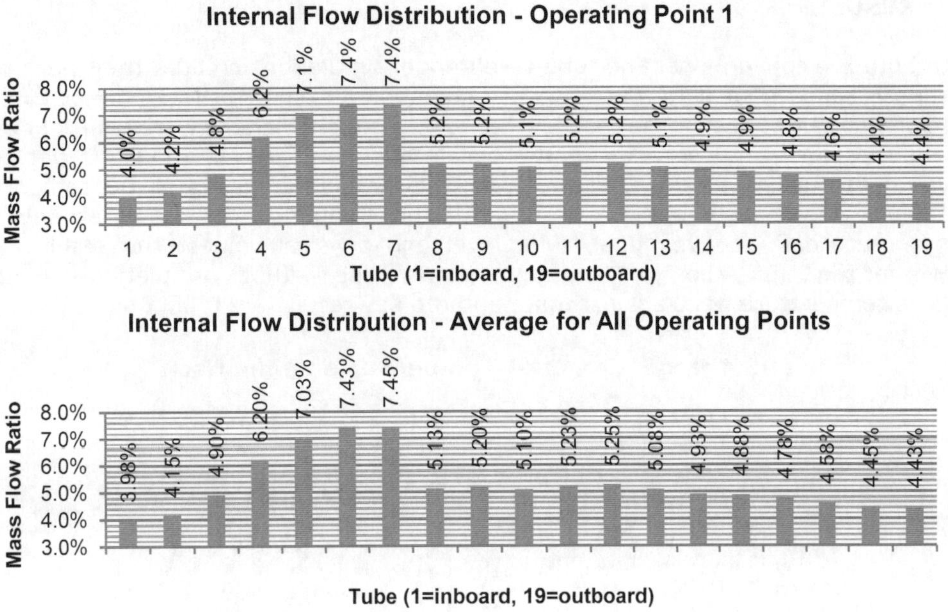

Figure 11 - CAC Tube Internal Flow Ratios

4.4 Tube Level Model Creation & Simulation

With the ratios between each tube known from the CFD, the GT 1D model could be created for the transient simulation. The model was created by taking the conventional method CAC model and separating it in to nineteen smaller CACs, each assigned with its individual flow rate as a percentage of the overall charge air mass flow entering the CAC inlet, but with other overall geometry maintained. The conventional and tube level models from GT Suite are shown in Figure 12. The feed and outlet for the nineteen tubes were combined to represent the pipework as installed in the vehicle.

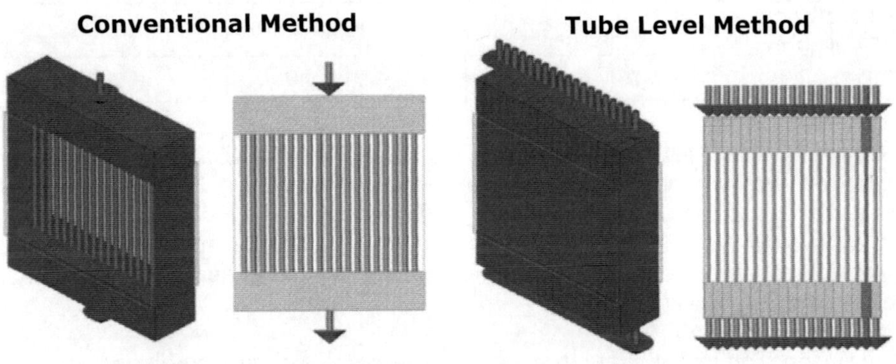

Figure 12 - Conventional & Tube Level CAC Models

The external flow distribution was applied in the same way as the conventional method so that the non-uniform flow would be included. This is done by creating an external airflow entry face on the CAC in GT suite and discretizing it to a fine resolution to which the flow distribution details from the 3D CFD are applied. The finished model represented the vehicle installation by containing both the external and internal non-uniform flow details, bringing the simulation in line with what the CACs would experience in car.

5 RESULTS

With both the conventional and tube level model simulations created they were run using the same input data and boundary conditions, except for the addition of the tube distribution ratios for the tube level model. Both simulations were transient and reproduced the ~220 seconds of the acceleration and decelerations as per the physical vehicle speed trace, allowing for the simulated results to be compared directly to the measured values. To ease the comparison of the three sets of temperature data (conventional, tube level and physical values) the results are shown for the LHS CAC only. The results for the three methods are plotted in Figure 13 as a temperature trace against time, and the key results are tabulated in Table 2.

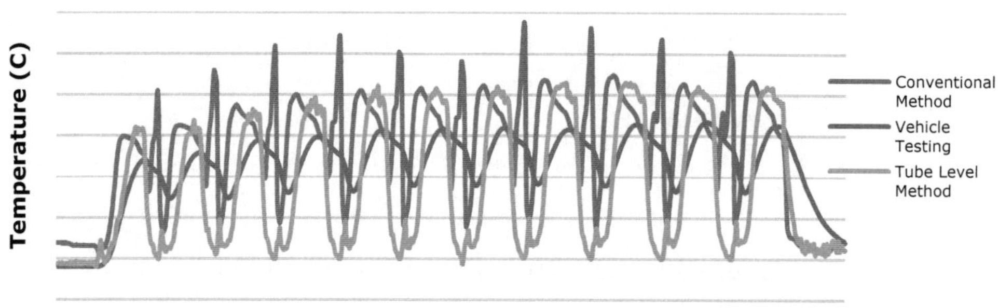

Figure 13 - Simulation and Physical LHS CAC Outlet Temperatures

Table 2 - Simulation Results

Method	Max Temperature	Percentage Error	Error – RMSE Average
Physical Testing	68.4 degC	NA	NA
Conventional	91.8 degC	34%	11%
Conventional Without Anomalies	81.2 degC	19%	10.42%
Tube Level	76.9 degC	14%	10.84%

5.1 Conventional Method Results

Looking at the results from the conventional method in Figure 13 it is clear to see that the outlet temperature from the LHS CAC is noticeably higher than that seen during the physical test, and that in between each acceleration run the temperature returns to lower value, closer to the initial stabilised condition, than the values recorded during the physical testing.

On further inspection of the simulated values, there are reoccurring erroneous values between each acceleration run that skew the temperature trace to show higher peak values when plotted. If these single values are ignored and the areas of the curve with a high density of data points are focused on, then the delta between the physical results and the conventional simulation results is reduced. The time at which the peak temperatures are achieved is slightly out of sync with the physical temperatures; the simulated peak values for the acceleration runs happen a few seconds before the physical values.

At the end of last deceleration, the physical temperature is returning to the initial condition but takes more time to get there than the simulation predicts. Looking at the peak temperature, the conventional method over predicts by 34% with a value of 91.8degC compared to a physical value of 68.4degC. Ignoring the anomalies this comes down to 81.2degC peak and an over prediction of 19%.

Overall the average percentage delta between the conventional method and physical values, excluding the anomalies, is 10.42% RMSE, which is comparable to the deltas seen in the CAC simulations from VTMS 12 (1), which gave an error of 10.3% for low speed and 6.2% for high speed driving conditions, when the highly dynamic nature of this drive cycle is considered. With the anomalies included this increases to ~11% RMSE.

5.2 Tube Level Method Results

Moving to the tube level simulation removes the anomalies and brings the occurrence of the peak temperatures in phase between the simulated data and the physical values; however the drop back to the initial condition is still present and actually goes closer to the initial condition temperature than the conventional method. The over prediction of the peak temperatures is also still visible, and is not the behaviour predicted when creating the method.

The average percentage error between the tube level simulation and the physical results is 10.84% RMSE, which compared to the average error of the conventional method, is an increase in error of 0.42%. If the peak temperature at the end of the cycle is focused on, as this is pass criteria for the test, then the peak value from the physical testing of 68.4degC is predicted as 76.9degC in the simulation, this is an over prediction of 13.8%, which is a 20% improvement over the conventional method.

6 DISCUSSION

From looking at the temperature plots in Figure 13 the simulated results from the tube level method show a much stronger match to the physical testing, both in the trend of the curve and the peak values of the results. The spikes present in the conventional method are removed, but the tube level method still shows an error when compared to the physical testing, showing an opportunity for further improvement.

6.1 Areas for Improvement

By looking back at the method, and the results achieved, areas of possible improvement for future application were identified. These areas of improvement are explained in detail below:

6.1.1 Data Processing

To assess the impact of the peaks in a similar way to removing the anomalies in the conventional method, and to see if it moved the tube method simulation results closer to the physical testing results, a 5 second moving average was taken across the temperature trace. The resulting curve is shown in Figure 14, and the new percentage error compared to the physical values is 7.9% RMSE; a notable improvement. One thought on why the improvement is so large from averaging the results is the frequency of the simulated results which were processed at 10Hz to keep the file size down, but with such rapid changes in velocity during the acceleration and deceleration cycles this frequency could be too low, which would allow some of the detail to be missed and cause the spikes between data points. The average curve gives a much smoother profile which is comparable to the trend of the physical results.

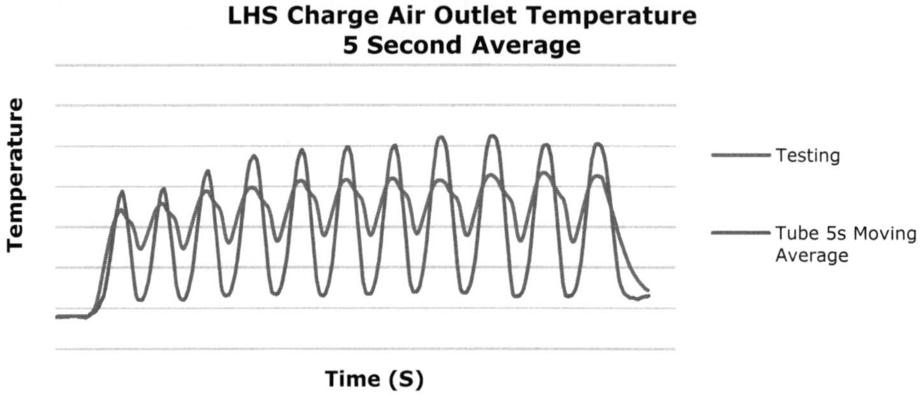

Figure 14 - Time Averaged Tube Method Results

The use of a automated script throughout the process would overcome the housekeep requirements of the larger data files this would generate, and still allow the tube level method to be time efficient for the level of accuracy it delivers.

6.1.2 Operating Point

With the averaged curve, the next most visible error is the under predicting of the charge air temperatures at the end of the deceleration, which as mentioned in Section 5 nearly return to the initial condition each time. The feed for this data is the convex hull from the operating points, which accurately captures the vehicle speed and fan speed seen during the cycle, but perhaps does not capture enough information between the minimum and maximum temperatures to allow for accurate interpolation. By looking at operating points in Table 1, the range between the values selection for simulation is quite small. This could be overcome by increasing the number of operating points selected, which would provide more data sets to interpolate between in the 1D model along with additional points to carry out the internal and external 3D CFD simulations. With this additional information it is hoped that the accuracy would improve, further work is underway to evaluate this theory.

A second suggestion for improving the results with regard to the operating points would be to change the axis on the convex hull depending on which heat exchanger is being investigated. While vehicle speed and fan speed are both good indicators for the mass flow through the cooling pack, it might be beneficial to use internal air temperatures or charge air pressure against vehicle speed for future CAC simulations, as this would hopefully capture a larger spread of data between which the GT Suite 1D model can interpolate.

6.1.3 Measurement Equipment

A question around the refresh rate of the thermocouples was raised when viewing the results, as it could be possible that the peaks seen in the simulation are experiences on the physical car, but the thermocouples do not react quickly enough to reach the minimum and maximum temperatures before the next part of the cycle begins.

The larger error between the lower temperatures in the simulation results and the physical testing results could be due to the thermal inertia of the thermocouples, which retain heat as the internal mass flow rate reduces, which in effect reduces the cooling air passing over the thermocouple.

To understand if these thoughts could appear in the results as an error, the thermocouple charge air outlet temperatures were compared against the engine intake manifold temperatures, which are recorded by the ECU as an input for fuel metering and spark timing. There was not a noticeable difference between the two sets of temperatures, while this does not rule it out as an error, it indicates that the thermocouples are within the response rate required by the engine, and therefore capture sufficient data for the intended application of this method.

6.1.4 Temperature Probe Position

To improve the accuracy of the physical measurement for the CAC outlet air temperature and rule it out thermocouple positioning as a source of error, more thermocouples could be used and their readings averaged. The current set up is a single thermocouple placed in the CAC outlet, which with the internal flow conditions known as non-uniform, might not be recording the maximum and minimum values seen in the airstream. The use of multiple thermocouples configured equally around the CAC outlet and then averaged would eliminate this as a source of error in future testing.

6.1.5 Component Characterisation Data

A second input that comes from physical testing is the component characterisation, which doesn't provide much detail for the low mass flow condition due to a low number of test points at these conditions, compared to higher mass flow rate conditions. The lowest data point from the component characterisation completed on the flow bench is 8x higher than the lowest flow condition experienced on the car. This could negatively impact the accuracy of the low flow conditions seen during the deceleration portion of the cycle due to interpolation errors when compared to the accuracy of the higher flow conditions during the acceleration, which have many more data points to interpolate between, and show a higher correlation between the tube method simulation and the physical results.

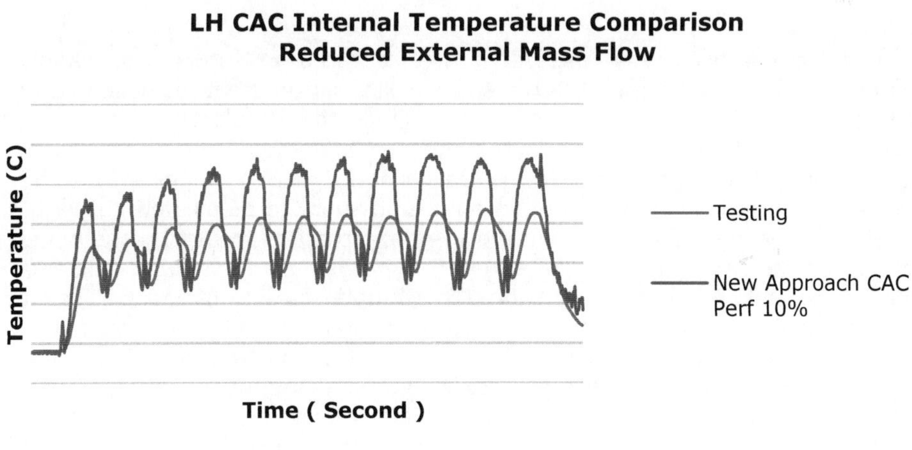

Figure 15 - Tube Level Method with Reduced External Heat Transfer

To understand the impact the characterised data has on the accuracy of the CAC outlet temperature predictions, an additional 1D simulation was run with the heat reject rate reduced by 10%. The results from this additional simulation are shown in Figure 15 against the physical testing results.

Adjusting the heat reject can clearly be seen to bring the overall trend of the curve closer to the test data, most notably the temperatures at the end of the deceleration sections of the drive cycle. From this result, it is safe to predict that if the level of detail obtained from the characterisation of the CAC and is used as an input for the new simulation method, then the accuracy of the result should improve.

6.1.6 Cooling Airflow Recirculation

When reviewing the simulated results against the physical testing results, recirculated airflow was suggested as a possible source of error that might not be captured in the simulation. To confirm if this was occurring; the temperature of the air entering the external face of the CAC was compared to the ambient air temperature, but with no noticeable difference so this was ruled out as a source of error.

6.2 Application

With any improvement in the accuracy of the charge air outlet temperature, the accuracy of the engine model will be improved, allowing for a closer representation of in-cylinder temperatures. While an improvement in the charge air temperature is always desirable, in the early stages of a project it is the accuracy of the simulated results that is most valuable, as they will be used for sizing hardware to meet the performance targets. Errors at this stage can be presenting in component sizing, which can either lead to an engine failing to meet its targets due to underspecified components, or the opposite where a component will be over specified. Either situation is undesirable as cost and project timing will increased through additional development loops to meet the targets or through part prices for over performing components.

The improvements in charge air temperature accuracy found in this method, while not as large as hoped, will allow for robust decision making when defining hardware, which in turn will support the powertrain as a whole in reaching its performance targets. Based on the comments from the VTMS 12 paper (1), the high load – high speed dynamic performance experienced in the Transient CAC System Performance Drive Cycle are the most challenging to simulate, so improvements yielded by the new tube level simulation method will hopefully increase for less challenging cycles and produce a lower percentage error when looking at the complete series of test cycles.

If the suggestions from this discussion are captured in future applications, a catalogue of characterised components and results will be created, which can then be used as a reference for future projects. Once this catalogue is in place the turnaround time will be reduced and the tube level method can be applied more frequently. It is possible to scale the performance of a component in the 1D simulation, which will allow for a component sizing study to be carried out with only a few sets of characterised data, creating the opportunities to use the new method maximise vehicle packaging. Another use of method could be the validation of existing components in a new application, for example carrying over a CAC from one model variant to another. The tube level simulation method could provide temperature data as an input to an engine model. If this is done to validate the use of a carryover heat exchanger, the project cost saving would be in excess of €800K from the design, development and tooling activities alone.

7 CONCLUSION

The tube level method of simulating captures additional details of the physical components and their behaviour at a level of detail not seen in the conventional method, most notably the closer correlation between the simulate temperature curve and test results when compared to the conventional method. However the accuracy improvement this additional detail delivers is lower than predicted. By smoothing the simulated results by using a time average in the post-processing stage, the accuracy of the simulated temperatures does come closer to the physical values recorded during a vehicle test cycle than the conventional method. This improvement can be applied when sizing components and assessing component and vehicle performance against a set of targets.

The repetitive nature of the test cycles lends themselves to the selection of operating points rather than simulating each and every data point, however care needs to be taken to make sure that the operating points accurately capture the component and vehicle behaviour, and not just the extremes.

Overall, with the addition of some post processing to the simulated results from the new method, it is possible to reproduce the outlet temperature of the charge air from the CACs, which has proven difficult in the past (1), to within 8% of the physical values over a transient simulation of 220 seconds. This level of accuracy, while not an exact match, will allow for the component and vehicle performance to be predicted in the very early stages of a project, reducing the reliance on physical testing and hopefully reducing the costs incurred during the development of a new vehicle due to hardware changes. This will give an increased accuracy overall and the method can be applied with the VTMS 12 method (1) to predict the cooling packages performance against both Vmax and transient conditions.

8 REFERENCES

1. **Price, A, et al., et al.** *Prediction of Transient Cooling Package Behaviour with Multiple Air Paths.* s.l. : IMechE, 2015.
2. **Jansen, Wilko.** *Drive Cycle Simulation of A Tiered Cooling Pack Using Non-Uniform Boundary Conditions .* s.l. : IMechE, 2015.
3. **Cengel, Yunus and Boles, Michael.** *Thermodynamics: An Engineering Approach.* s.l. : McGraw-Hill Education, 2014. ISBN 0073398179.
4. **Bentley Motors Ltd.** Cooling System Sign-off Procedure & Requirements. 2014.

HEAT EXCHANGE

Mechanism of oxide film destruction in fluxless brazing of aluminum under an inert atmosphere

T Yamayoshi, Y Itoh, A Fukumoto
UACJ Corporation, Research and Development Division, Japan

ABSTRACT

The addition of magnesium or lithium to aluminum brazing sheets was investigated with the aim of the clarifying the mechanism of oxide film destruction during fluxless brazing. The surface of the filler alloy after brazing was observed by scanning and transmission electron microscopy. The presence of complex oxide particles of magnesium or lithium with aluminum were confirmed on the surface of the filler alloy after brazing. The results indicate that the oxide film was broken into oxide particles by magnesium or lithium during brazing, and a new surface of molten filler alloy was exposed between the oxide particles, allowing fluxless brazing.

1 INTRODUCTION

Aluminum automotive heat exchangers are generally manufactured by brazing. In order to braze aluminum, it is necessary to destroy a strong and stable oxide film present on the material surface. For this reason, methods using a flux or heating in vacuum without using flux have been developed. Currently Controlled Atmosphere Brazing (CAB), destroying the oxide film by using a non-corrosive flux, is a mainstream method. However, it requires fluxless brazing in an inert atmosphere since flux residue in the heat exchanger can become a cleanliness problem.

Materials with added magnesium are known to be brazable without flux in an inert atmosphere. Also, it is reported that fluxless brazeability is improved by adding a wetting element such as bismuth (1, 2). A brazing sheet with lithium has also been developed (3). Magnesium and lithium added in the fluxless material have lower oxide formation free energy than aluminum. Therefore, it is assumed that these elements destroy the oxide film (Al2O3) by a reduction reaction. However, research on the nature of the oxide film after brazing does not seem to have been investigated.

In this study, the oxide film after brazing of the material containing added magnesium or lithium was observed by Scanning Electron Microscopy (SEM) and Transmission Electron Microscopy (TEM) to clarify the mechanism of oxide film destruction.

2 EXPERIMENTAL PROCEDURE

2.1 Materials

Table 1 shows the chemical composition of the brazing sheet provided in this study. As the base composition, the filler alloy contained 10 mass% of silicon, and the core alloy contained 1.2 mass% manganese. For No.1 material 0.6 mass% of magnesium was added to the filler alloy and for No.2 material 0.6 mass% of magnesium was added to the core alloy. The No.3 material contained 0.02 mass% of lithium in the filler alloy. In order to improve brazeability, for No.1 to No.3 materials 0.02 mass% of bismuths was added in the filler alloys. The No.4 material was a comparative material having no added magnesium and lithium, and had a flux of 3g/m^2 applied before brazing.

For producing these brazing sheets, the filler alloy was cladded on one side of the core alloy by hot rolling, being rolled down to a thickness of 0.4 mm. The temper is O (annealed), and the cladding ratio of the filler alloy was 10%. The brazing sheets were then degreased with acetone. Finally, the oxide film formed during manufacturing the material was removed by dipping in 2% nitric acid-1% hydrofluoric acid at room temperature for 90 seconds.

Table 1 Chemical composition of brazing sheets(mass%).

No.	Filler alloy					Core alloy			Remarks
	Si	Fe	Mg	Li	Bi	Cu	Mn	Mg	
1	10	0.17	**0.61**	—	0.02	0.16	1.2	—	
2	10	0.17	—	—	0.02	0.16	1.2	**0.62**	
3	10	0.17	—	**0.02**	0.02	0.16	1.2	—	
4	10	0.17	—	—	—	0.16	1.2	—	Flux application 3g/m^2

2.2 Brazing conditions

Figure 1 shows the clearance filling specimen used for evaluating the brazeability. For the clearance filling specimen, a vertical plate of AA3003 alloy was positioned on a flat brazing sheet, while one end of the plate was raised on a stainless steel rod to provide clearance between the plate and the sheet. Brazing was performed in a nitrogen gas atmosphere having an oxygen content of 5 ppm or less and a dew point of -70 °C or less, and the specimen temperature was held at 597 to 600 °C for 5 minutes. After brazing, the specimen was rapidly cooled to 300 °C in a nitrogen gas atmosphere having an oxygen content of 5 ppm or less to prevent oxidation during cooling.

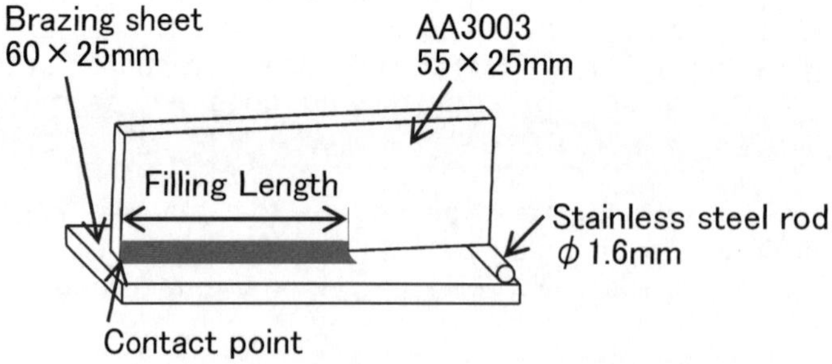

Figure 1 Schematic of clearance filling specimen.

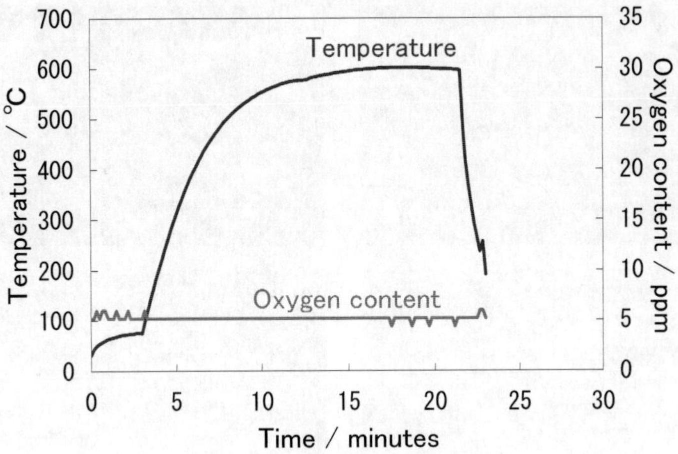

Figure 2 Heating profile.

2.3 SEM observation and TEM Analysis

To provide specimens of the filled clearance after brazing, the brazing sheets were cut out and the surfaces of the filler alloy were observed by SEM. In addition, the oxide film on the filler alloy after brazing was analyzed by TEM-Energy Dispersive X-ray spectrometry (EDX). Since lithium is a light element, it can not be detected with EDX. Therefore, lithium was analyzed by TEM-Electron Energy Loss Spectroscopy (EELS). A sample for TEM analysis was prepared by dipping the brazing sheet in a saturated iodine methanol solution, leaving only the oxide film.

3 RESULTS AND DISCUSSIONS

3.1 Brazing test results

Figure 3 shows the results of the clearance filling test. No.1 to 3 materials formed a fillet even without flux. The filling length of the brazing sheet containing magnesium in the core alloy (No.2), at 31mm, was longer than that in the filler alloy (No.1), at 21mm. However, it was not as large as the filling length using the CAB method (38mm). On the other hand, the material containing lithium in filler alloy (No.3) had a filling length of 36mm which is comparable to that of the CAB method. Furthermore, as shown in Figure 4, the No.3 material formed a uniform fillet, which was equivalent to CAB method. Cross-sections at a position 5 mm from the contact point are shown in Figure 5. All the specimens had a fillet shape essentially identical to that of the CAB method. These results indicate that lithium is an effective element for improving the fluxless brazeability.

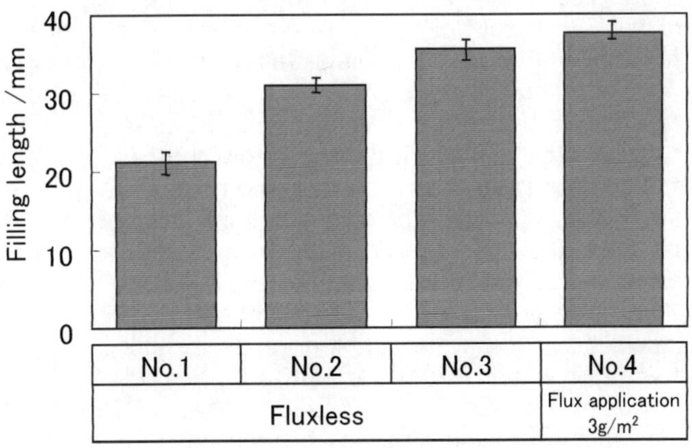

Figure 3 Results of clearance filling test.

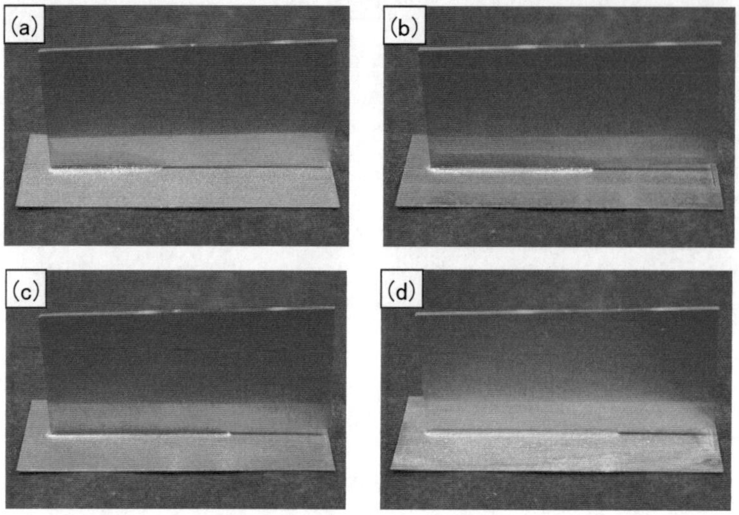

Figure 4 Appearance of clearance filling specimens. (a) No.1, (b) No.2, (c) No.3, (d) No.4 (with flux).

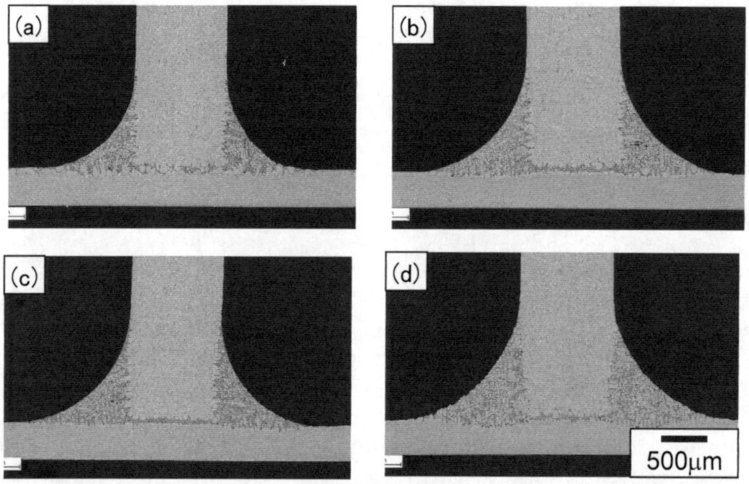

Figure 5 Cross-sections of the clearance filling specimens at the 5 mm from the contact point. (a) No.1, (b) No.2, (c) No.3, (d) No.4 (with flux).

3.2 SEM observations

Figure 6 shows the secondary electron images of the filler alloy surface after brazing for materials No.1 to No.3 and the surface of a representative sample before brazing for material No.1. Fine particles with white or grey contrast were observed on all brazed filler alloy surfaces, and the size of the particles was about 50 to 200 nm. It appears that the oxide film was broken into fine particles during brazing. In addition, a black region with smooth low contrast was observed between the fine particles. The area ratio of this black region was obtained by image analysis for the low magnification image shown in Figure 7. Brazing sheet containing magnesium in the filler alloy (No.1) had a black area ratio of only 13%. However, brazing sheet containing magnesium in the core alloy (No.2) had a ratio of 48%. The highest black area ratio was brazing sheet containing lithium in the filler alloy (No.3), which was 67%. The material with good brazeabilty had a high black area ratio. From this result, it was presumed that the black region is the part where the new surface of molten filler alloy is exposed. Based on these SEM observation results, we surmise that the oxide film was broken into fine particles, and then the new surface of the molten filler alloy was exposed between the particles.

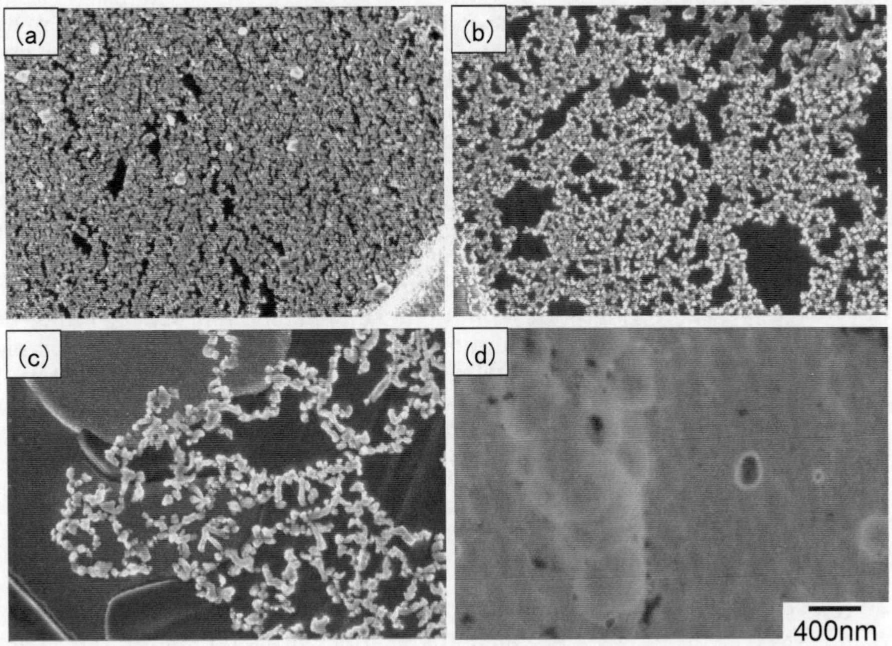

Figure 6 Secondary electron images of the filler alloy surface (a) No.1, (b) No.2 (c) No.3 after brazing and (d) the surface of a representative sample of material No.1 before brazing (high magnification).

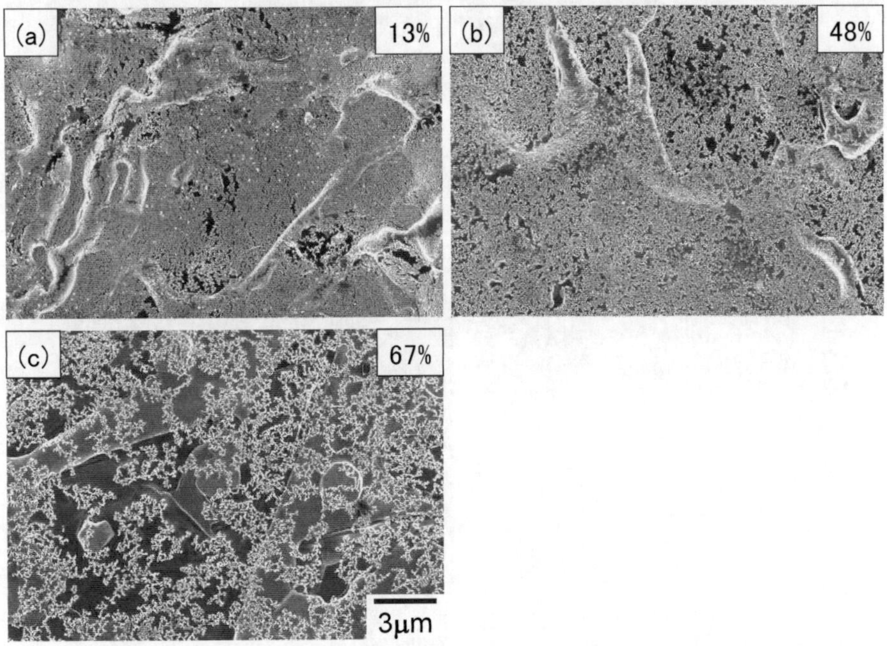

Figure 7 Secondary electron images of the filler alloy surface after brazing (low magnification) and black area ratio. (a) No.1, (b) No.2, (c) No.3.

3.3 TEM analysis

TEM analysis was carried out to identify particles observed in the SEM for No.2 and No.3 materials with good brazeability.

Figure 8 shows the TEM-EDX analyses of particles on the surface of No.2 material after brazing. The particles were observed with black contrast in the TEM image. Aluminium, magnesium, oxygen were detected from all particles. From the diffraction pattern shown in Figure 9, these particles were identified as $MgAl_2O_4$. Note that MgO was not found in this analysis.

Figure 10 shows the TEM-EELS analysis of particles on the surface of No.3 material after brazing. The particle was observed with white contrast in the STEM image. A strong lithium signal was detected from the particle. Aluminum was detected from around the particle and was also detected in it. In the aluminum region of the spectrum, two peaks were clearly observed. These peaks do not appear if aluminum is a metal, but are characteristic of an oxide. We concluded that lithium-aluminum based oxide particles are formed on the filler alloy surface of No.3 material after brazing.

We thermodynamically verified the validity of observations of a complex oxide with aluminum after brazing. The reaction formula of magnesium or lithium and Al_2O_3 are shown in Table 2. Although all the free energies of formation are negative, the free energy of formation of either complex oxide with aluminum is more negative than formation of a single oxide such as MgO or Li_2O. Therefore a reaction to form a complex oxide occurs thermodynamically. This corresponds with the result that the particles observed in this study were a complex oxide with aluminum.

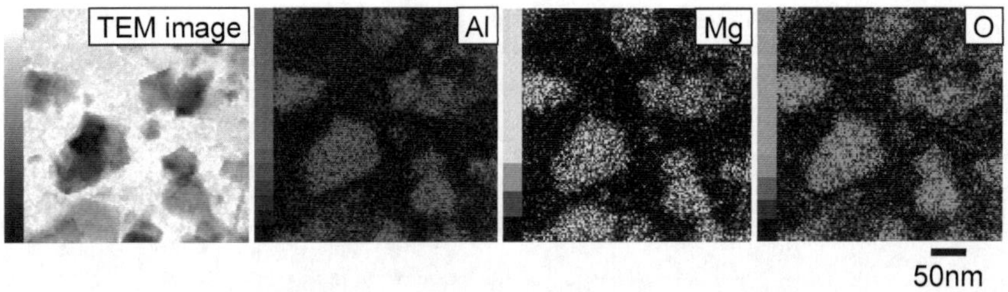

Figure 8 TEM-EDX analyses of particles on the surface of No.2 material after brazing.

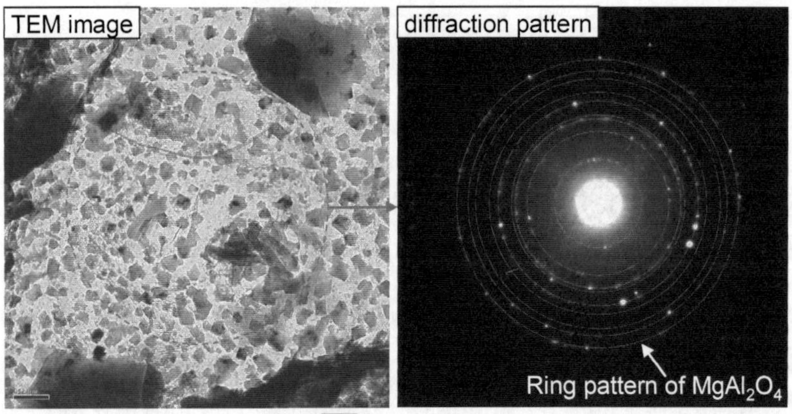

Figure 9 TEM image and diffraction pattern of particles on the surface of No.2 material after brazing.

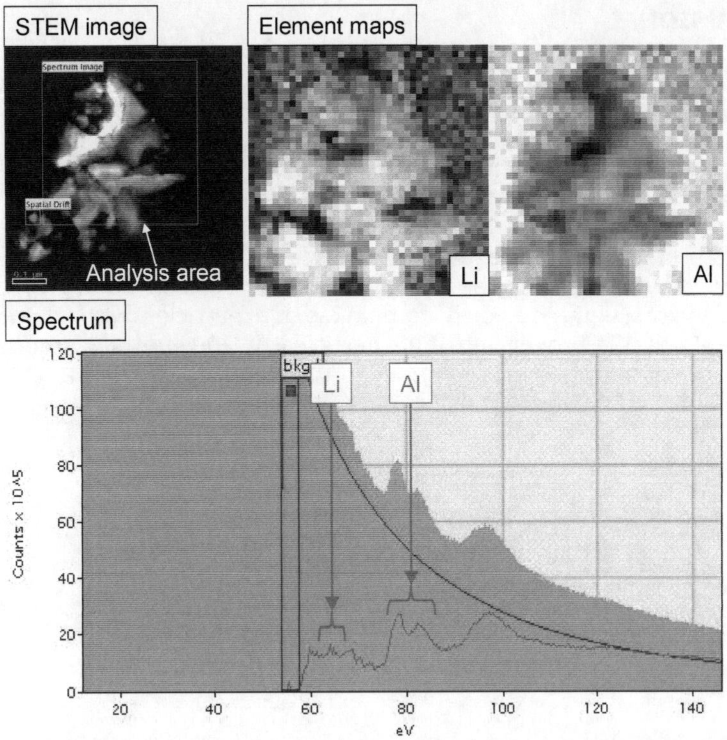

Figure 10 TEM-EELS analysis of oxide film of filler alloy after brazing No.3 material.

Table 2 Free energies of formation for reactions of Mg or Li with Al_2O_3.

No.	Reactions	⊿G (kJ) at 600°C	Remarks
1	$Mg + 1/3Al_2O_3 \rightarrow MgO + 2/3Al$	-45.1	
2	$Mg + 4/3Al_2O_3 \rightarrow MgAl_2O_4 + 2/3Al$	-73.9	complex oxide
3	$2Li + 1/3Al_2O_3 \rightarrow Li_2O + 2/3Al$	-17.7	
4	$Li + 2/3Al_2O_3 \rightarrow LiAlO_2 + 1/3Al$	-69.9 (at 500°C)	complex oxide

3.4 Mechanism of oxide film destruction

From the results of this study, the mechanism of oxide film destruction in fluxless brazing is thought to proceed as follows. Although the oxide film exists over the entire surface of the filler alloy before brazing, it reacts with magnesium or lithium during brazing and is broken into complex oxide particles by that reaction. Then, the molten filler alloy is exposed between the oxide particles. As a consequence, brazing is possible.

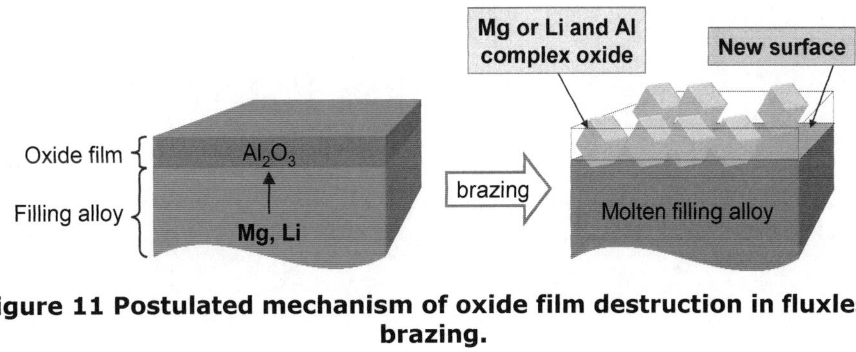

Figure 11 Postulated mechanism of oxide film destruction in fluxless brazing.

4 CONCLUSION

Fluxless brazing of aluminum is possible by the use of magnesium or lithium. It is effective to add magnesium to the core alloy than to the filler alloy. Lithium is a preferred element for improving the fluxless brazeability over magnesium. A brazing sheet containing 0.02% lithium in the filler alloy had a brazeability approaching that in the CAB method.

The mechanism of oxide film destruction is thought to be as follows. The additive element (magnesium, lithium) that diffused to the surface during brazing reacts with the oxide film (Al_2O_3) to form complex oxide particles with aluminum. As a result, the oxide film is broken into fine particles of 200 nm or less in size, and a new surface of filler alloy is exposed between the particles. Finally, a fillet is formed.

5 REFERENCES

1. W Schultze, H Schoer: Welding Journal, (1973), October, 644.
2. A Wittebrood, A Burger, S. Kirakham: VTMS, (2011), 35.
3 EP2835211: UACJ Corporation (2014).

Flux residues of aluminium brazing and engine coolants

L Orman [1], H W Swidersky [1], D Lauzon [2], U Seseke-Koyro [1]
[1] Solvay Fluor GmbH, Germany
[2] Solvay Fluorides, LLC, USA

ABSTRACT

Since the early 1980s, K-Al-F type fluxes have been used in aluminium brazing operations (i.e. CAB furnace, flame, and induction) without changes to the chemical composition. Over the same time frame, several changes in the cooling loop environment took place, which in combination with flux residues have led to certain problems with the coolant stability.

The discussion about flux residue remaining on internal surfaces of brazed heat exchangers has become an industry topic, which historically had not occurred prior to the changes mentioned above. Post Brazed Flux Residue [PBFR] is slightly soluble in water and in currently used commercial coolants. It has been shown in earlier publications that interactions can occur between dissolved flux and certain coolant additives which can lead to formation of gel-like substance. Also it has been determined that below a certain limit of PBFR in solution no deterioration of coolant is observed, thus having a flux type with low solubility could potentially eliminate the risk of gel formation.

Parameters affecting the level of dissolved flux include:

- Residue quantity (flux load and total brazed internal surface area)
- Time and temperature
- Coolant formulation and water content

This paper summarizes results of a test series using modified ASTM D4340 equipment for evaluating solubility characteristics and interactions between residues of various fluxes and commercially available coolants. Special focus is on recent flux developments with reduced residue solubility.

Keywords: Aluminium Brazing, K-Al-F Type Fluxes (NOCOLOK® Flux), Post Braze Flux Residue, Flux Residue Solubility, Engine Coolants, Coolant Additives, Coolant Stability, New Flux Formulations with Reduced Residue Solubility

1. INTRODUCTION

Potassium aluminium fluorides are standard fluxes for brazing of aluminium (i.e. controlled atmosphere brazing (CAB = furnace; flame; and induction). They consist of eutectic K-Al-F complex compounds, and their chemical composition remained unchanged since the introduction of this technology more than 30 years ago.

For the last 20 years, however, several changes in the thermal management system took place - of which the most important ones are:

- Coolant chemistry
- Heat exchanger design - incorporating more internal brazed surfaces with internal fluxing
- More heat exchangers in the system loop
- Reduced coolant volumes in smaller and more efficient heat exchangers
- Increased fluid flow velocities
- Increased local engine temperatures
- Introduction of low temperature radiator without degassing systems

All those changes are responsible for tougher working conditions for coolants. It is known that in extreme cases deterioration of the coolant can take place in systems where no brazing fluxes were present. In some cases the problem with stability of the coolant was attributed to the presence of Post Brazed Flux Residue (PBFR) [1, 2]. Another publication [3] showed that the influence of PBFR depends on the type of coolant, and that the addition of novel additives in formulations improves compatibility with CAB Al materials, resulting in:

– Increased pH stability
– Significant reduction of nitrite and silicate consumption
– Improved aluminium corrosion protection performance

As established by [4] there is a threshold value of maximum content of PBFR in solution for each coolant formulation. That value is expressed by weight of flux residue in solution per litre of coolant. This finding resulted in the determination by different auto makers a minimum value of PBFR inside a given heat exchanger. Thus one could say that the simplest method to reach that threshold would be to lower the flux load level before brazing. This however proved to be not so simple. Due to different brazing process limitations the level of flux load which would meet the threshold value in most cases is not possible to reach without high rejects level.

For the above reason Solvay Flux GmbH has adopted a different approach to the issue of the PBFR threshold level. Having a flux with a solubility limit of PBFR level below the threshold value would secure having low level of dissolved flux in the coolant regardless the PBFR load, thus securing safe operation of the cooling system.

In order to check if such approach could be justified in a series of experiments at Solvay's NOCOLOK® Tech Centre a modified ASTM D4340 test bench has been used for evaluating solubility characteristics and interactions between residues of various new fluxes and commercially available coolants.

2. EXPERIMENTAL

In a single run coolant mixture (500 ml) was filled from the top into a glass cylinder of the ASTM D4340 test bench. Heat-treated flux powder - "artificial" PBFR (i.e. flux which was brought to a molten state, cooled and solidifies and then milled into a powder) was added from the top of the testing cylinder. In each case the concentration of the added flux was 2g/l.

The coolant temperature was set to 135°C which resulted in keeping the real temperature of the coolant in the range of 125°C to 145°C. The coolant was heated up during 2 hours and then for the next 5 hours the test was run at temperature within the ranges mentioned above. The pressure correspondingly varied between 0.14 bar and 0.83 bar. The heat was turned off over night. Coolant samples (25ml) for chemical analysis were collected after each day of testing. The chemical analyses of the coolant samples were done by ICP.

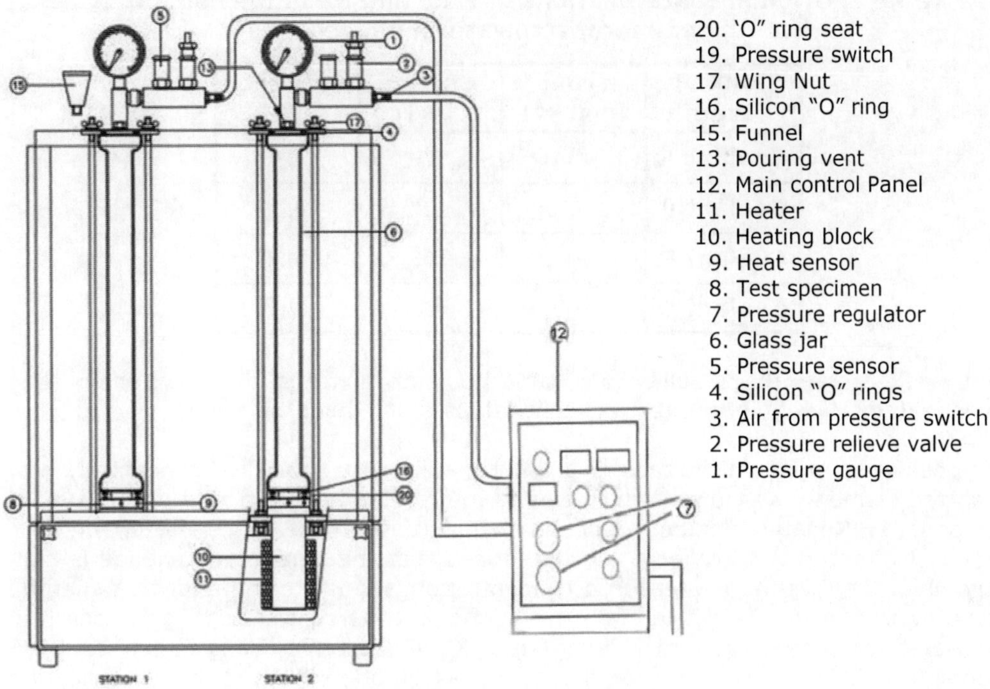

Fig. 1: ASTM D4340 test bench

At first the stability of pure coolant-deionized water 1:1 mixtures were tested. Then different fluxes were tested with one coolant. That allowed for choosing three fluxes with the lowest solubility which subsequently were tested with different coolants. The matrix of all experiments is shown in table 1 below.

Table 1: Matrix of experiments

Flux type	Coolants			
	Coolant A	Coolant B	Coolant C	Coolant D
Pure 1:1 mixture	X	X	X	X
Standard Nocolok Flux	X	X	X	X
$KAlF_4$	X			
$KAlF_4 + BaF_2$	X			
$KAlF_4 + SrF_2$ (A)	X			
Nocolok Li + BaF_2 (A)	X			
$KAlF_4 + SrF_2$ (B)	X			
Nocolok Li	X			
Nocolok + CaF_2	X	X	X	X
Nocolok Li + BaF_2 (B)	X	X	X	X
Nocolok Li + CaF_2	X	X	X	X

3. RESULTS OF EXPERIMENTS AND DISCUSSION

The original concentration of K, Si and Na in the coolant samples is presented in Table 2.

Table 2: Original concentrations of K, Si and Na in the coolant 1:1 demineralized water mixtures

Coolant type	K conc. [mg/kg]	Si conc. [%]	Na conc. [%]
Coolant A	134	127	3730
Coolant B	6830	107	457
Coolant C	7	106	4980
Coolant D	9	94	2780

In the first step the stability of water coolant mixtures over five days was investigated. The end of the test concentrations of K, Si and Na were analysed.

The positive numbers in the table 3 present results where the final concentration of analysed elements was lower, which can be interpreted as interaction of the coolant components. Negative numbers represent an increase in the concentration which is difficult to interpret; however it can be noticed that the percentage change is high only when the level of as-measured concentration is close to the analytical method accuracy and when the concentration is higher the percentage change is small in the worst case being equal to 215 mg/kg of K^- with relative change equal to 4.7% (coolant B). This observation should allow for assumption that the results indicating an increase in concentration are due to the inaccuracy of the analytical procedure. Having the above in mind one can notice that all the tested pure coolant mixtures shows good stability over the test period and temperature, though we could suspect that the coolant B is prone to minor changes, which is manifested by a relatively large change in the concentration of Na.

Table 3: Change of ion concentration of pure coolant mixtures after 5 days of testing at high temperature

Coolant type	ΔK conc. [mg/kg]	ΔK conc. [%]	ΔSi conc. [%]	ΔSi conc. [%]	ΔNa conc. [%]	ΔNa conc. [%]
Coolant A	-11	-8,2	-3	-2,3	-45	-1,2
Coolant B	-215	-4,7	11,5	10,7	91	19,9
Coolant C	1	14,2	-1,5	-1,4	145	2,9
Coolant D	-2	-22,2	-5	-5,3	-50	-1,8

Potassium levels* are used as the most reliable indicator for flux solubility [5]. Elemental content of K in NOCOLOK® is roughly equal to 30% which means that for calculation of total level of flux in solution the concentration of K^- was multiplied by a factor 3.3. For the other type of fluxes the weight ratio of K to the total mass of flux is slightly different. That difference on the basis of flux composition was taken into account when calculating the amount of flux in solution.

The results presented in Fig 2 and 3 show extreme cases. The highest solubility which was characteristic for coolant A and the lowest solubility was experienced by NOCOLOK® Flux + CaF_2 and NOCOLOK® Liflux + CaF_2 in combination with coolant D. It should be also pointed out that all pure NOCOLOK® PBFR was dissolved in coolant A coolant during the test.

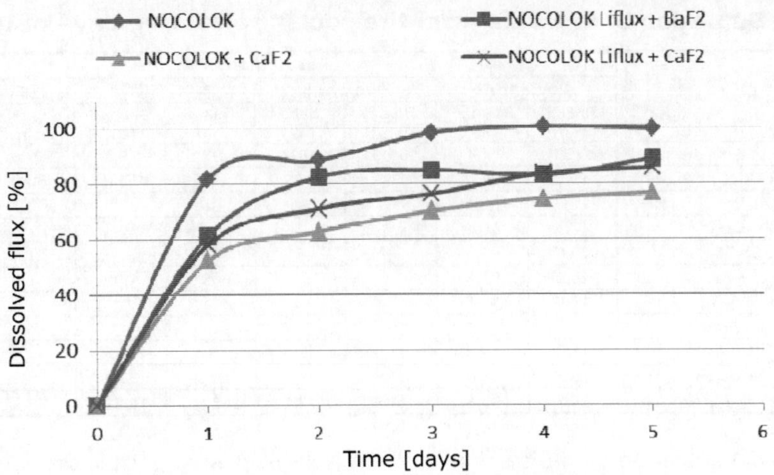

Fig. 2: Percentage of dissolved fluxes as function of time in the ASTM D4340 test; tested with Coolant A

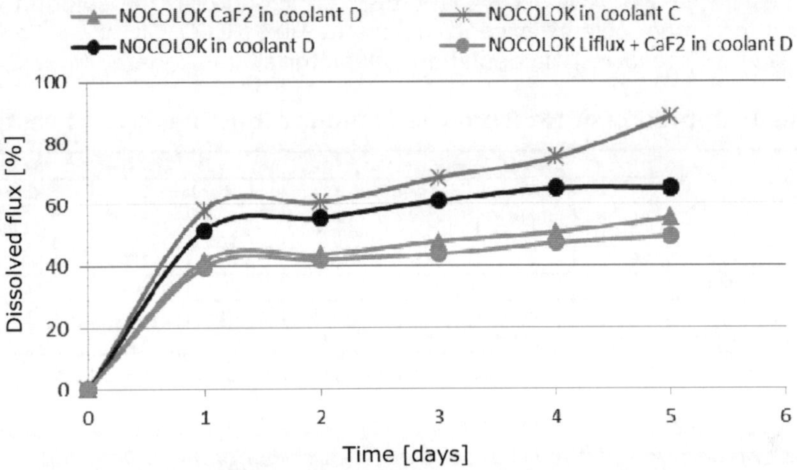

Fig. 3: Percentage of dissolved fluxes as function of time in the ASTM D4340 test; tested with Coolant C and D

The stability of the coolant inhibitors (interaction with dissolved flux) can be shown as a change of corrosion inhibitors ion concentration.

Analyses of the whole data indicated that for Si levels the main change of the concentration took place during the first 5 hours of testing, thus only the final results (after five days) are presented in this work.

Looking at the results presented in Table 4 one could conclude that pure coolants were stable at the test duration except a relatively small depletion of Si levels from Coolant A. The negative values should be rather contributed to inaccuracy of the testing methodology.

The first and the most obvious result is such that in case of Coolant A depletion of Si levels is in the range of 85% for all of the tested fluxes. The final level of Si in mg/l is below the limit determined by the car makers. Smaller depletion is observed for the coolant B though not all the fluxes were tested in this combination. The best result is obtained by flux NOCOLOK® Li Flux + CaF_2, and coolant D.

Table 4: Depletion of Si levels from the coolant after five days of testing

Flux type	Si depletion							
	Coolant A		Coolant B		Coolant C		Coolant D	
	Final Si [mg/l]	Change [%]	Final Si [mg/l]	Change [%]	Final Si [mg/l]	Change [%]	Final Si [mg/l]	Change [%]
Pure 1:1 mixture	130,0	-2,4	95,5	10,7	107,5	-1,4	99,0	-5,3
Standard NOCOLOK® Flux	19,0	85,0			49	53,8	50,0	46,8
$KAlF_4$	17,0	86,6						
$KAlF_4 + BaF_2$	17,5	86,2						
$KAlF_4 + SrF_2$ (A)	20,0	84,3						
NOCOLOK® Li + BaF_2 (A)	20,0	84,3						
$KAlF_4 + SrF_2$ (B)	18,0	85,8						
NOCOLOK® Liflux	18,5	85,4						
NOCOLOK® + CaF_2	15,5	87,8	23,0	78,5	53,5	49,5	49,0	47,9
NOCOLOK® Liflux + BaF_2 (B)	16,5	87,0	24,5	77,1	51,5	51,4	42,5	54,8
NOCOLOK® Liflux + CaF_2	15,5	87,8	24,5	77,1	54	49,1	55,0	41,5

Change in concentration of Na is presented in table 5. It seems that similarly to the changes of Si concentration in pure coolant mixtures all the coolants were quite stable at high testing temperature except Coolant B which showed depletion of Na in solution by about 20%. Looking at the results of testing with flux mixtures we can observe that coolant D showed the lowest depletion of Na in solution for all tested fluxes. In case of NOCOLOK® Li + BaF_2 depletion of Na in solution was also at low level for other coolants except coolant C. Flux NOCOLOK® Li + CaF_2 except coolant B showed the lowest Na depletion values for all investigated fluxes.

Table 5: Depletion of Na from the coolant after five days of testing

Flux type	Coolant A		Coolant B		Coolant C		Coolant D	
	Final Na [mg/l]	Change [%]	Final Na [mg/l]	Change [%]	Final Na [mg/l]	Change [%]	Final Na [mg/l]	Change [%]
Pure 1:1 mixture	3775	-1,2	366	19,9	4815	2,9	2830	-1,8
Standard NOCOLOK® Flux					4440	10,5	2550	8,3
NOCOLOK® + CaF_2	3480	6,7	378	17,4	4340	12,5	2580	7,2
NOCOLOK® Liflux + BaF_2	3640	2,4	450	1,6	4405	11,2	2665	4,3
NOCOLOK® Liflux + CaF_2	3935	-5,5	390	14,7	4905	1,1	2855	2,7

**All K concentration levels presented in this paper are averages from parallel tests run in two cylinders minus the K concentration of the original coolant mixture*

4. CONCLUSIONS

1. Solubility of PBFR depends on both the type of coolant and the type of flux used. The current data indicates that the best results are obtained for combination NOCOLOK® Liflux+ CaF_2 and coolant D. In this case solubility was lowered by about 50% as compared to standard NOCOLOK®.
2. Dissolved PBFR interacts with some coolant additives thus giving rise to potential "gel" formation. This interaction will be less when solubility of the flux in a given coolant is lower. Thus it seems plausible to state that fluxes with lower solubility in the given coolant mixture should be less prone to "gel" formation. This hypothesis requires further verification by real parts testing.
3. Coolant A - due to high depletion of Si - can be considered more critical when used in heat exchangers that contain some Post Brazed Flux Residue inside the cooling loop. Also the coolant B - due to some instability at higher temperature - may not seem to be the best choice.

LITERATURE

[1] B. Yang, A.V. Gershun, F.J. Marinho, and P.M. Woyciesjes, "Effect of Fluoride on Corrosion of Cooling System Metals in Ethylene Glycol-Based Antifreeze/Coolants", in Engine Coolant Technologies: 5th Volume, W. N. Matulewicz, Editor, ASTM STP 1491, pp. 77-86, ASTM (2008)

[2] C. Jeffcoate, M. Ranger, J. Grajzl, B. Yang, P. Woyciesjes, and A. Gershun, "Investigation of Interaction Between Coolant Formulations and flux Loading/Compositions in Controlled Atmosphere Brazed (CAB) Aluminum Surfaces in Heat Exchanger Applications", in Engine Coolant Technologies: 5th volume, W. N. Matulewicz, Editor, ASTM STP 1491, pp. 69-76, (2008)

[3] S. Claeys, "Improved Aluminum Passivation for Heavy-Duty Coolants" D15 – Sixth Symposium on Global Testing of Extended Service Engine Coolants and Related Fluids, J.W Marriott Hotel, Indianapolis, May 23rd, 2013

[4] P. Garcia, H.W. Swidersky "Investigation of Interactions between Silicate-Based Coolants and K-Al-F Brazing Fluxes", Solvay Fluor GmbH, Internal Report, 2010

[5] Solvay - Technical Information: "Post Braze Flux Residue and Engine Coolants", November 2011

WASTE HEAT RECOVERY

Extending the benefits of exhaust gas recirculation (EGR) in gasoline direct injection (GDI) engines with on-board fuel reforming

D Fennell, C Hergueta, M Bogarra, A Tsolakis, M Wyszynski, M Herreros
Mechanical Engineering, University of Birmingham, UK

ABSTRACT

Exhaust gas fuel reforming can improve modern Gasoline Direct Injection (GDI) engine fuel economy, and therefore reduce CO_2 emission in addition to controlling other pollutants such as NO_x and PM. Furthermore, it can potentially reduce pumping loses due to extended dilution. This technology relies on achieving energy recovery from the exhaust stream by endothermic catalytic reforming of hydrocarbons and a portion of exhaust gas. The hydrogen-rich reformate gas obtained during the process has higher enthalpy (up to 30%) than the hydrocarbon fuel utilised in the reforming catalyst. The obtained reformate is then re-circulated back to the engine's intake as REGR (reformed exhaust gas recirculation) using the EGR loop. The effects of REGR composition on engine performance and exhaust temperature have been investigated in this work, showing that REGR improves efficiency as well as gaseous and particulate emissions, and that variation in CO concentration in reformate has only a minor influence on combustion with a reduced range of useable ignition timings.

1. INTRODUCTION

In recent years, the automotive sector has incorporated of a wide variety of new technologies in order to meet stringent emission legislation standards. To achieve the targets, hydrogen has been seen as a very promising energy vector and has been researched widely for use in internal combustion (IC) engines in addition to fuel cells [1]. The hydrogen properties which make it suitable for IC engines are the low minimum ignition energy, which provides more stable ignition, and its high laminar flame velocity that speeds up early flame development, improving combustion stability and increasing indicated efficiency [2]. Its high diffusion coefficient, that may improve fuel-air mixing, and shorter flame quenching distance provide a more complete combustion by oxidizing hydrocarbons closer to the combustion chamber walls, and thus reducing the emission of total hydrocarbons (THC), carbon monoxide (CO) and, PM in both mass and number [3]. Hydrogen combustion can also raise the maximum in-cylinder pressure compare to gasoline due to its rapid combustion, leading to enhanced PM, CO and THC oxidation [4] but simultaneously increasing NO_x emissions [5] owing to the increased flame temperature. However, the use of hydrogen in IC engines has been a challenge for researches due to its drawbacks. Firstly, H_2 is still mostly produced from fossil sources using traditional techniques that lead to net energy losses during its transformation [6]. Secondly, its storage on-board of a vehicle is also a technological barrier because of its low energy density in terms of volume, which has deemed its use in road transport vehicles as impractical [7].

Research initiated at The University of Birmingham suggests the storage of hydrogen on-board in the form of hydrocarbon fuels or other energy vector such as ammonia-NH_3 [8]. These fuels can then be converted into hydrogen-rich gas through a catalytic fuel reforming process known as Reformed Exhaust Gas Recirculation (REGR). This technique is based on a heat recovery strategy where the thermodynamic benefit of the exhaust gas fuel reforming depends on the dominance of two main endothermic reactions: 1) steam reforming reaction (SR) and 2) dry reforming reaction (DR). REGR technique is justified when the hydrogen-rich gas enthalpy is higher than the hydrocarbon energy injected in the reformer, although a fuel efficiency benefit can be obtained with reformer process efficiencies lower than 100% [9]. Endothermic reactions occur between the CO2 and water contained in the exhaust with the hydrocarbons injected are produced during the reforming process. Any oxygen presence in the exhaust gases will be consumed by full or partial oxidation (POX). These exothermic reactions will only marginally affect the process efficiency given its low contribution in relation to the endothermic reactions (assuming low O_2 concentration). However, the exothermic reactions can be beneficial under certain engine conditions to increase the local catalyst temperatures and thus raise the total products yield. The mildly exothermic reaction water gas shift (WGS) reaction normally occurs in the last stage of the reforming process to further increase of H_2 yield [10].

REGR has been extensively studied for diesel and Port Fuel Injection (PFI) engines through simulation analysis [11, 12] and plasmatron reformer [13]. More recently, research has also been focused on GDI engines [14] comparing the benefits of simulated REGR to conventional EGR [15]. The EGR technique has been applied to GDI engines as it can decrease NO_x emissions and improve fuel economy by reducing pumping losses [16-18]. Furthermore, significant reductions in CO [19] and PM [20] have been also observed under EGR conditions. Therefore, the combination of exhaust thermochemical recovery, hydrogen and REGR can further extend the benefits of EGR in terms of fuel economy and emissions, while maintaining the combustion stability. In the literature, significant improvements in indicated engine efficiency, reduced NOx and PM have been reported. As references, simulated reformate gas, with a fixed hydrogen/CO ratio of 3:1 has improved indicated efficiency as the dilution range can be extended [15]. However, an increase in THC was observed. Other studies have shown 6% reduction in the engine fuel consumption [21] and 11% hydrogen peak yield [9] with the application of reformate gas using a full-scale prototype gasoline reformer integrated in a multi-cylinder GDI engine.

The aim of this work in using simulated gas is twofold: i) to establish the effect of different hydrogen and CO concentrations on the engine's efficiency and combustion performance including the influence of the dilution and pumping losses ii) guide the development of the already available fuel reformer as the current prototype presents reduced performance at low loads [9]. Furthermore, engine performance has also been investigated with REGR, across a wide range of engine speed/load conditions. The results are compared to equivalent data for baseline and EGR conditions.

2. EXPERIMENTAL SET UP

An air-guided four-cylinder turbocharged GDI engine has been used in this study. The specifications are depicted in Table 1 and the schematic of the engine and experimental setup is shown in Figure 1.

Table 1: Engine specifications

Engine Specifications	
Compression Ratio	10:1
Bore×Stroke	87.5 x 83.1 mm
Turbocharger	Borg Warner k03
Rated Power	149kW at 6000rpm
Rated Torque	300Nm at 1750-4500rpm

The engine has a dual scroll turbocharger and side-mounted solenoid injector with a centrally located spark plug. The aftertreatment consists of a conventional three-way catalyst (TWC), and consequently, the engine uses a homogeneous charge strategy and stoichiometric combustion conditions where the oxygen concentration is controlled by a Heated Exhaust Gas Oxygen (HEGO) sensor. A camshaft driven high pressure pump feeds the fuel rail and varies the fuel pressure with the engine operating conditions. Originally, the engine does not use an external EGR system; instead the OEM's calibration strategy reduces pumping losses through valve overlapping to increase the residual exhaust gas, known as internal EGR. For the experimental studies, an external EGR system was installed and it is composed of two different types of EGR loop, defined as High Pressure (HP) and Mixed Pressure (MP). For this study, the HP-EGR loop was used to enable the comparison between EGR and REGR. The EGR is passively cooled before being introduced into the intake manifold. A variable area flow meter was used to measure the flow of pre-mixed hydrogen and carbon monoxide into EGR stream to generate the desirable reformate compositions.

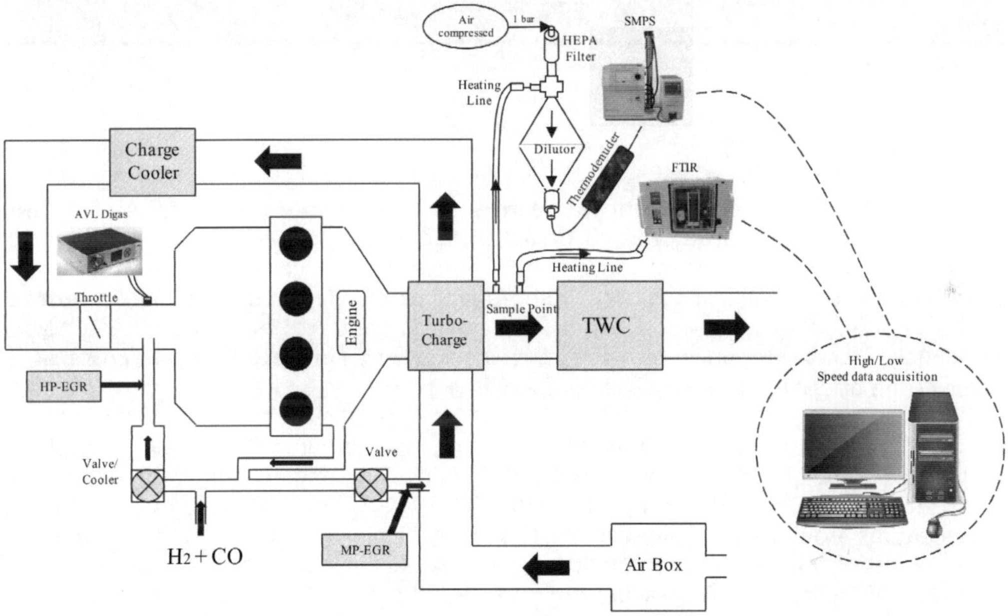

Figure 1: Schematic of engine and instrumentation setup

For in-cylinder pressure measurements, an AVL miniature piezo-electric pressure transducer coupled with a Baumer magnetic encoder every 0.5 crank angle degree (CAD) and 200 cycles were considered for the results. A LabView in-house program controlled the acquisition of in-cylinder pressure data.

A Fourier Transform Infrared Spectroscopy (FTIR) 2100 MKS was used to measure NO_x, the sample was previously filtered to avoid potential damage to the optical lenses. The charge dilution rate was obtained using an AVL Digas 440 non-disperse infrared analyser. It was determined by eq (1):

$$Charge\ Dilution\ Rate = \frac{CO_{2(intake)}}{CO_{2(exhaust)}} \cdot 100\ (\%) \qquad (1)$$

Particle size distribution measurements were carried out using a TSI scanning mobility particle size (SMPS) composed by a series 3080 electrostatic classifier, a 3081 Differential Mobility Analyser (DMA) and a 3775 Condensation Particle Counter (CPC). The samples were then diluted with air at dilution ratio (DR) of 10, using an ejector diluter system, fitted with a high efficiency particulate arrestance filter to precondition the sample. The system was previously heated at 190°C to avoid hydrocarbons and water condensation.

A standard EN228 gasoline with 5.1 % (v/v) ethanol content was injected into the combustion chamber. Hydrogen and carbon monoxide were obtained from industrial cylinder gas bottles in order to obtain the desired reformate composition. The fuel properties are shown in Table 2:

Table 2: Fuel properties [2]

Properties	Hydrogen	Carbon monoxide	Gasoline
Molecular weight	2.016	28	83.22[a]
Low Heat of combustion (kJ/Kg)	120000	10100	42200[a]
Limits of flammability in air (%)	4-75	12.5-74.6	1.1-6
Quenching distance (mm)	0.64	--	3.5
Adiabatic flame Temperature (K)	2390	2383	2370
Laminar burning velocity 360K (cm/s)	290	16.52	51

[a] **provided by shell**

3. TEST CONDITIONS

A sequence of two different tests was carried out to explore the benefits of REGR. For the first test, the target was to establish the effects of different hydrogen and CO concentrations in REGR on the engine's efficiency and combustion performance in addition to the effect of ignition timing advance on PM emissions. Based on the results obtained, the second test was carried out to estimate the engine performance characteristics with the most convenient simulated REGR across of a wider range of engine speed/loads conditions.

For the first test, the ignition timing sweep ranged between 60° and 20° CA before top dead centre (bTDC) at increments of 2° CA. The engine condition used was 80Nm/5.7 bar IMEP at 2000rpm and, 21% EGR/REGR. The two different reformate compositions simulated in this test are shown in Table 3. A fixed 3:1 hydrogen/CO ratio was used for the first composition (C1), based on published [22] results. The flow rate of hydrogen/CO gas mixture was adjusted so that the total volumetric combustible gas fraction in the REGR was 10%. Therefore, the H_2 and CO concentration in the REGR stream will be 7.5% and 2.5% respectively.

Table 3: Simulated REGR compositions

Composition	Simulated REGR
C1	H_2 and CO mixture with 10% total concentration, in a fixed 3:1 ratio flow of
C2	7.5% hydrogen, having equivalent hydrogen content to C1

In the second part of the study, a fixed of 7.5% hydrogen concentration was supplied in order to replicate REGR. It was decided that the CO fraction in REGR would be excluded. This decision was based on the results obtained in part one regarding the relative effect of CO; in general, compositions C1 and C2 showed comparable performance and the extended test matrix (detailed below) would require a large quantity of pre-mixed hydrogen/CO gas, being economically unfeasible. At each test point the maximum EGR or REGR rate was used, defined either by a combustion stability limit of 5±1% COV of IMEP; or by the physical flow limit of the EGR system. The speed/load map of interest was 30 to 110 Nm and 1500 to 3000 rpm at increments of 20 Nm and 500 rpm. For baseline condition the valve timing remained at the standard calibration settings; while for EGR and REGR valve timing was fixed at a setting resulting in low trapped residual (IVO 11 CAD bTDC and EVC 8 CAD aTDC). The ignition timing was modified such that combustion was phased for mass fuel burned MFB50% at 8°±2 aTDC for comparison purposes. The charge air temperature at the inlet port was controlled to 35°C at baseline condition. With the use of high recirculation rates it was not likely to maintain this temperature, so higher set points of 45°C and 50°C were used for the EGR and REGR conditions.

4. RESULTS AND DISCUSSION

4.1 Effect of reformate on engine indicated efficiency and PM emissions

The indicated efficiency and combustion initiation results are plotted in Figure 2 for 21% EGR and the baseline condition included for comparison purposes to simulated reformate. It is observed that the optimum ignition timing was practically the same for C1 compared to C2 so that the CO presence in the reformate composition had low influence on combustion initiation. However, the combustion stability was maintained for longer as ignition timing was retarded with C2. This effect is observed in the abrupt decrease in indicated efficiency for C1 at 35° bTDC ignition timing, which indicates the lower benefits of CO compared to hydrogen combustion, even with equivalent hydrogen concentration in the simulated reformate.

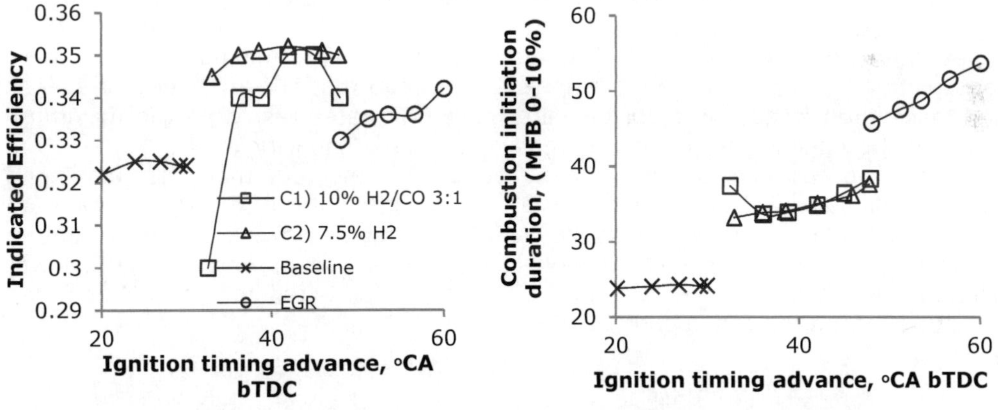

Figure 2: (a) Effect of REGR composition on indicated efficiency and (b) combustion initiation at 5.7bar IMEP, 2000rpm and 21% REGR

The changes in performance for each REGR composition were insignificant compared to 21% EGR and the baseline conditions, as achieved higher indicated efficiency than both EGR and the baseline in most cases. These results indicate that, for a fixed hydrogen fraction concentration the optimum ignition advance is determined by its strong influence on combustion rate. However, CO concentration reduced the range of useable ignition timings where the combustion is stable compared to REGR composition without CO, but it does not have relevant effect on initiation combustion rate.

The results in Figure 3 shows a general trend for increasing PM mass with advanced ignition timing; particularly for undiluted combustion at baseline conditions. This was attributed to the reduction of time for fuel-air mixture; leading to an increase in PM formation. The addition of REGR induced a high PM reduction with respect to baseline conditions. Composition C2, which uses 7.5% hydrogen in REGR, resulted in a 91% reduction in PM mass compared to baseline at optimum ignition timing. The gradient of the PM growth with ignition timing advance is significantly lower for combustion with REGR due to the replacement of some gasoline with gaseous fuel, which results in less liquid fuel presence in the cylinder. Consequently, the reduced time for charge mixing is less critical than for gasoline baseline. There was a negligible difference between C1 and C2, (both contain the same hydrogen concentration) which indicates that as the ignition timing is retarded, CO effect cease to be significant in PM formation. However, it was observed a slightly increase in PM for advanced ignition timings as the reduction of time for fuel-air mixture is also shortened for REGR compositions.

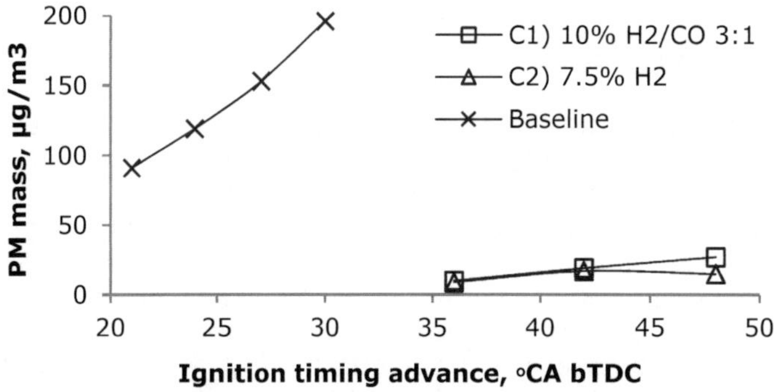

Figure 3: PM mass concentration against ignition timing advance

4.2 REGR performance across wider range of engine operating conditions

The recirculation rates used across the speed/load range for the EGR and REGR cases are shown in Figure 4. The limiting factor determining the maximum recirculation rate varied across the speed/load range map. At lower engine speeds and loads the combustion stability under high EGR rates restricted the maximum rate. As the engine and/or load increased, the flow capacity of the EGR system hindered the introduction of higher rates. It observed that the maximum (combustion stability limited) recirculation rate was 34% for REGR at low/moderate loads, when the manifold pressure ratio was high. However, the maximum achievable with EGR was 26% at high load and low speed, which was 8% lower dilution rate, indicating the beneficial effect of H_2 for extending the dilution rate without affecting the combustion quality.

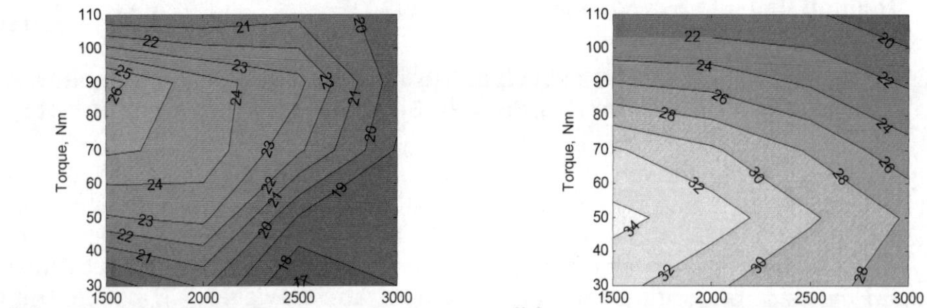

Figure 4: Recirculation rate (%) maps for (a) EGR and (b) REGR

The combustion stability is characterised by the COV of IMEP across the tested range for the REGR case in Figure 5 (a). When the COV is greater than 4%, the test point can be considered as limited by combustion stability. Numbers below 4% indicate the regions in which the recirculation rate was determined by the EGR system and the manifold pressure ratio; and there was insufficient dilution present to cause a significant deterioration in combustion stability. It can be seen a small area where COV of IMEP is 6%, indicating 32% of REGR (Figure 4) at that point exceed the limiting constraint of COV of IMEP. An outstanding increase is obtained at the lowest speed and load as the recirculation rate can be increased from 18% to 32% with REGR.

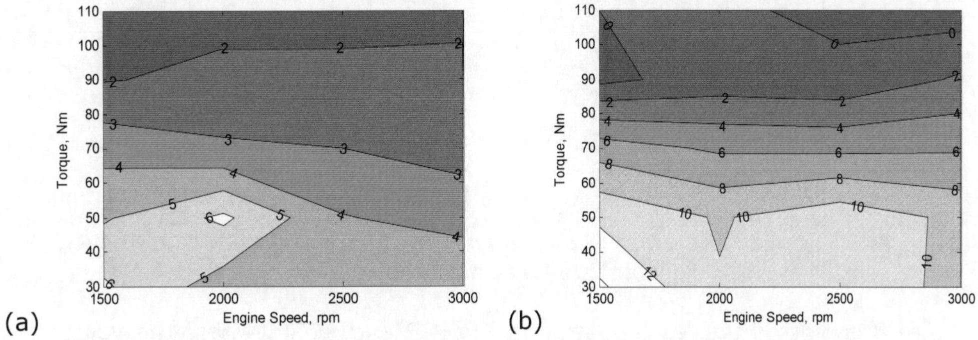

Figure 5: (a) Combustion stability with REGR (COV of IMEP) and (b) Increase in recirculation rate with REGR relative to EGR %

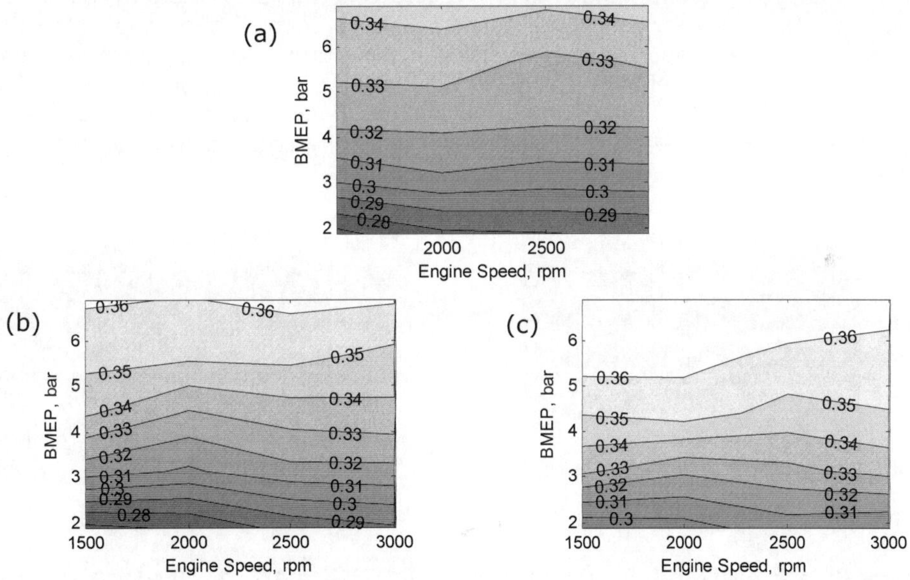

Figure 6: Indicated efficiency (%) performance maps for (a) Baseline, (b) EGR and (c) REGR cases

Indicated efficiency maps for baseline, EGR and REGR cases are compared in Figure 6. A general increasing trend for efficiency was observed across the map with EGR and further improvements with REGR. These results are clearly shown in Figure 7 (a). Conventional EGR is unable to offer significant efficiency improvements at low loads and speeds. However, EGR provided a uniform benefit of 4-7% in the upper half of the map. REGR substantially improves the low speed/load performance relative to both EGR and baseline due to the largely increased recirculation rates while maintaining combustion performance by hydrogen addition. Furthermore,

REGR provided a consistent efficiency improvement of between 2% and 3% beyond EGR, reaching 7-9% improvement compared to baseline across a large portion of the map. Particularly, the lack of improvement induced by EGR at low loads and speeds was attributed to slower combustion rates, which increases the time available for the flame front to be influenced by in-cylinder motion and gas expansion, resulting in lower combustion stability efficiency [23].

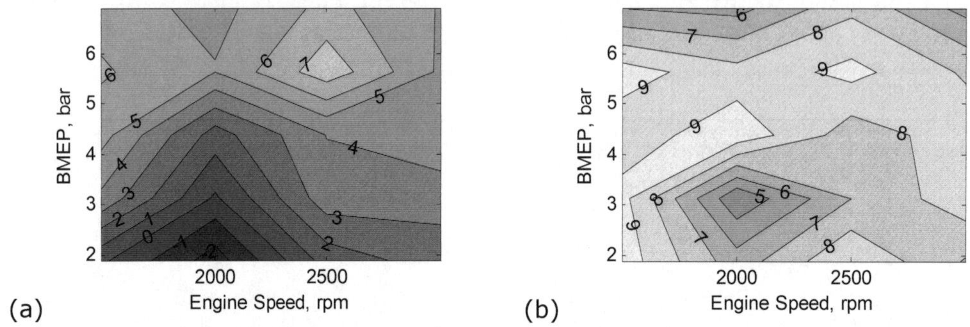

Figure 7: Indicate efficiency (%) improvement for (a) EGR and (b) REGR relative to the baseline

In Figure 8 is plotted the NO_x emissions with EGR and REGR, where the values are normalised against the baseline GDI engine NO_x emissions. The plots clearly indicate lower NO_x across the range, with between 40% and 90% reduction achieved by EGR (Figure 8 (a)). NO_x were reduced markedly by REGR, particularly for low loads due to extended dilution rates, and more consistently than EGR. Engine-out NO_x was between 82% and 96% lower across the map with REGR. However, it has been reported that at the same dilution rate, REGR emitted higher concentration of NO_x relative to EGR [15], since the higher adiabatic flame temperature and faster combustion of hydrogen compared to gasoline result in greater in-cylinder temperatures, leading to slightly increased NO_x formation rate for REGR relative to EGR.

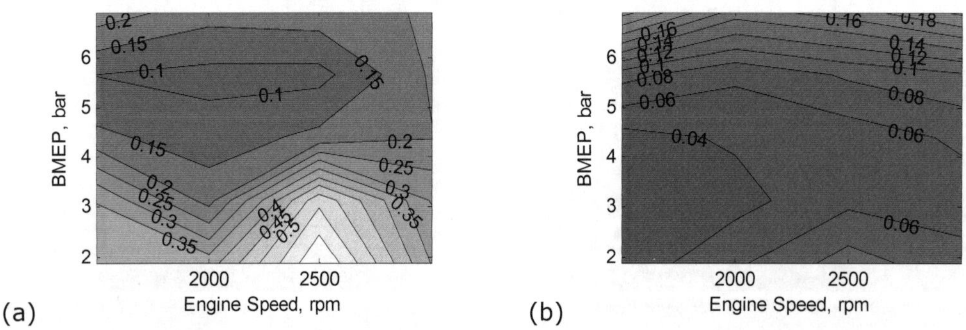

Figure 8: Normalised NO_x emissions for (a) EGR and (b) REGR relative to the baseline

Maps of engine-out temperatures for the baseline and REGR conditions are presented in Figure 9. Under EGR conditions, lower exhaust temperatures across the whole operating range were measured due to the higher heat capacity of the inert species contained in the combustion charge with REGR [23]. According to the results, the fuel reformer could be implemented closer to the exhaust port in order to utilise the highest exhaust stream temperature, which would be highly beneficial to carry out the endothermic reforming reactions maximising the reformer performance. The reformer could potentially be redesigned and mounted close to the exhaust ports but there are many additional considerations for the reformer re-design and engine integration. These include i) the effect of increasing exhaust back pressure on engine performance ii) high thermal loading of relatively intricate (thin walled) reformer components iii) packaging in this location would also be detrimental to turbocharger performance, since fuel reforming reactions would reduce the exhaust enthalpy and physical restriction would result in a pressure drop, reducing the turbine inlet pressure.

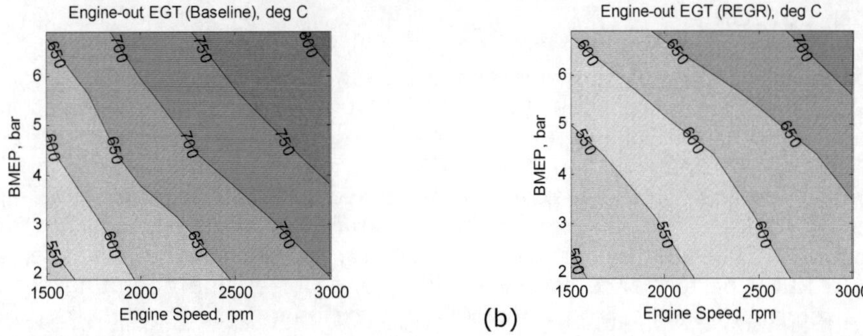

(a) (b)

Figure 9: Engine-out EGT map temperatures (°C) for (a) baseline and (b) REGR

The effect of REGR on post-TWC exhaust temperature is presented in Figure 10 as reference of the conditions achieved for the fuel reforming reactor fitted in this location. The baseline condition post-TWC showed a substantial reduction in temperature when compared to the engine-out temperature, mostly due to the expansion across the turbine and heat losses from of the exhaust system to the surroundings. However, the post-TWC temperature, when REGR is used, is comparable with the baseline condition as there is not significant drop in temperature. It is thought that the slightly enlarged engine-out HCs with high REGR condition resulted in exothermic oxidation reactions over the TWC; therefore, the temperature drop is insignificant. Even so, the post-TWC EGT was below 600 °C for the low/medium load speed range, which means that the reformer efficiency will growth as the speed and load are increased because of the endothermic reforming reaction would be more favoured at these conditions.

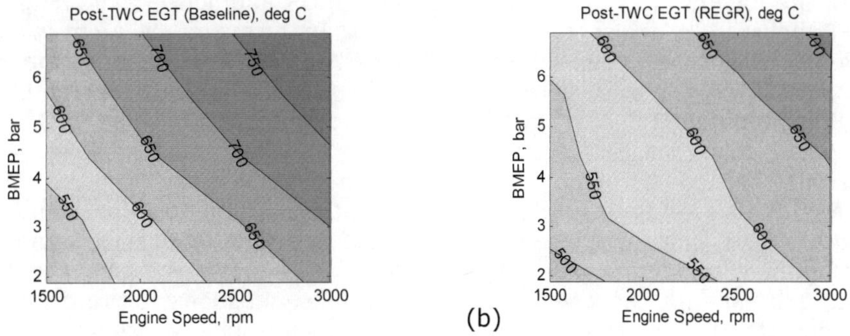

(a) (b)

Figure 10: Engine Post-TWC EGR map temperatures (°C) for (a) baseline and (b) REGR

Improvements in reformer performance are required at low load [9] in order to enhance engine efficiency (Figure 6), suggesting that a reformer re-design may be useful to achieve greater heat transfer under conditions when there is less waste heat available. Smaller size reactor or alternative catalyst geometries (e.g. corrugated cross flow) would enable to promote turbulent flow and as a result higher mass and heat transfer [24]. Therefore, improved heat transfer rates may be obtained from the engine bulk exhaust gas to the reforming catalyst during the reforming process. This would be favourable for endothermic reforming reactions at lower temperatures. The reduced size would also have benefits in terms of cost and integration within the engine exhaust and vehicle. Additionally, a reduction of catalyst thermal mass can minimise the warm-up time and reach the maximum performance in shorter time.

5. CONCLUSION

The extended benefits of reformed exhaust gas recirculation on a GDI engine with on-board fuel reforming has been demonstrated in this investigation. The investigation consisted of two experiments using hydrogen, or hydrogen and CO, added to EGR in order to simulate REGR and generate reformate with realistic, achievable composition. The experiments aimed to: 1) Determine the effect of different hydrogen and CO concentrations in REGR on the engine's efficiency, combustion and PM emissions as a function of ignition timing advance, and 2) establish REGR performance across a wider range of engine operating conditions, using a reformate composition where CO was excluded based on results previously obtained.

The effect of CO in the reformate composition showed that the optimum ignition timing was practically the same for composition C1 compared to C2, which indicates that CO presence in the reformate had low influence on combustion initiation. However, CO content in C1 reduced the range of useable ignition timings, since it showed an abrupt decrease in indicated efficiency for C1, even with equivalent hydrogen concentration. Additionally, the simulated REGR induced a 91% reduction of PM emissions at optimum ignition timing with respect to baseline conditions. The replacement of some gasoline with gaseous fuel (H_2 and CO) led to less liquid fuel presence in-cylinder and, as a result, lower particle formation.

Engine performance maps showed that REGR increased indicated engine efficiency between 6-9% across most of the engine range compared to baseline combustion, and 2-3% compared to combustion with EGR. The improvement was led by the beneficial physicochemical hydrogen properties for extending the dilution rate without affecting the combustion stability. Therefore, REGR enabled to increase the dilution rate for REGR up to 6% with respect to EGR addition, maintaining the COV of IMEP below 5% across the engine range. In terms of NO_x emissions, EGR achieved reductions between 40% and 90%, while NO_x were reduced markedly by REGR, accounting for 82%-96% lower across the map due to extended dilution rates. Engine-out EGTs were found to be significantly cooler with REGR, but the reduction of post-TWC temperature relative to baseline was not as large due to exothermic reactions across the TWC. The temperature maps produced provide a guide for fitting a reformer prototype, the most favourable location being close to the exhaust port due to the elevated temperature that favours the endothermic reforming reactions. However, the practical challenges of locating the reformer close to the exhaust has given way to propose suggestions for re-designing the reforming catalyst and achieve high reforming performance across of the full range of engine loads.

ACKNOWLEDGMENTS

The work was funded by an industry-academia collaboration project, CO_2 Reduction through Emissions Optimisation (CREO: ref. 400176/149), which is funded by the Innovate UK (Technology Strategy Board). Daniel Fennell also received a postgraduate scholarship from this project. Ford Motor Company is acknowledged for supplying the base engine and associated parts, Johnson Matthey and

Cambustion Ltd are also acknowledged for advice with the experimental part and analysis of the results. Internally, Carl Hingley and his team of technicians are acknowledged for their assistance during the test cell construction. Cruz Hergueta would like to thank the University of Birmingham for providing the fully funded scholarship.

REFERENCES

1. W. Mohamed, W.A.N. and M.H. M. Kamil, Hydrogen preheating through waste heat recovery of an open-cathode PEM fuel cell leading to power output improvement. *Energ.Convers.Manage*, (2016).**124**: p. 543-555, DOI: http://dx.doi.org/10.1016/j.enconman.2016.07.046.
2. He, Z., et al., Effects of H2 and CO enrichment on the combustion, emission and performance characteristics of spark ignition natural gas engine. *Fuel*, (2016).**183**: p. 230-237, DOI: http://dx.doi.org/10.1016/j.fuel.2016.06.077.
3. Zhou, J.H., C.S. Cheung, and C.W. Leung, Combustion, performance, regulated and unregulated emissions of a diesel engine with hydrogen addition. *Appl. Energ*, (2014).**126**: p. 1-12, DOI: http://dx.doi.org/10.1016/j.apenergy.2014.03.089.
4. Karvonen, M., et al., Technology competition in the internal combustion engine waste heat recovery: a patent landscape analysis. *J. Clean. Prod.*, (2016).**112, Part 5**: p. 3735-3743, DOI: http://dx.doi.org/10.1016/j.jclepro.2015.06.031.
5. Verma, G., et al., Experimental investigations of combustion, performance and emission characteristics of a hydrogen enriched natural gas fuelled prototype spark ignition engine. *Fuel*, (2016).**178**: p. 209-217, DOI: http://dx.doi.org/10.1016/j.fuel.2016.03.022.
6. Hajjaji, N., A. Houas, and M.-N. Pons, Thermodynamic feasibility and life cycle assessment of hydrogen production via reforming of poultry fat. *J. Clean. Prod.*, (2016).**134, Part B**: p. 600-612, DOI: http://dx.doi.org/10.1016/j.jclepro.2015.12.018.
7. Rossetti, I., M. Compagnoni, and M. Torli, Process simulation and optimisation of H2 production from ethanol steam reforming and its use in fuel cells. 1. Thermodynamic and kinetic analysis. *Chem. Eng. J.*, (2015).**281**: p. 1024-1035, DOI: http://dx.doi.org/10.1016/j.cej.2015.08.025.
8. Wang, W., et al., Ammonia as hydrogen carrier for transportation; investigation of the ammonia exhaust gas fuel reforming. *Int. J. Hydrogen Energ.*, (2013).**38**(23): p. 9907-9917, DOI: http://doi.org/10.1016/j.ijhydene.2013.05.144.
9. Fennell, D., et al., Thermochemical recovery technology for improved modern engine fuel economy - part 1: analysis of a prototype exhaust gas fuel reformer. *RSC Advances*, (2015).**5**(44): p. 35252-35261, DOI: 10.1039/c5ra03111g.
10. Leung, P., et al., Raising the fuel heating value and recovering exhaust heat by on-board oxidative reforming of bioethanol. *Energ. Environ. Sci.*, (2010).**3**(6): p. 780-788, DOI: 10.1039/b927199f.
11. Tsolakis, A., et al., Dual Fuel Diesel Engine Operation Using H2. Effect on Particulate Emissions. *Energ. Fuel*, (2005).**19**(2): p. 418-425, DOI: 10.1021/ef0400520.
12. Abu-Jrai, A., A. Tsolakis, and A. Megaritis, The influence of and CO on diesel engine combustion characteristics, exhaust gas emissions, and after treatment selective catalytic reduction. *Int. J. Hydrogen Energ.*, (2007).**32**(15): p. 3565-3571, DOI: http://dx.doi.org/10.1016/j.ijhydene.2007.02.014.
13. Bromberg, L., et al., Emissions reductions using hydrogen from plasmatron fuel converters. *Int. J. Hydrogen Energ.*, (2001).**26**(10): p. 1115-1121, DOI: http://dx.doi.org/10.1016/S0360-3199(01)00049-0.

14. Fennell, D., et al. GDI Engine Performance and Emissions with Reformed Exhaust Gas Recirculation (REGR). in SAE 2013 World Congress & Exhibition United States, 2013, SAE International.
15. Fennell, D., J. Herreros, and A. Tsolakis, Improving gasoline direct injection (GDI) engine efficiency and emissions with hydrogen from exhaust gas fuel reforming. *Int.J.Hydrogen.Energ*, (2014).**39**(10): p. 5153-5162, DOI: http://dx.doi.org/10.1016/j.ijhydene.2014.01.065.
16. Fontana, G. and E. Galloni, Experimental analysis of a spark-ignition engine using exhaust gas recycle at WOT operation. *Appl.Energ*, (2010).**87**(7): p. 2187-2193, DOI: http://dx.doi.org/10.1016/j.apenergy.2009.11.022.
17. Wei, H., et al., Gasoline engine exhaust gas recirculation – A review. *Appl.Energ*, (2012).**99**: p. 534-544, DOI: http://dx.doi.org/10.1016/j.apenergy.2012.05.011.
18. Luján, J.M., et al., Influence of a low pressure EGR loop on a gasoline turbocharged direct injection engine. *Appl.Therm.Eng*, (2015).**89**: p. 432-443, DOI: http://dx.doi.org/10.1016/j.applthermaleng.2015.06.039.
19. Alger, T., T. Chauvet, and Z. Dimitrova, Synergies between High EGR Operation and GDI Systems. *SAE Int. J. Engines* (2008).**1**: p. 101-114, DOI: 10.4271/2008-01-0134.
20. Lattimore, T., et al., Investigation of EGR Effect on Combustion and PM Emissions in a DISI Engine. *Appl.Energ*, (2016).**161**: p. 256-267, DOI: http://dx.doi.org/10.1016/j.apenergy.2015.09.080.
21. Bogarra, M., et al., Study of particulate matter and gaseous emissions in gasoline direct injection engine using on-board exhaust gas fuel reforming. *Appl.Energ*, (2016).**180**: p. 245-255, DOI: http://dx.doi.org/10.1016/j.apenergy.2016.07.100.
22. Fennell, D.A., *Exhaus Gas Fuel Reforming for Improved Gasoline Direct Injection Engine Efficiency and Emissions* in *Mechanical Engineering*. 2014, University of Birmingham: Birmingham.
23. Ladommatos, N., S. Abdelhalim, and H. Zhao, The effects of exhaust gas recirculation on diesel combustion and emissions. *Int. J. Engine Res.*, (2000).**1**(1): p. 107-126, DOI: 10.1243/1468087001545290.
24. Mucha, P., G. Gaiser, and M. Kuehnle, Corrugated Cross Flow Catalyst with Permeable Walls. *SAE Int. J. Fuels Lubr*, (2008), DOI: 10.4271/2008-01-0809.

Thermo-economic optimization during preliminary design phase of organic Rankine cycle systems for waste heat recovery from exhaust and recirculated gases of heavy duty trucks

L Guillaume, A Legros, R Dickes, V Lemort
Thermodynamics Laboratory, University of Liège, Belgium

ABSTRACT

Waste heat recovery (WHR) Organic Rankine Cycle (ORC) system is a very promising technology for reducing fuel consumption and consequently the CO2 emissions of future heavy-duty trucks (HDT). Nonetheless, the adoption of this technology in the automotive domain requires specific R&D activities going from the system definition to the on-board integration. This study focuses on the preliminary design phase of ORC systems recovering the heat wasted from two of the sources available on a HDT: the exhaust and recirculated gases. From these heat sources and their combinations, 6 possible architectures are identified. On the other hand, 4 volumetric expansion machine technologies are considered (scroll, screw, piston and vane Expanders). At the end, 24 topologies are therefore modelled considering the main components (Pump, Heat exchangers, Expansion machines). A three-step optimization method is proposed to identify the most promising system. First, the most suitable conditions are identified for the design of the ORC systems using a simple model of volumetric expansion machine. In a second step, the design phase, using more detailed models for the expansion machines, a thermos-economic optimization is performed. Finally, in a third step, the output power of the latter system models is maximized in off-design conditions, optimizing the evaporating pressure and the overheating degree.

Keywords: Organic Rankine Cycle, Waste Heat Recovery, Heavy Duty Trucks, Thermo-Economic optimization.

NOMENCLATURE

SYMBOLS

A	Heat transfer area [m²]
Bo	Boiling number [-]
C	Empiric coefficient [-] or Cost [€]
D	Diameter [m]
DC	Direct cost [€]
e	Plate thickness [m]
F	Coefficients in Martin's correlation [-]

SUBSCRIPTS

cd	Condenser
co	Corrugation
cond	Conduction
conv	Convection
cp	Compressor
ctrl	control
EG	Exhaust Gas

f	Fanning friction factor [-]
G	Mass velocity [kg/s.m²]
h	Specific enthalpy [J/kg]
IC	Indirect cost [€]
k	Thermal conductivity [W/m.K]
L	Plate length [m]
$\dot{M}$	Mass flow rate [kg/s]
m	Empiric coefficient [-]
mu	Dynamic viscosity [Pa.s]
n	Empiric coefficient [-]
N	Rotational speed [RPM]
Nu	Nusselt number [-]
P	Pressure [pa]
p	Pitch [m]
Pr	Prandtl number [-]
$\dot{Q}$	Heat flow rate [W]
Re	Reynolds number [-]
T	Temperature [°C or K]
SIC	Specific investment cost [€]
TIC	Total investment cost [€]
U, h	Heat transfer coefficient [W/m².K]
V	Volume [m³]
$\dot{W}$	Mechanical or electrical power [W]
x	Vapor mass fraction [-]

EGR	Exhaust Gas Recirculation
Eng	Engineering
eq	Equivalent
ev	evaporator
ex	Exhaust
exp	Expander
h	Hydraulic
instl	installation
l	Saturated liquid
liq	Liquid
LT	Low temperature
p	piping
pp	Pump
rot	rotational
s	swept or isentropic
su	Supply
syst	System
tp	Two-phase
v	Saturated vapor
vap	Vapor
wf	working fluid

GREEK SYMBOLS

β	Chevron angle [rad]
ϵ	Efficiency [-]
ρ	Density [kg/m³]

1. INTRODUCTION

It is now accepted that greenhouse gas emissions from human activities are enhancing the global warming effect keeping the earth at a temperature higher than it would otherwise be (1). In the last twenty years, the World energy consumption has increased by more than 30% (2) despite the adoption by the United Nations on Climate Change of the Kyoto Protocol (3). In 2012 (Figure 1) the transport sector was responsible of about one third of the European CO2 emissions. And it was pointed out that "*Heavy-Duty Vehicles (HDV) represent about a quarter of EU road transport CO2 emissions and some 6% of the total EU emissions"* (4). Despite recent improvements in fuel consumption efficiency, HDV emissions are still rising, mainly because of the increasing number of vehicles in traffic. Nowadays, HDV are the second biggest source of emissions within the transport sector, i.e. larger than both international aviation and shipping. Thus, the reduction of these CO2 emissions has become a strategic goal of the EU.

A very promising solution is the valorization of the waste heat, which is about 60% of the combustion energy, transforming it into mechanical or electrical energy to increase the overall vehicle energy efficiency and consequently reducing its CO2 emissions and fuel consumption. The heat re-use can be performed by means of a heat engine (e.g. organic or steam Rankine cycle systems) using the waste heat as energy source, as it is being adopted for large stationary applications. The adoption of such technology in the automotive domain requires specific R&D activities in order to identify the most appropriate system architecture and integration level so to achieve sustainable cost and the reliability requirements. A major part of these activities is dedicated to the selection of the working fluid and the expansion machine. The working fluid selection process has been widely investigated in several studies (6,7,8,9,10,11) but no universal optimal fluid is indicated since the choice is highly dependent to the target application. A detailed list of the guidelines and indicators that should be considered when selecting an organic fluid for power generation are reported in (9).

This study focuses on volumetric expansion machines and particularly on the scroll, screw, piston, and vane technologies for which experimental data is available. Numerous design steady-state models of ORCs are built. Using these models, a three steps optimization method is proposed starting with the selection of the design conditions. Then, for these conditions, a thermo-economic optimization is performed to design the systems. Finally, the performance of the systems is optimized and evaluated in off-design conditions, optimizing the evaporating pressure and the overheating degree.

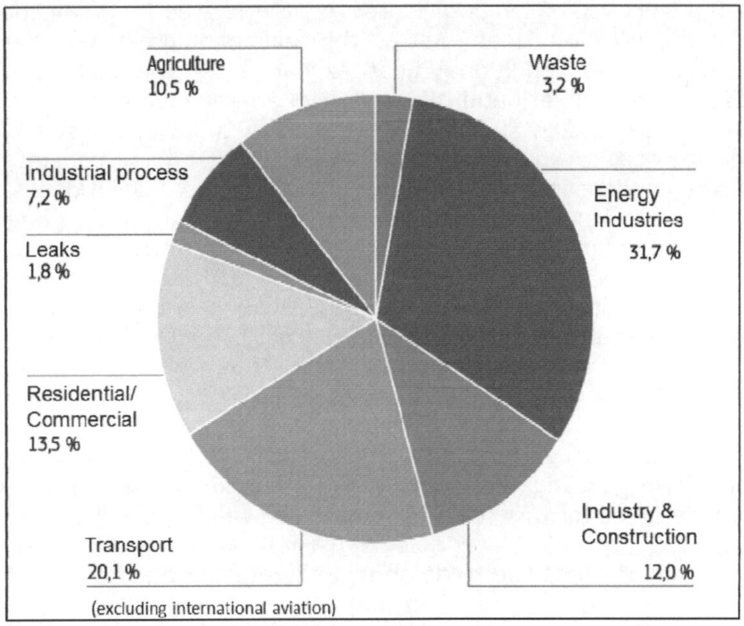

Figure 1: European CO2 emissions by sector in 2009 (5)

2. DEFINITION OF THE CASE STUDY

This paper attempts to address this problematic of selecting the architecture, the expander technology, and the working fluid for a waste heat recovery organic (or non-organic) Rankine cycle system devoted to be coupled to a truck engine. In a first step, the use case is defined in terms of investigated heat sources and heat sinks, working fluid candidates, technologies of the ORC components and engine operating conditions on which the system will be designed and its performance will be evaluated.

2.1 Heat sources and heat sinks

Multiple heat sources can be considered to recover the thermal energy wasted by the engine. A list of the potential heat sources available on a truck equipped with an Exhaust Gas recirculation (EGR) system is given in Table 1. In this table, the temperature levels of the heat sources and their capacity flow rates are compared. Because, increasing the number of sources will increase the level of complexity in terms of packaging and control and the cost of the system, it is proposed, in this study, to investigate only two heat sources. In (12), a second law analysis is performed and shows, based on the work performed in (13) and (14), that the most interesting sources are the exhaust gases, and the recirculated gases, to which correspond the highest temperature levels. The charged air is not considered since a modification of its temperature level might largely affect the performance of the engine.

The only heat sink available on a truck is air, the ram air created by the vehicle movement and the air pulsed by the fan. The vehicle cooling module, usually composed of the air conditioning (A/C) condenser, the charge air cooler (CAC), the main engine radiator and the cooling fan, uses this air as secondary fluid for the different heat exchangers. Because of space constraints, adding a new heat exchanger (direct condenser or a radiator dedicated to an indirect condenser) in the front end of the truck is challenging. Moreover, the limited heat rejection capacity margin of the cooling module as well as the additional pressure drops resulting from the insertion of an additional heat exchanger in the cooling module are constraining the heat recovery and might lead to an engagement of the fan, which consumes a lot of energy. A configuration without excessive fan power consumption, due to the ORC system, is the use of the engine coolant in order to feed an indirect condenser as studied in (14). Innovative solutions such as the increase of the heat rejection capacity using part of the body panels as heat exchangers (flat heat exchangers) functionalizing the aerodynamic under-body and of part of the side body panels were proposed (15) but might be more complex to implement. Finally, another possibility, the one retained in this study, is the use of a low temperature radiator dedicated to the Rankine condenser and placed between the charge air cooler and the engine radiator. This configuration leads to an increase in net output power of the WHR system of about 11% compared to the use of the engine coolant (12).

Combining the two investigated sources and the single heat sink, 6 possible topologies are identified for this study. They are shown in figure 2.

Table 1: Description of the heat sources available on the investigated truck*

Heat source	Temperature [°C]	Capacity flow rate [kW/K]
Exhaust gases	Mid to high (325)	High (0.16)
Recirculated gases	High (450)	Low (0.04)
Charged air	Low to mid (95)	High (0.2)
Engine coolant	Low (90)	High (1)
Oil	Low (90)	Low (N.A.)

*US10 11l 320 kW

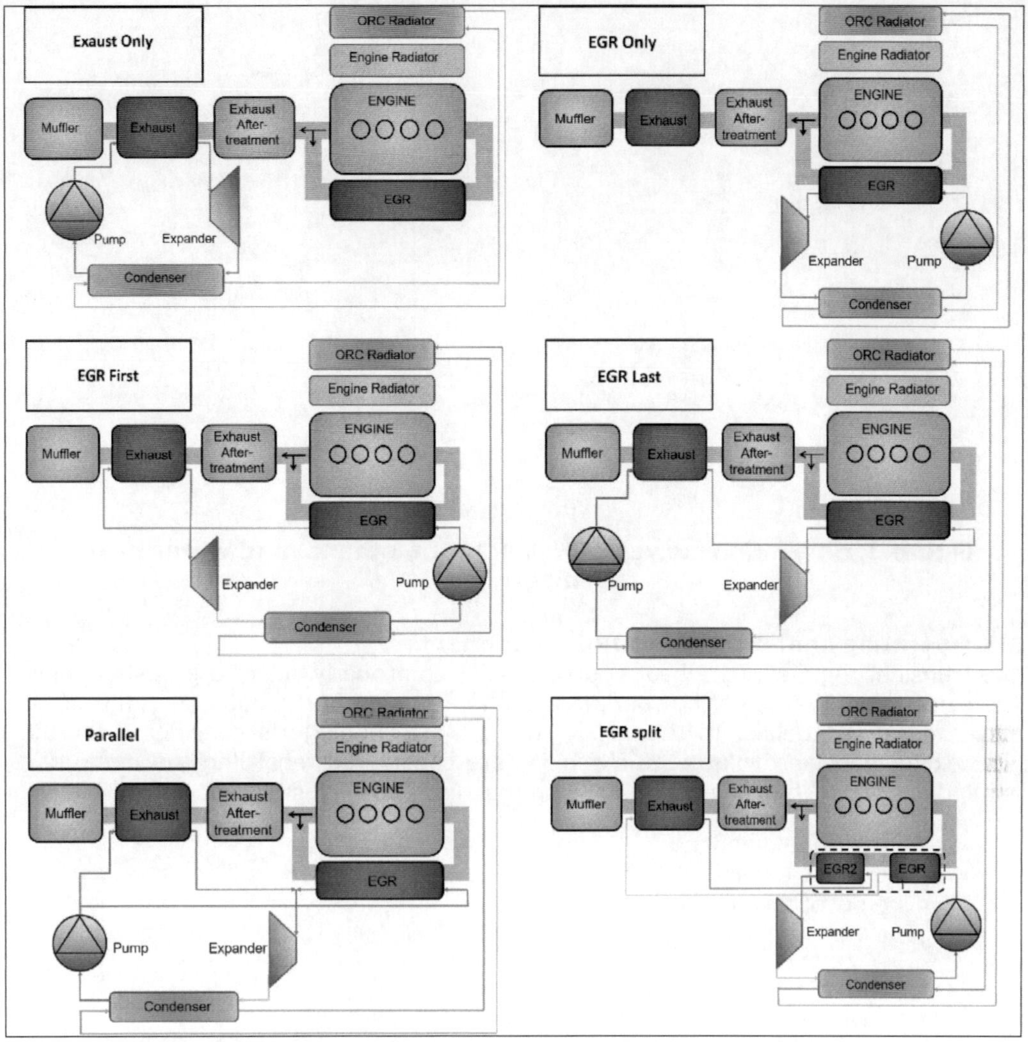

Figure 2: Description of the 6 investigated topologies (CAC not represented)

2.2 Working fluid candidates

Hundreds of working fluids are available on the market and the selection of working fluids has been treated widely in scientific publications. Most of the time, a comparison between a set of candidate working fluids in terms of thermodynamic performance and based on a thermodynamic model of the cycle is proposed. Quoilin (16) proposed some guidelines and indicators (thermophysical properties, thermodynamic performance, environmental impact, etc) when selecting the most appropriate working fluid. He also showed that very few studies include additional parameters considering the practical design of the ORC system, mainly because of the difficulty to define a proper function for the multi-objective optimization of the cycle. He concluded that economic considerations should drive the final choice. In this study, three working fluids, which are currently the most frequently investigated (17,18,19,20,21,22) and covering a wide range of critical temperatures, are selected for the thermodynamic and thermo-economic optimizations of the ORC systems: HFC-R245fa, water and ethanol.

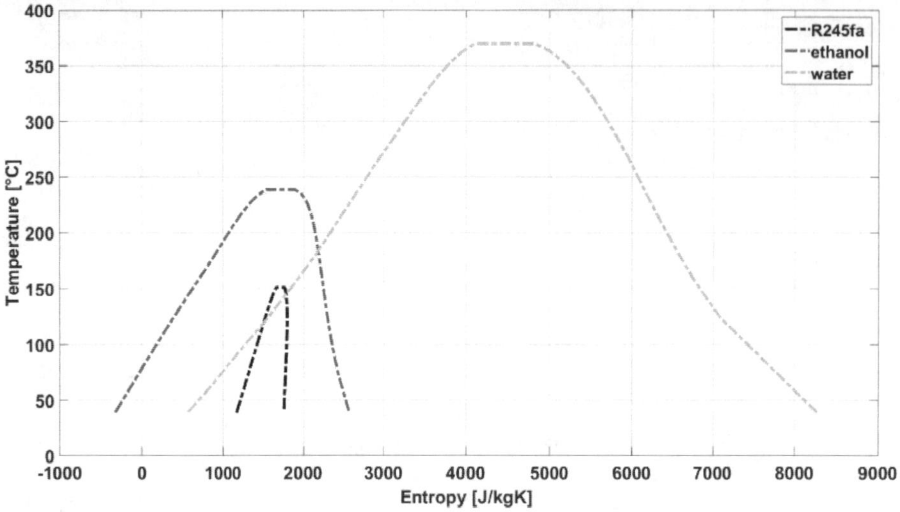

Figure 3: Saturation curves of HFC-R245fa, ethanol and water in T-s diagram

2.3 Operating conditions and ORC components

The transient regime of the heat source implies to properly select the design point. The selection of the operating conditions is the first step of the optimization process and is discussed below. In this study, the plate technology is considered for the heat exchangers and volumetric machines are considered regarding the pump and expander. Four technologies of expansion machines are investigated:

- The scroll expander
- The screw expander
- The piston expander
- The vane expander

3. MODEL OF THE WASTE HEAT RECOVERY SYSTEMS

The case study being defined, the models of the components used for these systems are briefly described in this section. All the models are implemented in the Matlab environment.

3.1 Models of the heat exchangers

The plate heat exchangers are modeled by means of the Logarithmic Mean Temperature Difference (LMTD) method for counter flow heat exchangers. They are subdivided into three moving-boundaries zones corresponding to the three phases in which the working fluid can be found. In addition, the two-phase region is divided in *N* finite volumes. Each of the single-phase zones and each cell of the two-phase region is characterized by a heat transfer area *A* and a heat transfer coefficient *U* whose product is known based on the exchanged power and the LMTD.

$$AU = \frac{\dot{Q}}{LMTD} \qquad (1)$$

The heat transfer coefficient U is calculated by considering two convective and one resistive heat transfer resistances in series (secondary fluid side, metal wall and refrigerant side).

$$\frac{1}{U} = \frac{1}{h_{wf,conv}} + \frac{1}{k_{wall,cond}/e_{wall}} + \frac{1}{h_{sf,conv}} \qquad (2)$$

The area of each zone can therefore be computed and the total heat transfer area of the heat exchanger is calculated as follow.

$$A_{tot} = A_{liq} + \sum_{i=1}^{N} A_{tp,i} + A_{vap} \quad (3)$$

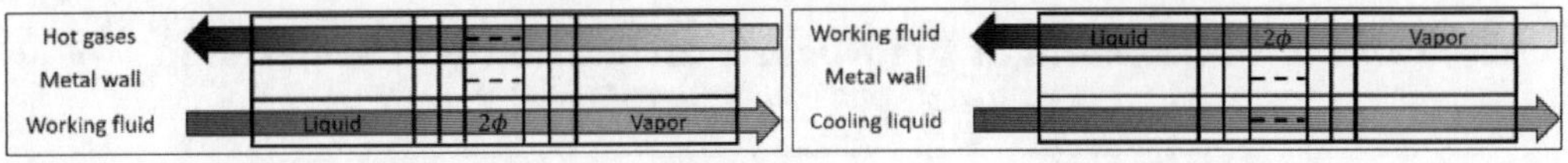

Figure 4: Description of the evaporator (left) and condenser (right) models

3.1.1 Single phase heat exchange coefficients and pressure drops

3.1.1.1 Working fluid side

Convective heat transfer coefficients (h_{conv}) and Fanning friction coefficients (f) are evaluated using Martin's correlation (23):

$$f = \left[\frac{\left(0.045 * \tan(\beta) + 0.09 * \sin(\beta) + \frac{F_0}{\cos(\beta)}\right)^{0.5} + \left(1 - \frac{\cos(\beta)}{\sqrt{3.8 * F_1}}\right)}{\cos(\beta)}\right]^2 \quad (4)$$

$$Nu = 0.205 * Pr^{\frac{1}{3}} * (f * Re^2 * \sin(2 * \beta))^{0.374} \quad (5)$$

with

- $if\ Re < 2000$:

$$F_0 = \frac{16}{Re} \quad (6) \qquad F_1 = \frac{149.25}{\mathrm{Re}} + 0.9625 \quad (7)$$

- $if\ Re \geq 2000$:

$$F_0 = (1.56 * \log(\mathrm{Re}) - 3)^{-2} \quad (8) \qquad F_1 = \frac{9{,}75}{\mathrm{Re}^{0.289}} \quad (9)$$

And where β is the chevron angle [rad] per Martin's definition, G is the mass velocity [kg/s.m2], D_h is the hydraulic diameter [m]. The friction pressure drops are computed with the following relation:

$$\Delta P = \frac{(2 * f * G^2)}{\rho * D_h} * L \quad (10)$$

where ρ is the mean density of the fluid [kg/m³] and L is the plate length [m].

3.1.1.2 Secondary fluid side

For the water-cooled condenser, the Martin's correlation is used as for the working fluid side. It shows good agreements when compared to the experimental data. On the other hand, for the gas heat exchangers (i.e., the evaporators), correlations found in literature lead to large errors on the heat flow rate prediction (up to 50 %). Because the heat transfer resistance is mainly located on the gas side, the heat exchange coefficient must be defined accurately. Thus, measurements were performed on an existing plate type heat exchanger built for an automotive ORC application and a Dittus-Boelter type correlation was derived to evaluate the Nusselt number and the heat exchange coefficient on the gas side (equation 11).

$$Nu = C * Re^m * Pr^n \qquad (11)$$

The C and m parameters were calibrated minimizing the error between the measured heat flow rate and the prediction of the model.

Table 2: Parameters of the Nusselt correlation on the gas side ($Re \in [10^4 - 3.8\ 10^4]$)

Parameter	Value
C	0.05
m	0.8
n	0.333

3.1.2 Two-phase heat exchanges and pressure drops

Convective heat transfer coefficients (h_{conv}) and Fanning friction coefficients (f) are evaluated using correlation for evaporation and condensation by Han and al. (24,25):

- Evaporation

$$Ge_1 = 2.81 * \left(\frac{p_{co}}{D_h}\right)^{-0.041} * (\beta)^{-2.83} \qquad (12) \qquad Ge_2 = 0.746 * \left(\frac{p_{co}}{D_h}\right)^{-0.082} * (\beta)^{0.61} \qquad (13)$$

$$Ge_3 = 64710 * \left(\frac{p_{co}}{D_h}\right)^{-5.27} * (\beta)^{-3.03} \qquad (14) \qquad Ge_4 = -1.314 * \left(\frac{p_{co}}{D_h}\right)^{-0.62} * (\beta)^{-0.47} \qquad (15)$$

$$Bo_{eq} = \frac{q}{G_{eq} * i_{fg}} \qquad (16) \qquad Nu = Ge_1 * Re_{eq}^{Ge_2} * Bo_{eq}^{0.3} * \mathrm{Pr}_l^{0.4} \qquad (17)$$

- Condensation

$$Ge_1 = 11.2 * \left(\frac{p_{co}}{D_h}\right)^{-2.83} * (\beta)^{-4.5} \qquad (18) \qquad Ge_2 = 0.35 * \left(\frac{p_{co}}{D_h}\right)^{-0.23} * (\beta)^{1.48} \qquad (19)$$

$$Ge_3 = 3524.1 * \left(\frac{p_{co}}{D_h}\right)^{-4.17} * (\beta)^{-7.75} \qquad (20) \qquad Ge_4 = -1.024 * \left(\frac{p_{co}}{D_h}\right)^{-0.0925} * (\beta)^{-1.3} \qquad (21)$$

$$Nu = Ge_1 * Re_{eq}^{Ge_2} * \mathrm{Pr}_l^{\frac{1}{3}} \qquad (22)$$

Where β is the chevron angle [rad] according to Martin's definition, p_{co} is the corrugation pitch [m], Bo_{eq} is the equivalent boiling number, q is the heat flux [W/m²] and G_{eq} and Re_{eq} are the equivalent mass velocity [kg/s.m²] and Reynolds number defined as a function of the fluid vapor mass fraction x and the densities of the saturated liquid ρ_l and vapor ρ_v:

$$G_{eq} = G * [1 - x + x * \left(\frac{\rho_l}{\rho_v}\right)^{0.5}] \qquad (23) \qquad Re_{eq} = \frac{G_{eq} * D_h}{\mu_l} \qquad (24)$$

The friction pressure drops are calculated as follow:

$$f = Ge_3 * Re_{eq}^{Ge_4} \qquad (25) \qquad \Delta P = \frac{2 * f * G_{eq}^2}{\rho_l * D_h} * L \qquad (26)$$

3.1.3 *Sizing of the heat exchangers*

For given plate spacing, width, and corrugations (amplitude, chevron angle, and enlargement factor) of the plate heat exchanger, two degrees of freedom are available when sizing the component: the length and the height. These two degrees of freedom are fixed by the heat exchange area requirement and the limitation on the pressure drop on the secondary fluid side.

3.2 Pump model

A model with constant values of global efficiency (equation 27) and volumetric efficiency (equation 28) is proposed for the pump. Based on these efficiencies, the mass flow rate, the outlet temperature, and the mechanical power consumed by the pump can be predicted using the following equations.

$$\epsilon_{g,pp} = \frac{\dot{M}_{wf,pp} * \left(h_{wf,ex,s,pp} - h_{wf,su,pp}\right)}{\dot{W}_{pp}} \quad (27) \qquad \epsilon_{v,pp} = \frac{\dot{M}_{wf,pp}}{\rho_{wf,su,pp} * N_{rot,pp} * V_{s,pp}} \quad (28)$$

For given inlet temperature and pressure conditions, mass flow rate and rotational speed of the pump, the displacement can be computed using equation 28.

3.3 Expander model

A semi-empirical model is proposed to predict the mass flow rate, the outlet temperature and the mechanical power produced by the investigated expanders. This model retains the most important physical phenomena inherent to the expansion machine and involve a limited number of parameters (around 10) which are identified on the basis of performance points. Studies in laboratory showed that these models can predict the performance of the expansion machines with good accuracy. Moreover, their semi-empirical nature enables to extrapolate the performance (26) of the machine for different operating conditions and design characteristics (displacement, sections of the inlet and exhaust ports, etc.). The model used as starting point is the one proposed by Lemort and al. (27) which has been validated for the scroll expander.

Based on the model, the global and volumetric efficiencies of the machine can be evaluated using equations 29 and 30:

$$\epsilon_{g,exp} = \frac{\dot{W}_{exp}}{\dot{M}_{wf,exp} * \left(h_{wf,su,exp} - h_{wf,ex,s,exp}\right)} \quad (29) \qquad \epsilon_{v,exp} = \frac{\dot{M}_{wf,exp}}{\rho_{wf,su,exp} * N_{rot,exp} * V_{s,exp}} \quad (30)$$

3.3.1. *Sizing of the expanders*

For given inlet temperature and pressure conditions, mass flow rate and rotational speed of the expander, the displacement can be computed using equation 30. As for the heat exchangers on the gas side, experimental data is used as starting point for the sizing of the expansion machines (26,28,29, 30). Indeed, semi-empirical models are used to describe each of the four investigated technologies and the parameters of the models must be defined. It is proposed to first calibrate these parameters using measurements performed on actual machines used as reference. Performance indicators are then defined for each source of losses (Filling factor, friction mean effective pressure, etc.) and are conserved during the design phase of the ORC system. Using these indicators, the corresponding parameters required by the semi-empirical model can be computed for the machine being currently designed.

Table 3: Sizing method used for the expansion machines

Reference expansion machine: Model parameters calibrated based on experimental data	**Reference expansion machine: definition of performance indicators on the design point**	**New expansion machine: new set of model parameters obtained conserving performance indicators on the design point**
3 Heat transfer parameters (supply exhaust and ambient): AU_{ref}	3 Heat transfer effectiveness (NTU method): ϵ_{th}	3 Heat transfer parameters (supply exhaust and ambient): AU_{new}
Internal leakage: $A_{leak,ref}$	Filling factor: Φ	Internal leakage: $A_{leak,new}$
Mechanical losses: α_{ref}, $\dot{W}_{loss,ref}$, $C_{loss,ref}$	Friction mean effective pressure: $fmep$	Mechanical losses: α_{new}, $\dot{W}_{loss,new}$, $C_{loss,new}$
2 Pressure drops parameters (supply and exhaust): d_{ref}	Pressure drop efficiency: $\epsilon_{dp} = \Delta P/P$	2 Pressure drops parameters (supply and exhaust): d_{new}

3.4 System model

The global model of the system is obtained by interconnecting each component model. The two main assumptions for the evaluation of the performance in design and off-design are the following:

- The sub-cooling degree at the outlet of the condenser is 5K.
- The fixed global efficiency of the pump is 50%.

4. SELECTION OF THE DESIGN CONDITIONS

The first step of the design process is the selection of the design conditions. To achieve this, the frequency distribution of the couples (engine torque, engine speed) over the driving cycle is mapped. 5 torque levels and 6 speed levels are considered in this study, which leads to 17 operating points. For each of these points, a system model including a simplified version of the expansion machine (including only losses due to the non-adapted volume ratio) are designed and optimized. The superheating degree and the built-in volume ratio of the expander are fixed and the evaporating pressure and the pinch-point of the heat exchangers of the 6 topologies are optimized for the 3 working fluids so to maximize the power produced. The mass of the heat exchangers is constrained to a limit of 60 kg while the pressure is limited to a maximum of 32 bar based on the information coming from a truck manufacturer. Then the power produced by each topology is maximized in off-design for the other 16 operating points, still optimizing the evaporating pressure. Finally, for each of the 17 systems designed, an average is performed on the power produced and weighted over the 17 points using the frequency distribution so to identify the most promising design conditions for each topology and each fluid. At the end, the point corresponding to the most important fuel consumption is leading to the maximal weighted average power and is selected for the design phase.

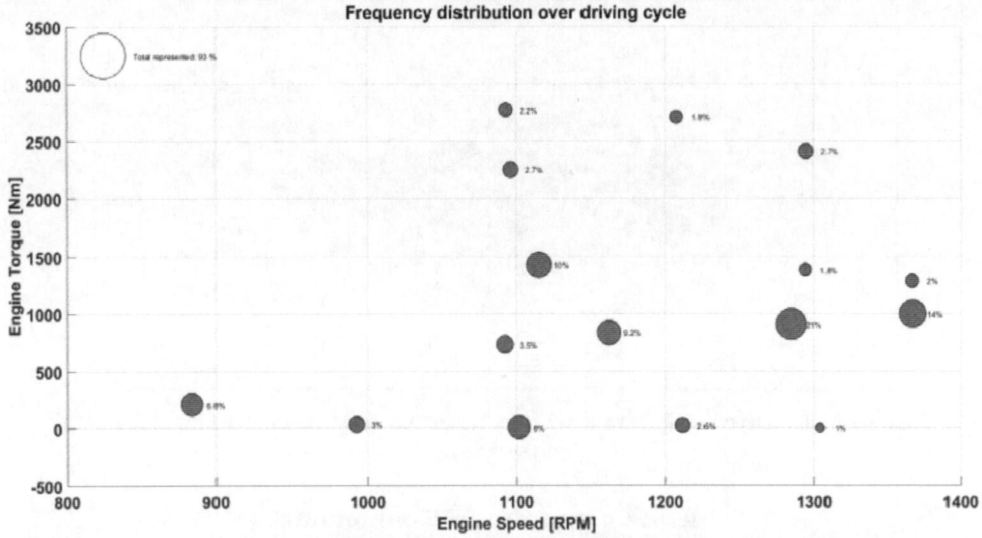

Figure 5: Frequency distribution of the couples (engine torque, engine speed) over the driving cycle

5. DESIGN: THERMO-ECONOMIC OPTIMIZATION

The design points being defined, the 24 systems including the detailed models of the 4 expansion machine technologies can be designed using the 3 working fluids (72 simulations models).

5.1 Costs

The objective of the design optimization is the minimization of the specific investment cost (SIC), ratio between the Total Investment Cost (TIC) and the power produced by the system. The TIC is the sum of the direct cost (DC) and Indirect Costs (IC). In the DC can be found the cost of the system components (C_{syst}), of the fluid (C_{wf}), of the piping (C_{piping}) and of the instrumentation and control of the system ($C_{instr,ctrl}$). The IC are mainly composed of the engineering (C_{eng}) and contingency costs (C_{cont}). Except for the cost of the system, assumptions are achieved on the DC and the IC. They are described in equation 31 to 39. The cost of the system is calculated based on an estimation of the cost of the components made from experience and discussion with manufacturers. These costs are listed in table 3. The costs of the expansion machine technologies are estimated using Solidworks. 3D sketches (Figure 6) were performed, or were directly used when available, to establish the list of the main parts of the machines and evaluate their cost as a function of the displacement of the expander assuming the material (stainless steel) and the fabrication process (molding) and a serial production of 20.000 machines per year.

$$C_{syst} = C_{pp} + C_{EGR} + C_{EG} + C_{EGR2} + C_{exp} + C_{cd} + C_{LT} \quad (31)$$

$$C_{instl} = 0.3 * C_{syst} \quad (32)$$

$$C_{piping} = 0.15 * C_{syst} \quad (33)$$

$$C_{instr,ctrl} = 0.2 * C_{syst} \quad (34)$$

$$DC = C_{syst} + C_{wf} + C_{instl} + C_{piping} + C_{instr,ctrl} \quad (35)$$

$$C_{eng} = 0.08 * DC \quad (36)$$

$$C_{cont} = 0.15 * C_{eng} \quad (37)$$

$$IC = C_{eng} + C_{cont} \quad (38)$$

$$TIC = DC + IC \quad (39)$$

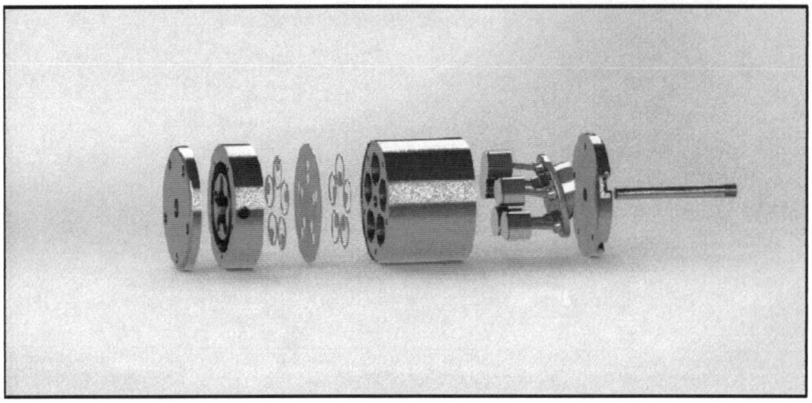

Figure 6: Example of 3D sketches used to estimate the cost of the expanders

Table 4: Cost of the ORC components

Component	Variable	Cost [€]
Heat exchanger	Mass (kg)	$250 + 5 * Mass$
Expander	Displacement (cm³)	
Scroll **Screw** **Piston** **Vane**		$15.94 * V_{s,cp}^{0.667}$ $47.739 * V_{s,cp}^{0.4167}$ $14.223 * V_{s,cp}^{0.6582}$ $11.111 * V_{s,cp}^{0.8716}$
Low temperature circuit	/	200
Working fluid	Mass (kg)	
R245fa **Ethanol** **Water**		$25 * Mass$ $1.096 * Mass$ $0.124 * Mass$

5.2 Constraints and optimization variables

When selecting the design conditions, a simplified model of the expansion machine is used. The constraints of the optimization are also reduced. When considering the actual system, additional constraints should be considered. The rotational speed, the inlet temperature, the vapor mass fraction, and the pressure ratio are generally limited for each expansion machine technology. These validity ranges are summarized in Table 4 resulting from a literature review. On the other hand, the temperature of the exhaust gases at the outlet of the evaporator must be kept above a minimal value (around 100°C) to avoid the condensation of acids. The NOx reduction strategy should not be impacted by the WHR system and the recirculated gases should be kept below a maximal temperature (around 120°C). These constraints as well as the optimization variables during the design are summarized in Table 5.

Table 5: State of the art performed for the expansion machine technologies

Expander technology	Scroll	Screw	Piston	Vane
Rotational speed range [RPM]	< 10000	< 25000	$500 - 6000$	< 4000
Maximal inlet temperature [°C]	250	490	> 500	< 165
Range of built-in volume ratio [-]	$1.5 - 4.9$	$2.5 - 6$	$3 - 14$	$2 - 4.5$
Range of pressure ratio [-]	25	50	*same as in ICE*	*N.A.*

Table 6: Constraints and variables of the thermo-economic optimization

Optimization variables	**Economic constraints**
Pinch point in the heat exchangers	Mass of the heat exchangers
Evaporating Pressure	Materials
	Constraints for WHR
	Outlet temperature of the exhaust gases Temperature of the EGR Heat rejection capacity
	Technical constraint (expander)
Expander rotational speed Overheating degree Built-in volume ratio	Limited speed range Maximal inlet temperature Maximal vapor mass fraction Maximal pressure ratio

5.3 Results

This study does not have the pretention to give the exact figures of what would be the performance of the waste heat recovery system installed on a truck but rather to give a methodology and the trends. This is a prediction work which is obviously limited by the assumptions made. Being more accurate would require more complex and deterministic models of the ORC components (CFD). However, it would also lead to unacceptable calculation times at this stage because of the number of topologies to investigate and the number of optimizations to perform. Here a good compromise between accuracy and calculation time is expected because of the use of validated but still relatively simple semi-empirical models (0D).

The active-set algorithm through the Matlab optimization toolbox is used to performed a global minimization of the SIC of each system. Results are shown in Figure 7. It can be observed that topologies 2 to 5, including the EGR as heat source are the most promising whatever the working fluid. The topology with the split of the EGR heat exchanger is on the other hand more expensive while the power produced by the systems is limited when using the exhaust gases only. In general, ethanol seems to be the most appropriate working fluid. Also, because of the low temperature of the heat source, water should be avoided using topology 1. Moreover, water is not well adapted to the vane expander. This is explained by the constraint on the maximal inlet temperature with this technology. The high built-in volume ratio of the piston expander makes it not suitable to be used with R245fa. Finally, the most appropriate solution, based on the assumptions made here is the use of the screw expander integrated in the parallel topology and using ethanol as working fluid.

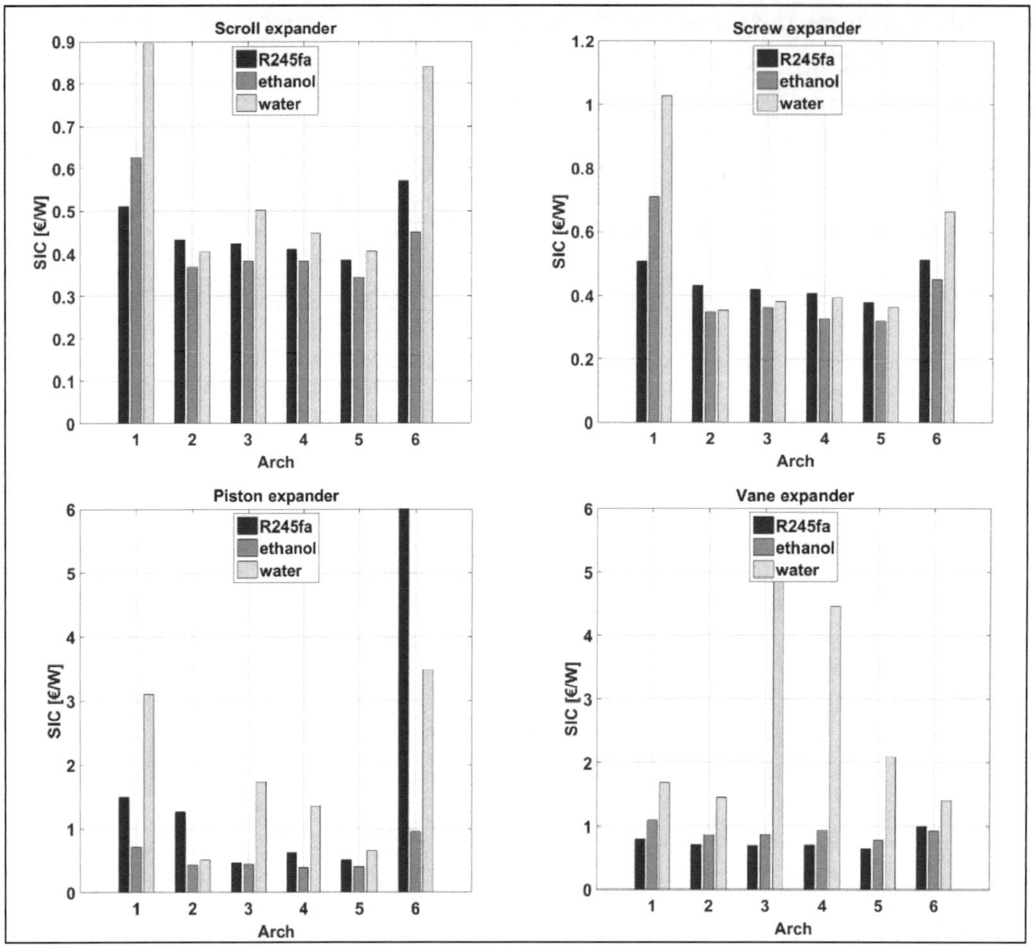

Figure 7: SIC of the 4 expansion machines, 6 topologies and 3 working fluids

6. OFF DESIGN PERFORMANCE: OPTIMIZATION

Finally, the performance of the 4 most promising architectures (2 to 5) and using ethanol as working fluid are maximized optimizing the evaporating pressure and the superheating degree for each of the 16 off-design operating points of the frequency distribution map (Figure 5). A weighted average based on these frequencies is performed on the power produced by the systems. Results are shown in Figure 8. Again, the parallel architecture using the screw expander seems to be the most appropriate solution for the WHR system. However, results are close between the scroll, the screw and the piston expander and performance and cost cannot be the only selection criteria at the end. Thus, for instance, the parallel architecture will lead to a more complex control strategy since two mass flow rates must then be regulated and specially the flow cooling the recirculated gases.

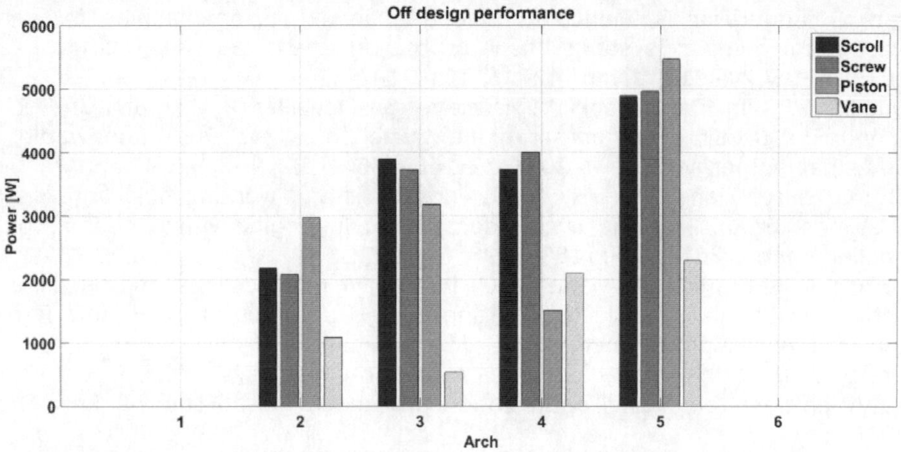

Figure 8: Off-design performance of the most promising topologies: Average power produced by the systems and weighted based on the frequency distribution of the engine operating points

7. CONCLUSION AND PERSPECTIVE

In conclusion, a 3-steps optimization methodology has been proposed for the selection of the heat sources, working fluid and expansion machine of an ORC recovering the heat wasted from the engine of a heavy-duty truck. 6 topologies, 3 working fluids and 4 expansion machines technologies are considered. First an optimization is performed to select the design condition. Then a thermo-economic optimization is achieved to design the systems. Finally, the performance is maximized in off-design. Based on the results, it is shown that ethanol is the most appropriate working fluid and the use of the screw expander leads to the best average performance. However, results are close between the scroll, the screw, and the piston expanders. Quasi-static simulations could be performed along driving cycles of the truck to evaluate the energy recovered by the systems and thus the fuel consumption reduction. To be more accurate, this would also require evaluating the power consumption of the fan when it is engaged. Finally, criteria other than performance and cost should be considered (complexity of the system and of the control strategy, safety, etc.).

REFERENCES

1. Kokic P, Crimp S, Howden M, A probabilistic analysis of human influence on recent record global mean temperature changes, Climate Risk Management 2014;3:1-12.
2. IEA, Key world energy statistics, http://www.iea.org/publications/freepublications//publication/KeyWorld2014.pdf 2014.
3. UNFCCC, Kyoto protocol to the united nations framework convention on climate change, http://unfccc.int/resource/docs/convkp/kpeng.html, 1992.
4. European comission, Towards a strategy to address co2 emissions from HDV, http://ec.europa.eu/clima/policies/transport/vehicles/heavy/index.htm, 2013.
5. European comission, Climate action, http://europa.eu/pol/pdf/ipbook/fr/climate_action_fr.pdf.
6. Badr O, Probert SD, O Callaghan PW, Selecting a working fluid for a Rankine cycle engine, Applied Energy 1985;21:1-42.
7. Bao J, Zhao L, A review of working fluid and expander selections for organic rankine cycle, Renewable and Sustainable Energy Reviews 2013;24:325-342.

8. Latz G, Andersson S, Munch K, Selecting an expansion machine for vehicle waste-heat recovery systems based on the rankine cycle, SAE Technical Paper 2013-01-0552, 2013, doi:10.4271/2013-01-0552.
9. Quoilin S, Van Den Broek M, Declaye S, Dewallef P, Lemort V, Techno-economic survey of organic rankine cycle (orc) systems, Renewable and Sustainable Energy Reviews 2013;22:168-186.
10. Shu G, Liu L, Tian H, Wei H, Yu G, Parametric and working fluid analysis of a dual-loop organic rankine cycle (dorc) used in engine waste heat recovery, Applied Energy 2014;113:1188-1198.
11. Toffolo A, Lazzaretto A, Manente G, Paci M, A multi-criteria approach for the optimal selection of working fluid and design parameters in organic Rankine cycle systems, Applied Energy 2014;121:219-232.
12. Grelet V., Rankine cycle based waste heat recovery system applied to heavy duty vehicles: topological optimization and model based control, PhD Thesis, 2016.
13. Teng, H., Regner, G., and Cowland, C. Waste heat recovery of heavy-duty diesel engines by organic Rankine cycle part i: Hybrid energy system of diesel and rankine engines. SAE Technical Paper, 2007. SAE International.
14. Edwards, S., Eitel, J., Pantow, E., Geskes, P., Lutz, R., and Tepas, J. Waste heat recovery: The next challenge for commercial vehicle thermomanagement, 2012. SAE International Journal of Commercial Vehicles, 5:395–406.
15. NoWaste Project, http://nowasteproject.eu
16. Quoilin S, Sustainable energy conversion through the use of organic Rankine cycles for waste heat recovery and solar applications., Ph.D. thesis, University of Liege, Liege, Belgium, 2011.
17. Amicabile S. Lee J., KUM D., A comprehensive design methodology of organic Rankine cycles for the waste heat recovery of automotive heavy-duty diesel engines, Applied Thermal Engineering 87 (2015) 574e585.
18. Grelet V., Dufour P., Nadri M., Reiche T., Lemort V., Modeling and control of Rankine based waste heat recovery systems for heavy duty, IFAC-PapersOnLine 48-8 (2015) 568–573.
19. Grelet V., Reiche T., Lemort V. Nadri M., Dufour P., Transient performance evaluation of waste heat recovery Rankine cycle based system for heavy duty trucks, Applied Energy 165 (2016) 878–892.
20. Maciann V, Serrano JR, Dolz V, Sanchez J, Methodology to design a bottoming Rankine cycle, as a waste energy recovering system in vehicles. Study in a HDD engine, Applied Energy 2013;104:758-771.
21. T. Yamaguchi, Y. Aoyagi, N. Uchida, A. Fukunaga, M. Kobayashi, T. Adachi, et al. Fundamental study of waste heat recovery in the high boosted 6-cylinder heavy duty diesel engine SAE Int J Mater Manuf, 8 (2) (2015), pp. 209–226 http://dx.doi.org/10.4271/2015-01-0326
22. Zheng N, Zhao L, Wang XD, Tan YT, Experimental verication of a rolling-piston expander that applied for low-temperature organic rankine 485 cycle, Applied Energy 2012;112:1265-1274.
23. Martin H, A theoretical approach to predict the performance of chevron-type plate heat exchangers, Chemical Engineering and Processing 35 1996;301-310.
24. Han DH, Lee KJ, Kim YH, Experiments on the characteristics of evaporation of R410a in brazed plate heat exchangers with different geometric configurations, Applied Thermal Engineering 2003;23;1209-1225.
25. Han DH, Lee KJ, Kim YH, The characteristics of condensation in brazed plate heat exchangers with different chevron angles, Journal of the Korean Physical Society 2003;43,1;66-73.
26. Dickes R. Dumont O., Daccord R., Quoilin S., Lemort V., Modelling of organic Rankine cycle power systems in off-design conditions: An experimentally-validated comparative study., Energy 2017, In Press, Accepted Manuscript, Available online 30 January 2017.

27. Lemort V, Quoilin S, Cuevas C, Lebrun J, Testing and modeling a scroll expander integrated into an organic rankine cycle, Applied Thermal Engineering 2009; 29; 3094-3102.
28. Oudkerk JF., Development of a semi-analytical model of volumetric expander for system-level simulation 2016, in Proceedings of ECOS 2016.
29. Vodicka V., Guillaume L., Mascuch J., Lemort V., Testing and modeling a vane expander used in an orc working with hexamethyldisiloxane (MM) 2015, in proceedings of ASME ORC 2015.
30. Dumont O., Experimental investigation of four volumetric expanders, article in preparation, IV International Seminar on ORC Power Systems, ORC2017 13-15 September 2017, Milano, Italy.

Waste Heat Recovery Technologies: A technical comparative assessment of EHRS, TEG, ORC and TAC

S Uppuluri, H Khalane, A Naiknaware
Computational Sciences Experts Group, USA
D Sahoo, A Kotrba
Tenneco Inc, USA

ABSTRACT

This paper is a comparative assessment of four waste heat recovery technologies. Organic Rankine Cycle (ORC), Thermo-Electric Generators (TEG), Exhaust Heat Recovery System heat exchangers (EHRS) and Thermo-acoustic converters (TAC). All these technologies operate quite differently, capturing a different ratio of exhaust heat, operating under different engine points and suitable for different types of vehicles. The technical characteristics of each of these have been assessed using a vehicle level energy model for a light duty truck and a class 8 truck, as applicable. Results indicate that the percentage of waste heat captured with the each of these technologies is drastically different and does not correlate with their fuel economy impact.

1 INTRODUCTION

There have been significant efforts underway on waste heat recovery to capture some of the 40% fuel energy leaving the tailpipe and to improve the fuel economy of the fleet. Some of these technologies include Organic Rankine Cycle (ORC), Thermo-Electric Generators (TEG), Exhaust Heat Recovery System heat exchangers (EHRS) and Thermo-acoustic converters (TAC) among others. All these technologies operate quite differently, capturing a different ratio of exhaust heat, operating under different engine points and suitable for different types of vehicles. Though they have been numerous papers discussing the technical details of construction, design optimization and controls of each of these technologies, there has not a good comparison of these technologies from the macro perspective to understand and contrast them against the objective – efficiently recover as much of the 30%-40% of fuel energy being lost through the exhaust. This technical assessment is novel in its approach as it compares each of these technologies for the amount of exhaust heat recovered, energy generated, the benefits and penalties of each of them and the net fuel economy impact on the vehicle. In addition, a market study was also conducted by the technical market research team to understand the perception of these technologies by the OEMs. This market research complemented the technical study well as the research funding to explore such waste heat technologies and the mindshare by the ultimate customers is heavily based on the perception of these technologies as well and not just the technical and cost merits of them.

2 WHR TECHNOLOGIES CONSIDERED

Here is a brief description of the four waste heat recovery technologies that were considered in this study for comparison.

2.1 Exhaust Heat Recovery System

The exhaust heat recovery system is a heat exchanger between the exhaust and the engine coolant to recover some of the exhaust heat back into the engine during the cold warm-up portion of the drive cycle. This additional heat speeds up the warm-up of the engine, thereby reducing its friction and improving the fuel economy. The thermal energy captured by the EHRS is kept as thermal energy (as opposed to other WHR technologies where thermal energy is converted to electrical or mechanical energy). Figure 1 shows the schematic of a cooling system of a conventional powertrain vehicle using an EHRS system. Detailed information on the EHRS systems can be found in [1] and [2].

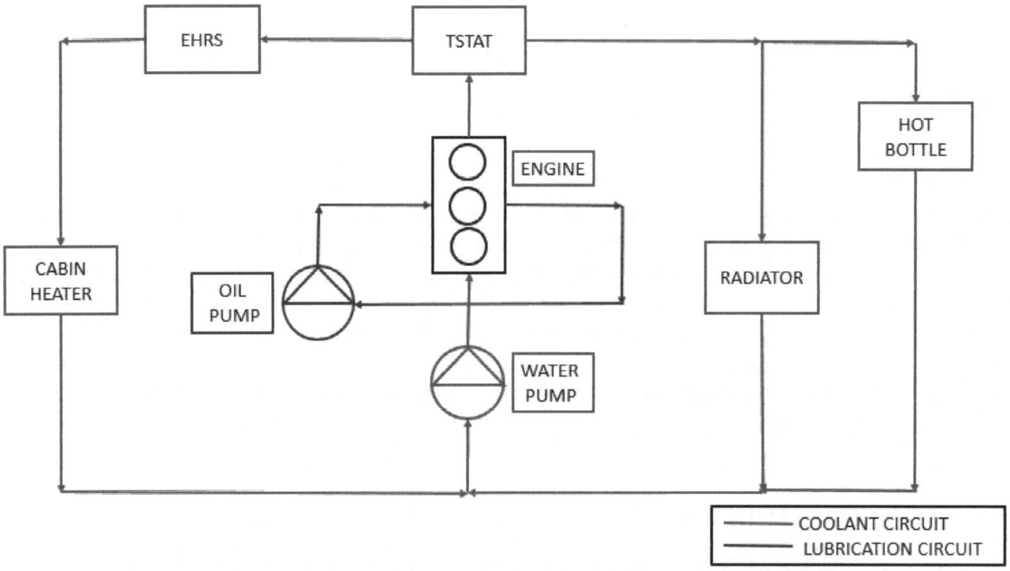

Figure 1: Simplified vehicle cooling system layout with EHRS system

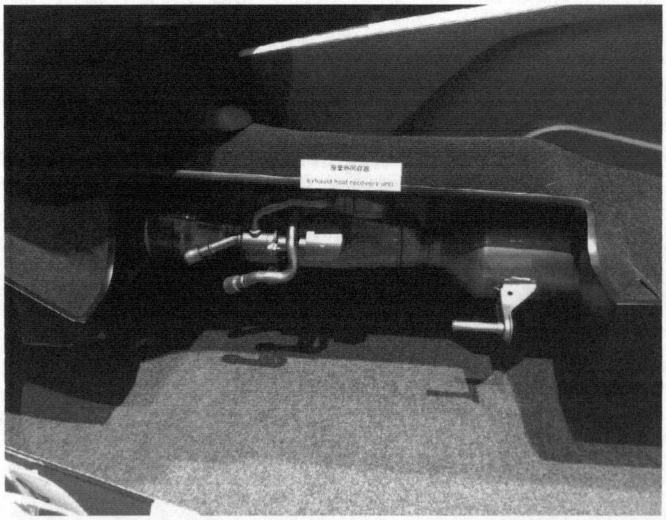

Figure 2: EHRS system installed underbody of a Toyota Prius

2.2 Thermo-Electric Generators

A Thermoelectric generator is a solid-state device that is typically made out of the rare earth metals and generates electricity from a temperature difference across it (Seebeck effect). The material must have a high electrical conductivity but a low thermal conductivity for it to serve as an effective thermoelectric material. A TEG module typically has multiple cards of TEG material arranged in a unique configuration to extract the maximum amount of electrical energy out of the exhaust. The hot side of the TEG is typically the exhaust and the engine coolant is the cold side. The TEG electrical power generated increased with increasing exhaust mass flow and increasing temperature differential across the TEG material. Figure 3 below shows how a TEG works. Additional technical details on how TEGs work and their current challenges are well documented in [3], [4] and [5].

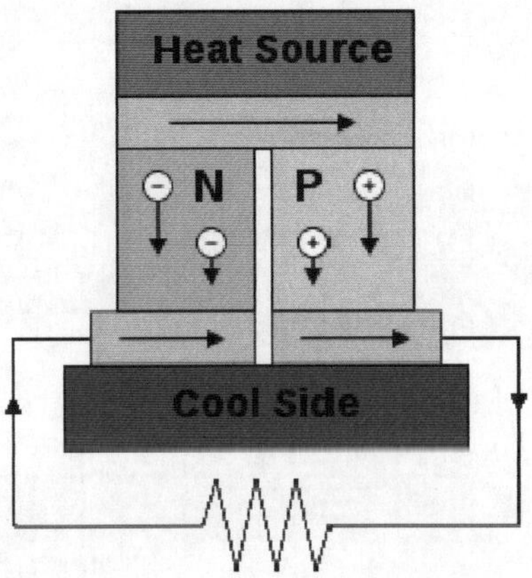

Figure 3: Thermo-electric generator (2)

2.3 Organic Rankine Cycle

In its simplest form, the Organic Rankine Cycle system is a two phase system comprising of an evaporator to capture the exhaust heat into the organic fluid, which then expands over expander to generate mechanical energy. The majority of this heat captured is rejected back out to the environment using a condenser. The operation of this system isn't unlike a two phase A/C system in the vehicle. The ORC provides a fuel economy benefit by coupling the mechanical energy back into the engine and by reducing the load operating point on the engine. The ORC system described here and the boundary conditions for operation were referenced from [6] and [7].

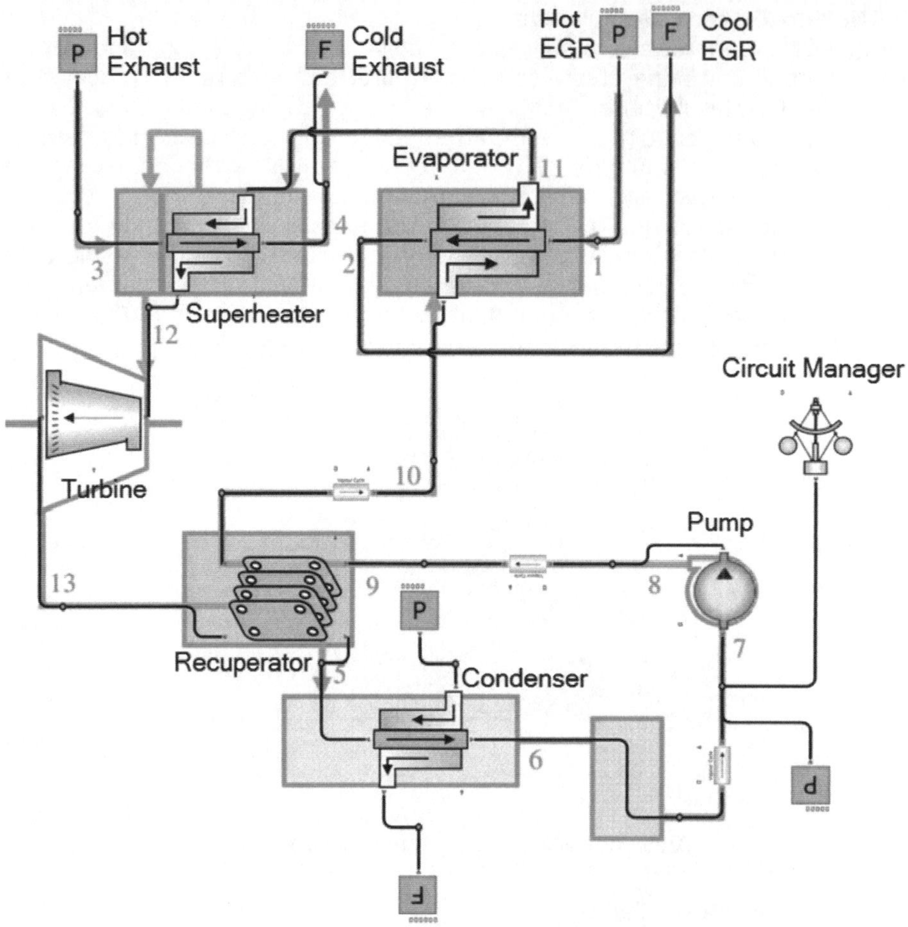

Figure 4: A schematic layout of the Organic Rankine Cycle system with a recupertator and a 2 stage heater

2.4 Thermo-acoustic generator

A Thermoacoustic converter TAC consisting of two modules, a thermal module and a generator module connected together by two acoustic ducts. The thermal module contains the regenerator sandwiched between hot and cold heat exchangers, with a thermal buffer tube abutting the hot heat exchanger, and functions as an acoustic amplifier of sound power flowing from the cold to the hot side of the regenerator. The generator module consists of a pair of back to back linear alternators connected to flexures that form the dividing walls between a pair of largely planar compression chambers and a shared expansion chamber. The generator module contains no sliding parts or seals and thus does not require any lubrication and does not suffer from any wear.

The motion of the flexure walls produces high intensity sound waves in the pressurized helium gas (working fluid) hermetically sealed within the TAC. This acoustic power travels from the compression chambers through a tuned acoustic duct to the cold side of the thermal module, where it is amplified by flowing through to the hot side. This amplified power is then returned to the generator module through a second tuned acoustic duct to the shared expansion chamber which is on the other side of the moving flexures15. The power delivered to the flexures is greater than that needed on the compression sides to produce the starting acoustic power and the extra power is converted to electricity through motion of the pair of attached electro-magnetic linear alternators.

In order to achieve reasonable output power density with the very limited stroke of the flexures compatible with infinite fatigue life the TAC operates at relatively high pressure and frequency. This high frequency has the additional advantage that the tuned acoustic ducts connecting the thermal module and generator module are relatively short allowing for a compact arrangement. The length of the ducts must be chosen to match the operating frequency and also to provide the correct phasing of the acoustic power at both the modules.

A 1 kW TAC system as outlined above has been built and successfully operated using both electrical heaters and a natural gas burner to deliver heat to the hot heat exchanger in the thermal module. The cold heat exchanger was kept cold with a circulating fluid.

Figure below shows the principle of operation for the TAC.

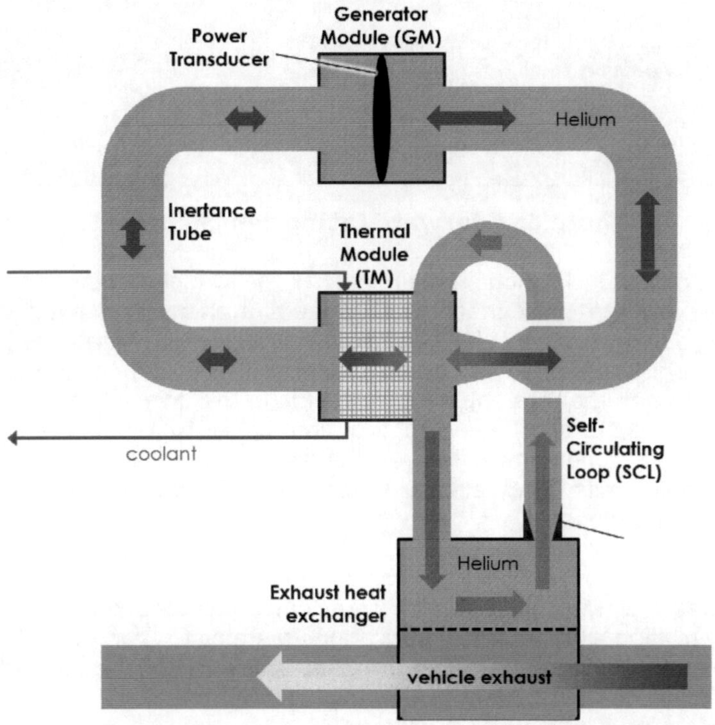

Figure 5: Principle of operation of the thermos-acoustic converter

3 COMPUATIONAL MODEL FOR FUEL ECONOMY ASSESSMENT

3.1 A vehicle modelling approach

A complete vehicle energy balance model was developed in Matlab/Simulink® and Stateflow using two separate modelling approaches, a forward looking model and backward (hybrid) looking model. The forward looking model includes a driver model which adjusts the throttle to affect the vehicle speed. The vehicle speed is in return a function of the torque generated by the engine. The driver model is a combination of a PI controller combined with a 1 sec, 2 sec and 3 sec look ahead. The forward looking model is more appropriate for scenarios where controls and fast-transient phenomenon are of interest in the study. Examples of this are a study of transmission shift map or a torque converter lock-up strategy. The EHRS and the TEG were quantified using the forward looking model. The schematic below show the architecture of the forward looking vehicle model.

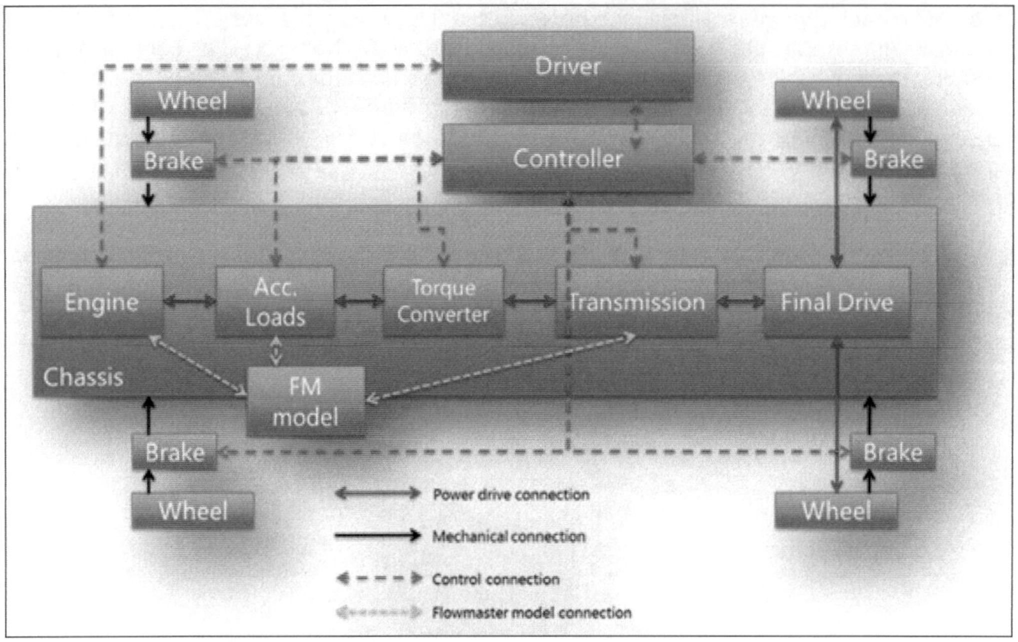

Figure 6: A forward facing vehicle model

A backward facing vehicle model assumes that the vehicle follows the given drive cycle perfectly and the appropriate wheel speeds, transmission shaft speed, gears and engine loads are back-calculated from it. It is however sometimes possible that the required vehicle velocity cannot be practically achieved by the vehicle either limited by the torque that the engine can provide in the given time or limited by the other factors. In such scenarios, a practical limit check was implemented to ensure that the operating conditions of any of these sub-modules did not exceed the maximum capable limits. The backward facing model requires less input data than the forward looking model.

In the current study, a forward looking approach was used for the vehicle model. This model was closely coupled with a transient underhood thermal model of the cooling system and the engine lubrication system to compute the warm-up performance and the difference in the warm-up times due to the technologies.

3.2 Vehicle model of a light duty truck

The a forward looking vehicle model of a light duty pick-up truck with ~3L gasoline boosted engine with a 6 speed automatic transmission was developed and correlated to UDDS cold start, UDDS hot start and US06 drive cycles. The vehicle was a conventional powertrain without start-stop feature. Figure 7 below shows the correlation of the vehicle model with measured dynamometer data.

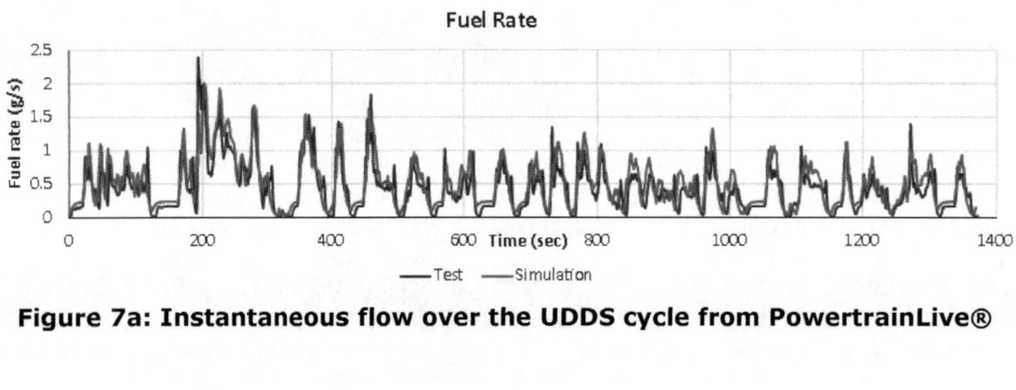

Figure 7a: Instantaneous flow over the UDDS cycle from PowertrainLive®

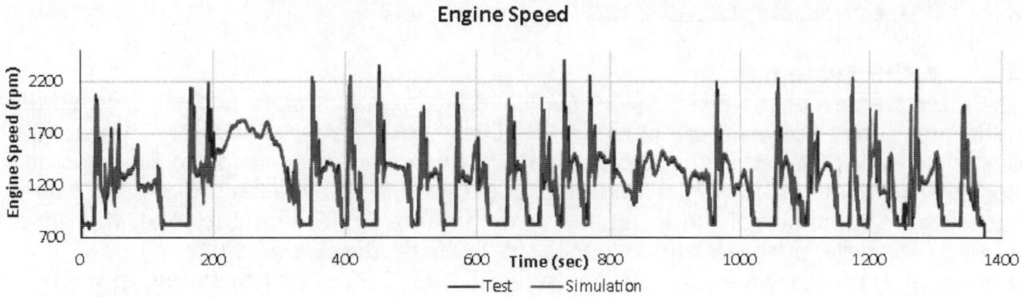

Figure 7b: Engine Speed over the UDDS cycle from PowertrainLive®

3.3 Engine thermal model

The primary purpose of the vehicle thermal model was to capture the engine, coolant and oil systems inertia and the change in warm-up due to the heat load from the WHR system. A baseline engine thermal model was developed and correlated with the vehicle warm-up data. The appropriate corrections for change in engine friction due to higher temperatures were are also captured. The friction model implemented here is fully illustrated in [8] and [9]. Figure below shows a simplified schematic of the major sub-systems of the underhood vehicle thermal systems modelled in Flowmaster. The model consists of the engine cooling, engine oil and the front-end airflow sub-systems.

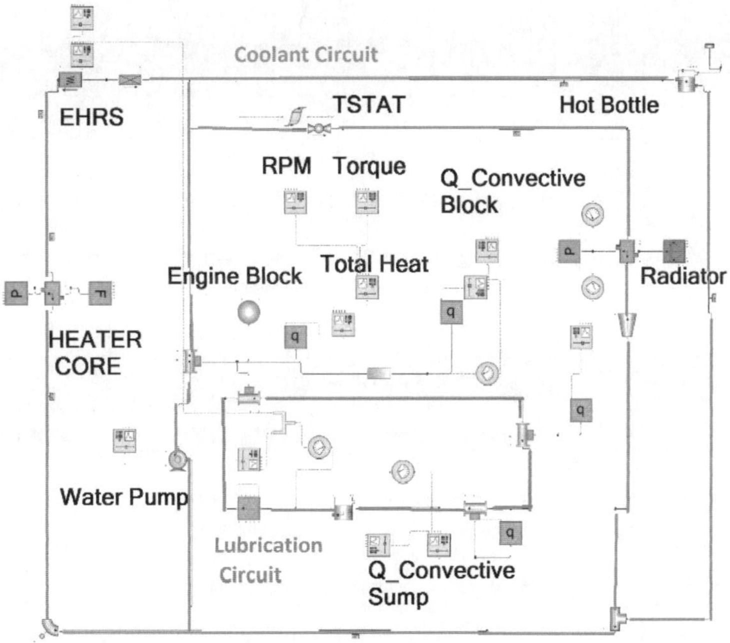

Figure 8: The simplified vehicle thermal inertia model in Flowmasterp

The engine speed and brake torque are used by the engine thermal model to look up the instantaneous combustion heat, frictional heat, coolant and oil flow rates and calculate the temperatures in various sub-systems. The key outputs of the model are the transient temperatures of the coolant and engine oil in addition to the thermal performance of individual components such as the radiator, engine oil cooler, etc. The full details of the model has been discussed in a previous work [8] and [11].

4 FUEL ECONOMY ASSESSMENT

4.1 EHRS System

The effectiveness of the EHRS system was ~85% cycle averaged and was turned on until the engine coolant temperature reached 90C. Above 90C, the thermostat opens indicating that there is excess heat in the system that needs to be rejected. So at this point, the EHRS is counter-productive and is bypassed. Figure 9 below shows the comparison of the exhaust energy and the EHRS heat duty over the drive cycle. The cycle average effectiveness of 85% of the EHRS system during the portion of the cycle when it was on (approx. first 250 sec) translated to a cycle average of 19% of total of exhaust heat capture rate over the UDDS cycle.

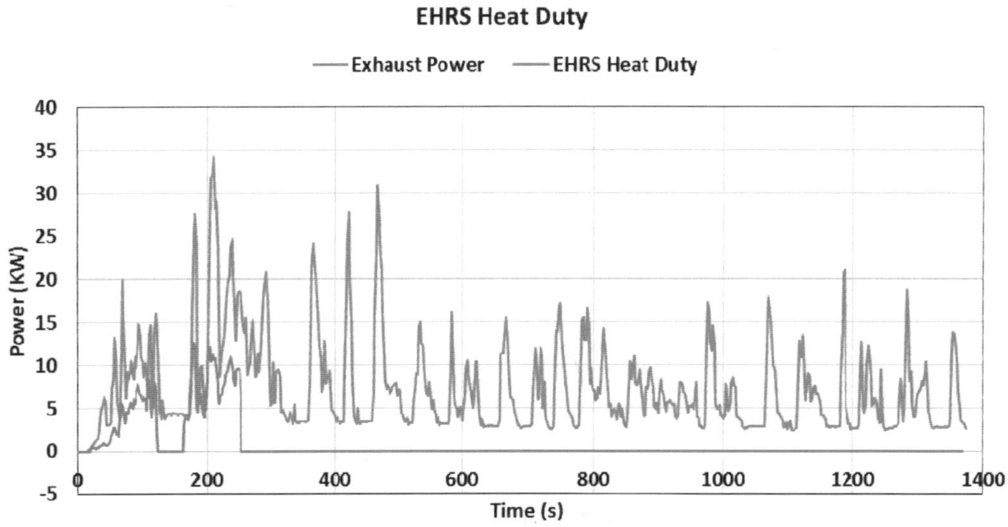

Figure 9: Ratio of exhaust heat captured by EHRS system

Vehicle level analysis indicates that the engine warm-up time reduces from ~1000sec to 770 sec with a fuel economy benefit of approximately 1% over the FTP75 cycle. See table below.

Table 1

	Baseline Case	With EHRS	% Improvement
FTP-75 Fuel Economy	**17.74 mpg**	**17.92 mpg**	**1.0%**
Oil Warm-up temp	1000 sec	772 sec	24%

4.2 TEG System

The benefit of the TEG system was also analysed on the 3L light duty truck. The performance of typical TEG system sized for a light duty truck is shown in the figure below. The TEG power generated is a function of exhaust mass flow and exhaust temperature. The authors are keenly aware that the performance of the TEG system can vary quite a depending on the TEG material used, arrangement of the TEG cards and the location of the TEG in the exhaust system. However, the purpose of this study was a comparative assessment with other WHR technologies, and configuration very similar to the one that is currently being tested by the industry is being considered.

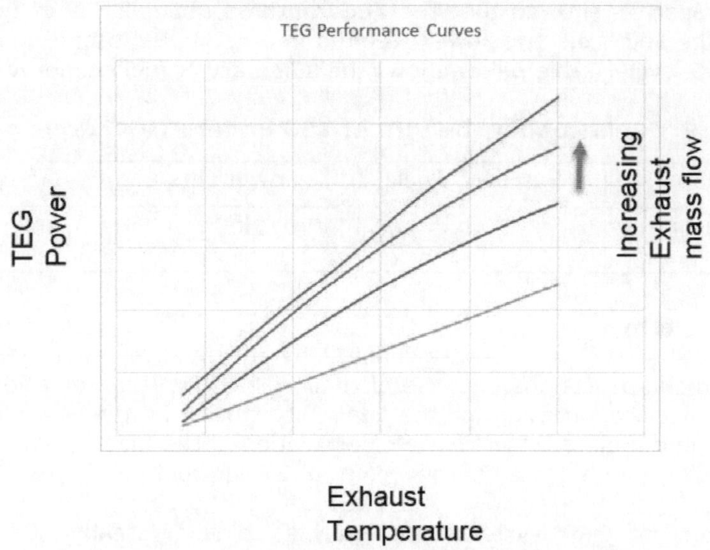

Figure 10: TEG performance characteristics

The TEG power generated in the UDDS drive cycle is shown in the figure below. Results indicate that the cycle average TEG power generated as around 103W with a peak in-cycle power above 200W. It can be observed that during cold start, the TEG power is quite low due to the exhaust temperature at the TEG inlet location. It takes 150-200sec for the TEG power to get up to standard operating efficiencies.

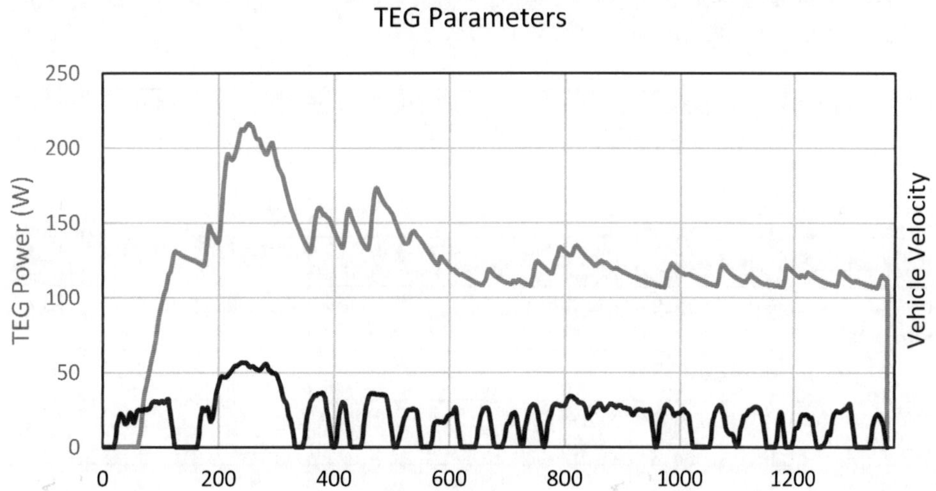

Figure 11: TEG power generated over the UDDS cold start drive cycle

The TEG module was integrated with the vehicle model for a light duty truck to evaluate the fuel economy impact of the TE module. It was assumed that the entire TE power will be used by the alternator. The fuel economy benefit will be realized by the reduction in alternator torque on the engine. Given that a typical vehicle with conventional powertrain will have anywhere between 200W-300W of load on the alternator at all times, this was a fairly acceptable assumption. A constant efficiency of 55% was assumed for the alternator. This is reasonably accurate assumption given the flatness of the efficiency curve of the alternator.

Vehicle level analysis showed that the TEG captured about 2.5% of total exhaust energy, and the 103W electric power resulted in a 0.7% fuel economy benefit over the FTP75 drive cycle. Table below shows the summary of fuel economy benefit.

Table 2: Fuel economy benefit of TEG under FTP75 drive cycle

	Baseline vehicle	(with TEG)	FE Benefit %
Fuel Economy (FTP75)	17.69	17.81	0.7

4.3 ORC System

The ORC systems that are being evaluated by the industry today are sized for commercial truck applications because of their size. Given the limitation of getting ORC system specification sized for the light duty truck, the fuel economy analysis was performed on a Class 8 truck. The configuration of the Class 8 truck was with a 13L diesel engine with a gross vehicle weight of 27,215kg.

The benefit of the ORC system was analysed over 3 steady state operating conditions (B25, B50 and B100). The configuration of the ORC system and its operating conditions in Park [7] and Teng [8].Table below shows the summary of the 3 operating points and the fuel economy benefit observed over these three points. Analysis indicates that about 7% of the exhaust heat was captured by the ORC system and the net fuel economy benefit of the system was 3.5%-4% for the three operating points.

Table 3: Operating conditions for the ORC system and class 8 truck

Parameters from ORC System analysis using Flowmaster®				
		B50	B75	B100
Total Exhaust Heat Captured	kW	67.7	84.9	121.2
% Heat Captured	%	59%	56%	64%
Condenser heat rejection	kW	60.0	74.6	105.4
Expander Power	kW	7.7	10.3	15.9
% of captured thermal energy converted to Mechanical Power	%	11%	12%	13%
% of Exhaust power re-captured	%	7%	7%	9%
Parameters from Vehicle Analysis using PowertrainLive®				
		B50	B75	B100
Engine Speed	rpm	1616	1616	1616
Engine Torque (Nm)	Nm	921.8	1382	1808
Engine Power	kW	156	234	306
Baseline Fuel Economy	mpg	6.24	4.33	3.35
Fuel Economy (with ORC)	mpg	6.47	4.49	3.49
Fuel Economy Improvement		3.8%	3.6%	4.1%

4.4 TAC System

Numerical simulations were performed for the 1kW TAC system. Following table shows the TAC simulation operating parameters. The TAC is interfaced with the exhaust post TWC as exhaust provides medium grade high temperature source to recover the energy. The recovered energy was estimated by modelling all the components of the system as explained in [7].

Table 4: Operating conditions for the TAC system

Mean Pressure (helium)	120 bar
Oscillating Pressure Amplitude	~6 bar
Coolant Temperature	80 °C
Design Frequency	500 Hz

The calculated heat transfer was then used as an input to the thermoacoustic model of the TAC to calculate the predicted power output of the system. The details of the placement of the thermal module and generator module in an actual exhaust system are not yet fixed or determined. Figure below shows the TAC power generated. It can be noted that the TAC power generated is significantly higher during the highway portion of the cycle.

The fuel economy benefit of this TAC system was evaluated on a light duty truck described above. It was assumed that all the power generated by the TAC was fed back to the alternator and helped offset the accessory load on the engine. The fuel economy benefit in just the city cycle was quite small (~0,5%), but there was a notable FE improvement when considering both city and highway due to the high TAC power generated in the highway portion of the drive cycle.

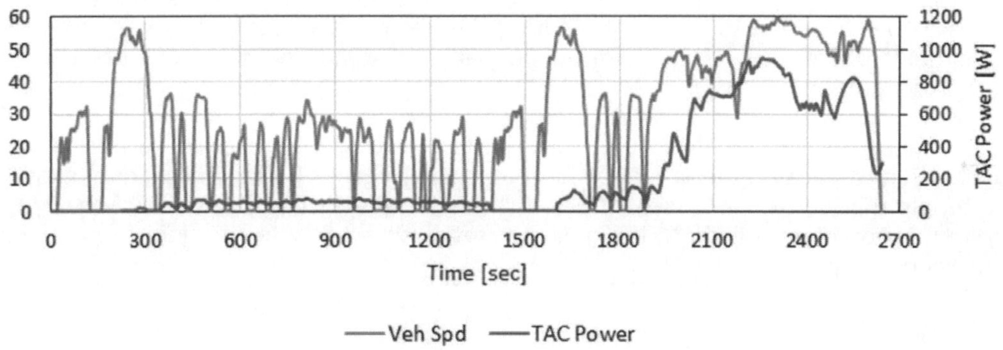

Figure 12: TAC power generated over the FTP75 + highway drive cycle

Table 5: Fuel economy benefit of TAC under UDDS CS and FTP75 + Hwy drive cycles

	UDDS CS		FTP+HWY	
	Fuel economy [mpg]		Fuel economy [mpg]	
	Baseline	With TAC	Baseline	With TAC
Simulation FE (mpg)	15.6	15.67	20.21	20.75
% improvement		**0.45%**		**2.67%**

5 DRAWBACKS AND ADVANTAGES OF THE TECHNOLOGIES

EHRS system

One of main advantages of the EHRS systems is its simplicity and the low cost compared to the other technologies discussed here. The fuel economy benefit for the technology as observed from the calculations comes from reduction in engine friction and fuel consumed during the warm-up portion. This benefit however is insufficient to justify the cost of the EHRS. In certain vehicles with conventional powertrain which have an idle bump strategy employed, the EHRS provides a much larger fuel economy benefit due to an early expiration of the idle bump strategy. The biggest benefactors of the EHRS system, however, are hybrid vehicles. During cold start, the engine for the hybrid vehicles turns on to provide cabin heating, and stays on until a certain cabin temperature (or coolant temperature depending on the manufacturer) has been achieved. An EHRS system helps speed up this warm-up, allowing the engine to turn off earlier, and provide a clear fuel economy improvement. The Toyota Prius and Chevy Malibu both have an EHRS in production vehicles.

The EHRS system is a great enabler of other technologies that will provide fuel economy benefit but likely does not justify its cost purely based on reduction in engine friction due to faster warm-up.

TEG system

One of the main advantages of the TEG system is that it has no moving parts (unlike an ORC system) and it generates electrical energy directly which can be used to off load the alternator load. Another benefit of the TEG system, which is not obvious, is the faster engine warm-up. The TEG system uses coolant to cool and maintain the cold end of the TEG at a target temperature. This means that the TEG system does add heat to the engine during the warm-up phase, reducing its friction and fuel consumption. This warm-up benefit complements its benefit of electric power generation.

The main disadvantages of the TEG system are cost, reliability and power generation capacity. The majority of the costs of the TEG comes from exotic materials that it uses to generate electricity. And this cost/power ratio of the TEG system is too high, but there is significant research activity in the field to improve this ratio.

ORC system

The main advantage of the ORC system is its power generation capability from the exhaust. It is the highest power generators of the waste heat technologies considered here. The ORC systems are currently in use by the power generation industry and are being miniaturized and adapted for the automotive market.

The disadvantage of the ORC is it complexity and size. The ORC system has all the complexities of an AC system as it works on a similar principle. It has many components, and offers challenges of individual component sizing and design, system control, packaging and additional weight. The ORC system also generates mechanical power which either needs a gearbox to bring the mechanical energy back to the engine, or a generator to convert that mechanical energy to electrical energy. In addition, the size of the ORC system in its current state of the art is quite bulky for passenger cars and is more appropriate for freight trucks and large pick-up trucks where the weight of the ORC system has a smaller impact and packaging is more feasible.

TAC system

The TAC system generates a significant amount of power once warmed up. The novel construction of the system (as the TAC does not have a pump) means that its has no moving parts either. The working fluid, high pressure helium, in the TAC is also available readily in the industry if and when the technology goes into larger scale production.

The main drawback of the TAC is its inertia. Preliminary modelling shows that the TAC takes almost 20-25min to warm-up and start generating power. This can be observed in Figure 12 shown in the earlier analysis. Once the TAC starts generating power, it produces enough power to make the cycle average power attractive. There is work underway on the TAC development to reduce the inertia and its response time.

6 SUMMARY AND CONCLUSIONS

Each of the four different technologies assessed had different ratios of exhaust energy capture and fuel economy benefit. A table below shows the summary comparison of the % exhaust energy captured, and the corresponding fuel economy benefit from it. The calculations indicate that there is no direct correlation between the amount of exhaust energy captured and the fuel economy benefit. The fuel economy benefit is very dependent on what form the energy is being generated, during what engine operating condition is this energy being generated and how it is being used in the vehicle. As an example, electrical energy generated has nearly 2X the benefit of the power they generated because of the low alternator efficiency (~55%).

Table 6: Summary of exhaust heat captured by the technologies

Technology	% of exhaust energy captured	% Fuel economy benefit observed
EHRS	**19%**	**1%**
TEG	**2.5%**	**0.7%**
ORC	**7%**	**4%**
TAC	**3.5%**	**2.7%**

REFERENCE LIST

1 Diehl, P., Haubner, F., Klopstein, S., and Koch, F., "Exhaust Heat Recovery System for Modern Cars," SAE Technical Paper 2001-01-1020, 2001, doi:10.4271/2001-01-1020

2 http://energy.gov/sites/prod/files/2014/03/f8/deer11_barrieu.pdf

3 Vijayagopal, R. and Rousseau, A., "Impact of TEGs on the Fuel Economy of Conventional and Hybrid Vehicles," SAE Technical Paper 2015-01-1712, 2015, doi:10.4271/2015-01-1712

4 Wijewardane, A. and Stobart, R., "Addressing the Heat Exchange Question for Thermo-Electric Generators," SAE Technical Paper 2013-01-0550, 2013, doi:10.4271/2013-01-0550

5 Yang, Z., Winward, E., Lan, S., and Stobart, R., "Optimization of the Number of Thermoelectric Modules in a Thermoelectric Generator for a Specific Engine

Drive Cycle," SAE Technical Paper 2016-01-0232, 2016, doi:10.4271/2016-01-0232

6 Park, T., Teng, H., Hunter, G., van der Velde, B. et al., "A Rankine Cycle System for Recovering Waste Heat from HD Diesel Engines - Experimental Results," SAE Technical Paper 2011-01-1337, 2011, doi:10.4271/2011-01-1337

7 Teng, H., Klaver, J., Park, T., Hunter, G. et al., "A Rankine Cycle System for Recovering Waste Heat from HD Diesel Engines - WHR System Development," SAE Technical Paper 2011-01-0311, 2011, doi:10.4271/2011-01-0311

8 Scott, T., Chang, F., Khandaker, M., and Uppuluri, S., "Transient Thermal Modeling of Power Train Components," SAE Int. J. Passeng. Cars - Mech. Syst. 5(2):962-973, 2012, doi:10.4271/2012-01-0956

9 Shayler, P., Leong, D., and Murphy, M., "Contributions to Engine Friction During Cold, Low Speed Running and the Dependence on Oil Viscosity," SAE Technical Paper 2005-01-1654, 2005, doi:10.4271/2005-01-1654

10 Sandoval, D. and Heywood, J., "An Improved Friction Model for Spark-Ignition Engines," SAE Technical Paper 2003-01-0725, 2003, doi:10.4271/2003-01-0725

11 Uppuluri, S., R. Khalane, H., Umbarkar, Y., and Naiknaware, A., "Cold-Ambient Warm-Up Predictions: A Novel Approach Using 1D Computational Models," SAE Technical Paper 2016-01-0198, 2016, doi:10.4271/2016-01-0198

Thermal and power analysis of a modularised thermoelectric power generator for passenger vehicles

K Huang [1], B Li [1], Y Yan [1,2], J Li [3]

[1] Fluids and Thermal Engineering Research Group, Faculty of Engineering, University of Nottingham, UK

[2] Fluids & Thermal Engineering Research Centre, University of Nottingham Ningbo, China

[3] China FAW Group Corporation R&D Centre, China

ABSTRACT

Recently, the thermoelectric power generation (TEG) technology emerges as alternative solution to the emission reduction challenge in this area. In this paper, we present an innovative pathway to an improved heat supply into concentric shape-adapted TEG modules integrating the heat pipe technologies. In order to enhance heat transfer for achieving higher efficiency, in our cases, the heat pipes are configured in the radial direction of the exhaust streams. Our analysis shows that potentials of power output are adequate for the limited space under the chassis of the passenger car. The efforts towards greater convective heat transfer also can be increased by adjusting the heat pipes at dual sides of the concentric TEG modules. Heat enhancement at the hot side of the TEG has an effective impact on the total power out of the TEG modules. However, such improvements can be offset by the adjustment from the coolant side. Furthermore, our results highlight the importance of heat transfer versus the TEG power generation under two possible configurations in the passenger car. Highest power output per repeat unit is achieved at 29.8 W per 0.45 Litre with ZT value 0.87 for Bi_2Te_3-based thermoelectric material in our studies.

Keywords: Thermoelectric power generation, Heat Enhancement, Exhaust heat, Temperature-dependent material properties, Heat pipe, Heat exchanger

1. INTRODUCTION

Previous decades have seen enormous developments in the renewable technologies by utilizing the alternative ways to obtain the usable energy. Thermoelectric Power Generator (TEG) technology stands out from the crowd through the exploration [1]. Unlike the traditional fossil fuel energy systems, TEG system converts waste heat directly into electricity through the Seebeck effect of the semiconductor materials whereby a temperature difference is maintained between the hot and cold side of the TEG module. TEG systems have the advantages of silent operation, high reliability, and they have no moving or mechanically complex components rendering them lasting for a very long time. They also operate at wider operational temperature range than most existing energy recovery systems [2, 3]. However, there are many challenges ahead of the large scale applications. Although one of the biggest challenges is relatively low heat to electricity conversion efficiency [4], recent decades have witnessed notable progress in thermoelectric materials, stimulated by new ideas in bulk materials [5]and nanostructures [6]. Nevertheless, there is still plenty of space for innovation and development of thermoelectric materials to take advantage of their solid-state nature, scalability, silent operation, and environmental friendliness [7]. It is highly anticipated from such material development that the TEG system will be very applicable in automotive waste heat industry shortly.

Even though the performance of thermoelectric materials has been improved significantly, there are still many challenges at the device level to prevent TEG devices from real applications. In most cases, it is poor heat transfer design that inhibits achieving the ideal efficiency. Liu et al. have performed CFD simulation and experimental study on the structural design of the heat exchanger to improve the efficiency of TEG system [8, 9]. HoSung Lee has studied the optimal design of thermoelectric devices with dimensional analysis [10]. The optimum designs including the power output and the efficiency on the external load resistance as well as the geometry of thermoelectric units.

In practice, there is limited space in the vehicle exhaust system which is suppressed by manufacturer for maximising the cabin room. Therefore, the effective area of the TEG modules is inevitably constrained by the space under the chassis of the vehicle, in particular for the modern passenger car. Moreover, the weight of thermoelectric generator is also a vital factor in fuel economy [11]. Therefore, the need of light weighted thermoelectric generator system should also be considered carefully for the development of vehicular TEG systems. Wang et al. have presented an innovative design of heat pipes-assisted heat exchanger structure to overcome the common conflicts between the needs for high heat flux into TEG system and the limitation of overall heat exchanger dimensions and weights [12].

In the current work, a novel design of a concentric cylindrical thermoelectric generator system is presented for use in the automotive exhaust system. Instead of using a bulky and heavy heat exchanger, the novel TEG system has a compact and lightweight heat sink which is assisted by heat pipes. As a reliable and efficient heat transfer component, heat pipes have been widely used for the thermal management of electronic devices. Heat pipes have the advantageous in that they have high heat conductivity, fast thermal response, and are light weight [13, 14]. The combination of heat pipes with heat exchanger is projected to reduce the weight of the TEG system and the whole vehicle as well, consequently improving the fuel economy.

The proposed design is capable of adapting to the shape of the exhaust pipe besides accommodating more TEMs in the same installation length. In this paper, the numerical studies about the heat transfer performance of this TEG system is conducted with different heat transfer coefficient conditions at hot and cold sides of the heat sink. Furthermore, this present study also investigates the impact of heat enhancement at the hot side on the total power out of the TEG modules and the influence of the arrangement of the coolant side.

2. MODEL DESCRIPTION

2.1. Schematic structure

As shown in Fig.1, the TEG system consists of a series of repeat units that segregated along with exhaust stream to adapt for the practical exhaust pipe. The repeat unit is made up of four concentric TEG modules, three hot plates and two cooling plates including twelve heat pipes in each plate as highlighted in blue and red colour in Fig. 1a and Fig. 1b. The exhaust stream interacts with heat pipes and transfers the heat into TEG in the radial direction of the exhaust stream. The channel spaces of both the exhaust and the coolant are tightly shaped by the heat pipes, the surface of the cooling plates and the shell of the heat exchanger. According to the demands of power outputs, this configuration can be easily adapted by exchanging the channels of exhaust and coolant flows and adjusting the number of repeat units. In reality, each unit has all different flow parameters regarding $T_{w,n}$ coolant temperature, $T_{e,n}$ exhaust temperature. These differences result in decreasing the temperature of the exhaust and increasing the temperature of the coolant. Therefore, the specific configuration of flow patterns will make great impacts on thermoelectric power generation.

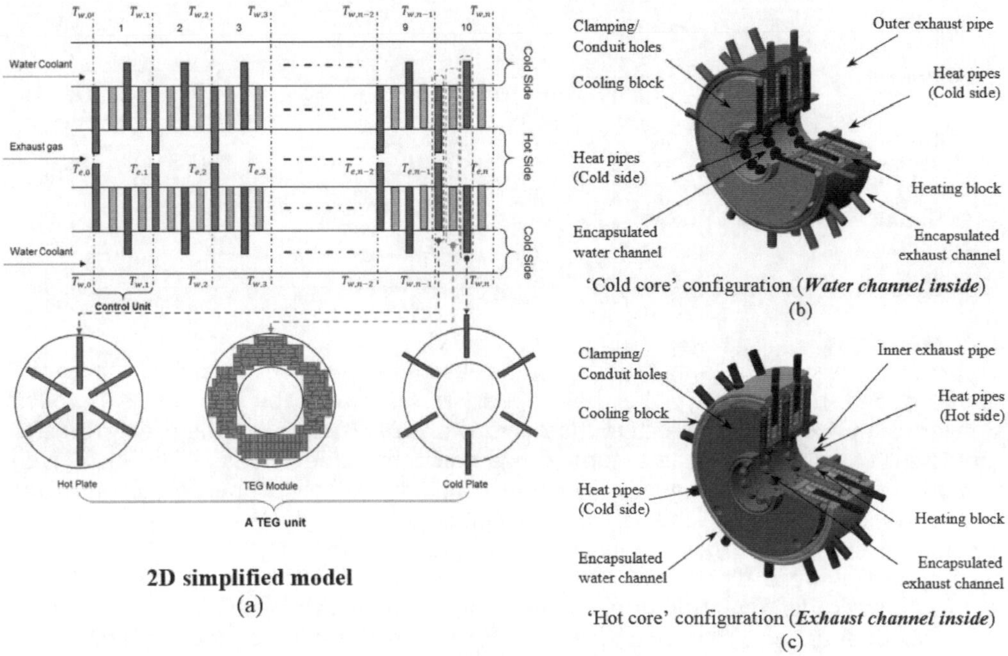

Fig. 1. A schematic of the (a) 2D TEG system model and the two 3D configurations of hot and cold sources: (b) 'Cold core' configuration and (c) 'Hot core' configuration

Comparing with some commonly used TEG modules, concentric TEG module is applied instead of square-shaped TEG modules, and the method of self-clamping by adjacent TEG modules is introduced to further reduce the need for the single clamp on each TEG modules. In this model, the primary assumption of constant temperature in the exhaust and coolant flow is made for simulation purpose. Also, the parallel link inside each repeat unit is set in which the total internal electrical resistance R_{in} is equal to a quarter of individual TEG module resistance R_{TEG}. The parameter analysis of matched load resistance is conducted for tracking the maximum power point when the external load is fluctuating. Providing the same power output of each unit, the total power generation of the whole system is calculated by multiplying the number of units used in the exhaust pipe.

Specifically, in Fig. 1b, the coolant flow can be allocated to an inner axial flow by replacing the exhaust flow to outer axial flow channel in respect of utilising the merit of the higher heat capacity of the water coolant. In addition, extruded fin stacks can be installed on heat pipes to enhance heat transfer convection in the exhaust streams. Unlike the conventional square shape TEG modules, the concentric and cylindrically shaped TEG can be easily accommodated with round exhaust pipe. Given that the alteration of clamping method in this newly design, the uneven thermomechanical stress on the components are expected to be minimised by replacing the individual clamps into four simple self-locking metal cables through the internal holes of TEG modules. Additionally, the geometric details of major TEG components and heat pipes are listed in the Table 1.

Table 1 Geometric details of major TEG components and heat pipes

Parameters	Thermo-electric leg	Copper connector	Heat pipes	Ceramic plate	Insulation ring	External copper plate	Internal copper plate
Length (mm)	2.55	5.8	6	-	-	-	-
Width (mm)	2.55	2.55	6	-	-	-	-
Height (mm)	2.55	0.3	40	0.7	16	23.1	7
Inner Diameter (mm)	-	-	-	50.8	40.8	50.8	32.8
External Diameter (mm)	-	-	-	101.6	50.8	115.6	101.6

2.2. Materials

In order to ease the simulations, some assumptions are made to save the computational time. Firstly, the thermal resistance caused by coolant h_w is much smaller than exhaust convective heat transfer h_e. Therefore, the heat transfer coefficient of cold-water h_w is assumed as a constant value in this study. Also given that the complexity of exhaust mixer and combustions, air properties are used as an approximation. And the error associated with neglecting the combustion products is usually no more than about 2% [15]. As shown in Table 2, the major assumption of heat pipes is the thermal conductivity which is assumed at 400 W/m·K in the radial direction of each component (same as copper) but 4000 W/m·K in the axial direction. This setting is widely used in the engineering for easing the simulation calculation with acceptable accuracies.

Table 2 Materials Properties of major components in the TEG system

Parameters	Copper plate	Copper connector	Ceramic Plate	Insulation material	Heat pipe
Thermal conductivity (W/m·K)	400	400	180	0.022	400 (Radial) 4000 (Axial)

The thermal conductivity of heating and cooling plates is set as copper material at 400 W/m·K. The dielectric material, Aluminium Nitride (AlN) ceramic, is designed concentrically to separate the copper plates and the TEG legs, and the thermal conductivity of high-grade AlN is assumed at 180 W/m·K. The MicroFoam® insulation material is installed to block the heat loss from the exhaust to the coolant, and the thermal conductivity is set as constant at 0.022 W/m·K due to its stable thermal performance across high temperature range.

Table 3 Temperature-dependent polynomials of thermoelectric materials properties

Thermoelectric Leg	Seebeck coefficient $\boldsymbol{A}$ (V/K)	Thermal conductivity $\boldsymbol{\Lambda}$ (W/m·K)	Electrical conductivity $\boldsymbol{\sigma}$ (S/m)
n-type	$2.0 \times 10^{-13}T^3 + 1.0 \times 10^{-9}T^2 - 9.0 \times 10^{-07}T + 8.0 \times 10^{-6}$	$-4.0 \times 10^{-8}T^3 + 5.0 \times 10^{-5}T^2 - 2.6 \times 10^{-4}T + 5.1$	$-2.6\times 10^{-3}T^3 + 4.60T^2 - 2.7\times 10^3T + 5.8 \times 10^5$
p-type	$-3.0 \times 10^{-14}T^3 - 3.0 \times 10^{-9}T^2 + 2.0 \times 10^{-6}T - 0.20 \times 10^{-4}$	$2.0 \times 10^{-7}T^3 - 0.20 \times 10^{-4}T^2 + 8.0 \times 10^{-4}T - 7.8$	$-2.6\times 10^{-3}T^3 + 4.60T^2 - 2.7\times 10^3T + 5.8 \times 10^5$

The thermometric material Bismuth Telluride Bi_2Te_3, which is widely available in common market, is electrically connected in series here. As the TEG thermal properties under changing temperature conditions are highly varied, all the relevant properties are set as the temperature-dependent variables as shown in Table 3. The interpolations are used here to estimate the Seebeck coefficient, thermal conductivities, and electrical conductivities of the n-type and p-type thermoelectric legs.

3. RESULTS AND DISCUSSION

3.1. System performance of repeated TEG unit

In our simulated cases, the best results of the repeated Bi_2Te_3-based TEG unit are shown in Fig. 2 (Parallel linked among four concentric TEG modules) in the last section. The major features of the TEG are matched load voltage, matched load current and matched load power as shown in U-I-P curves of Fig.2b, matched load voltage can be obtained at 4.5 V when the matched load resistance reaches to 0.7 Ω ($R_L/4$, a quarter of each modular resistance at 2.8 Ω). Meanwhile, the best output power can be achieved at 29.8 W for the same load resistance condition in which the match load output current is 6.62A through terminal connectors. The Water-TEG-Exhaust configuration referring to radial direction outperforms the Exhaust-TEG-Water configuration shown in Fig.2d regarding the power rating.

For the Exhaust-TEG-Water configuration, the match load power is only achieved less than 5 W in a total of 4 TEG modules although best convective conditions are applied. Limited heat transfer area could be the major reason that causes such undesired results. To compare Fig.2a to Fig.2c, the temperature at hot side heat pipe are kept at 555 K and 371 K respectively. There is still room for improvement to increase the hot side temperature of TEG legs. However, it should be noted that the fin stacks, which not simulated in our cases, will contribute to higher heat transfer coefficient at the interface of the exhaust and the heat pipes. The main obstacles to applying extra fin stacks are the limited space within the standard exhaust pipe. It is the advantage of the Water-TEG-Exhaust configuration that the fin stacks can be easily applied to the hot side heat pipe at the external channel. Therefore, a higher power rating could be anticipated.

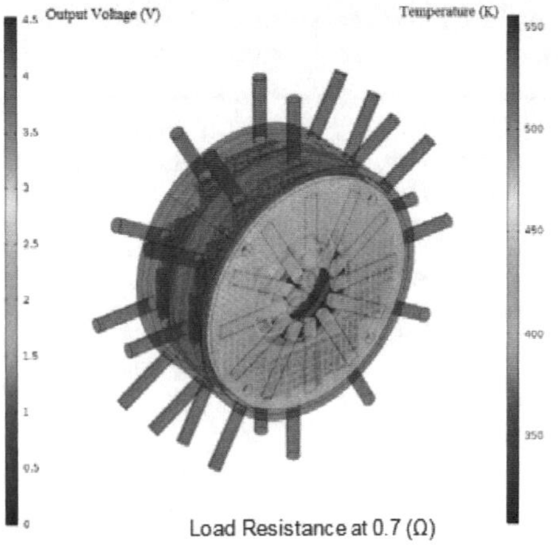

(a) Matched load output voltage and temperature distribution at external exhaust channel configuration

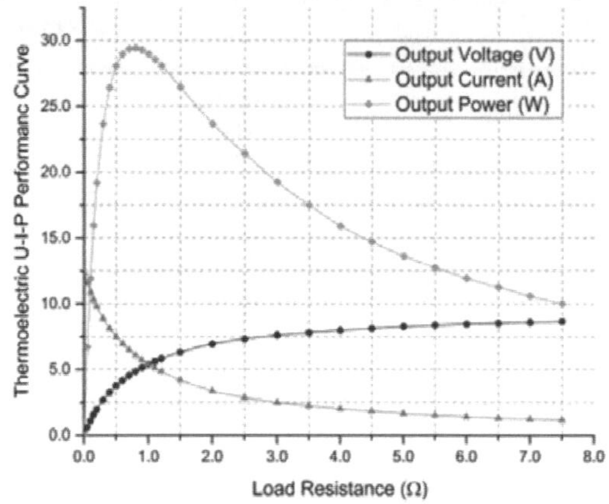

(b) Performance curve for external exhaust channel TEG unit

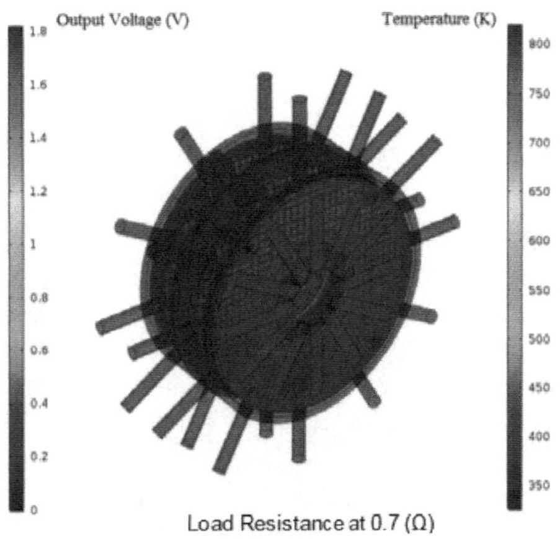

(c) Internal exhaust channel configuration

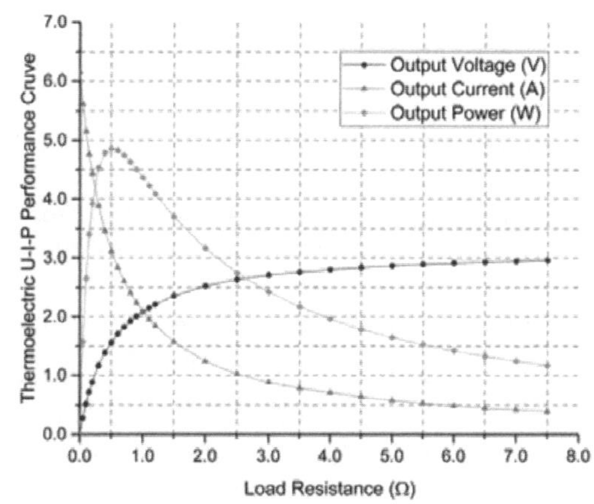

(d) Performance curve for internal exhaust channel TEG unit

Fig. 2. Repeated TEG temperature distribution simulation (a) and (c) and performance comparison between (b) External and (d) Internal exhaust channel configuration

To maintain a sustained temperature difference at both sides of the TEG, heat pipes are deployed in our cases by means of integrating them with the concentric cooling/heating copper blocks. The main reason for such deployment is to reduce the thermal resistance of heat sinks at both ends of the TEG surfaces. As shown in Fig.3a and Fig.3b, the cooling block can be maintained at 401.7 K while the heating block can be achieved at 555.7 K. Both copper blocks show excellent temperature distribution evenly which help to retain the temperature of TEG legs within each module. Therefore, a proper temperature distribution of TEG legs can be obtained as seen in Fig.3c that the temperature difference within the module is about 20 K in the radial direction of the exhaust flow.

In addition to the arrangement of heat pipes, it is worthy noted that proper insulations around the TEG modules also contribute to maintaining a bigger temperature difference as shown in Fig.3d. The thermal bridging may happen without sufficient thermal insulation at the junctions between heating blocks and cooling blocks. This issues may result in undesirable heat loss that bypasses the TEG into the water coolant. As seen in Fig.3d, the TEG modules are tightly sandwiched between stacked cooling and heating blocks. Therefore, thermal insulation material with high performance are required and carefully placed to solve such issues. In this case, the TEG module operates at its optimal temperature condition ranging from 473 K to 523 K. The figure of merit, ZT value is achieved at 0.867 which is acceptable at the current stage regarding the conversion efficiency.

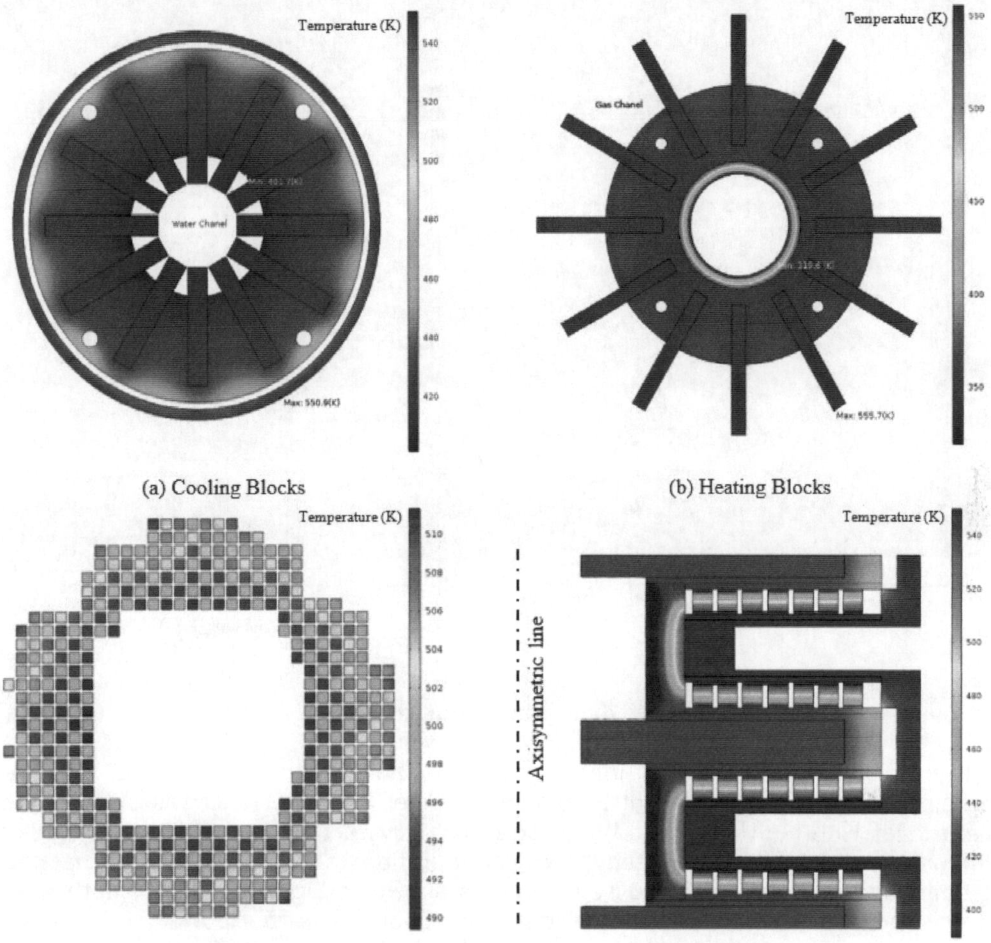

(a) Cooling Blocks (b) Heating Blocks

(c) Cross section at middle of TEG modules (d) Temperature across the cut out through axisymmetric line

Fig. 3. Thermal performance of thermoelectric power generator system at (a) Whole system simulation Image (b) Heating Block (c) Cooling Block (d) TEG Module

3.2 Comparison of different configurations (Compare Exhaust in centred pipe versus water in centre pipe) (U-I-P curves)

In the exhaust pipe of modern vehicles, heat transfer coefficient ranges from 100 $W/m^2 \cdot K$ to 1500 $W/m^2 \cdot K$ according to the practical engine performance. The exhaust flows steadily in the turbulent region. On the other hand, the heat transfer coefficient of the water coolant in vehicle radiators is in the region of 5000 $W/m^2 \cdot K$ to 15000 $W/m^2 \cdot K$ within the fin stacks. Nevertheless, the performances of TEG are closely related to the layout of the heat source and heat sinks, careful thermal designs of the TEG heat exchanger should be examined regarding the heat transfer performances.

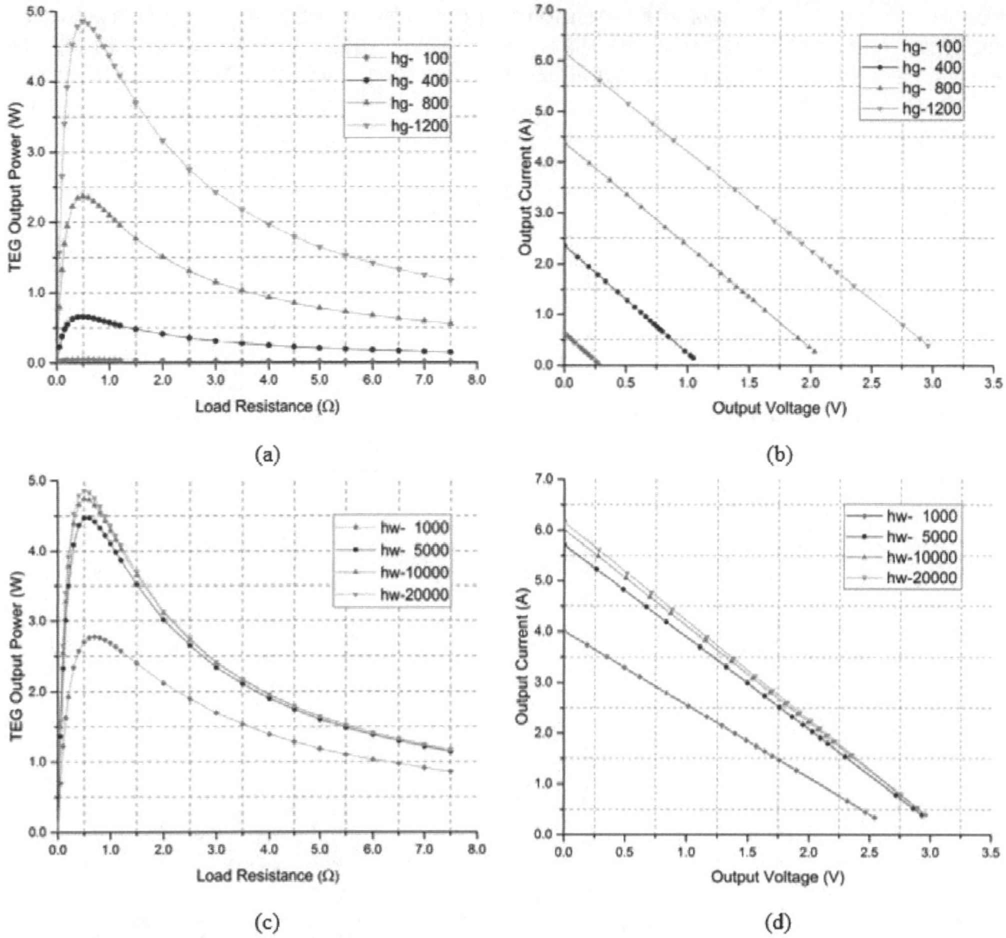

Fig. 4. Exhaust gas inside

In the radial direction, the heat loss of the conventional exhaust pipe takes place at the pipe surface to the ambient atmosphere. Given that the functions of TEG, the waste heat with high entropy, should be maximally recuperated along the path of exhaust stream flow. In our study, two configurations are both simulated that the comparisons of them are discussed here. As shown in Fig.4 and Fig.5, both heat source and cold source impose enormous impacts on the output power of the TEGs. However, the overall power rating of the Water-TEG-Exhaust configuration is much better than the Exhaust-TEG-Water configuration.

In Fig.4a, the output power increases significantly with higher heat transfer coefficient. And the U-I profile changes correspondingly under the same trend as seen in Fig.4b. Interestingly, the trend in heat transfer improvement at coolant side, which presented in Fig.4c, shows disproportional increase in output power for the Exhaust-TEG-Water configuration. Although the output power of TEG still increases, a weaker impact of higher heat transfer coefficient is noted in Fig.4c and Fig.4d. The growth of both output voltage and output current become inactive as shown in Fig.4d. The resistance of TEG, as presented as the slope of the curves in Fig.4d, seems affected by the enhanced coolant flow. The results of U-I curves in Fig.4d tend to be normalised to an upper limit when the heat transfer coefficient increases.

Compared with the results of Exhaust-TEG-Water configuration in Fig.4, it shows better TEG performance of output power, consistent with heat transfer coefficient growth for the Water-TEG-Exhaust configuration. It is mainly attributed to the larger surface area of the external channel. More heat flux can be transferred into the heating blocks under same heat transfer coefficient conditions. Another main benefit of this setting is that extra fin stacks can be configured on the heat pipes due to the ample space in the exhaust channel. It can be positively expected that the higher heat flux will be achieved under such development.

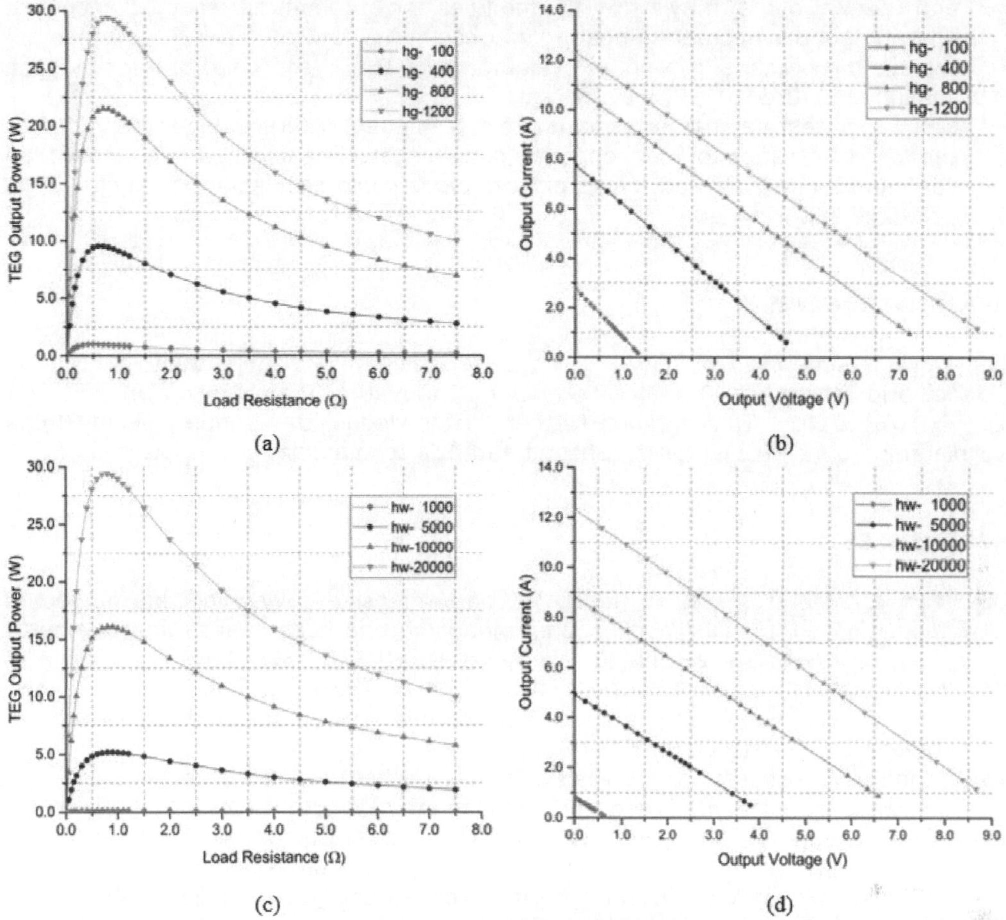

Fig.5. Exhaust gas outside

4. CONCLUSIONS

In this paper, a multiphysics model is developed and simulated in multiscale, in which the convective heat transfer of exhaust and coolant in channels, as well as the heat conduction and electrical conduction in thermoelectric materials, are conjugated and resolved to comprehensible levels. The model is based on an analytical solution of the nonlinear heat transport equation for thermoelectric, which enable it to take into account the temperature-dependent material properties of TE elements. The performance of a novel compact thermoelectric power generation system is studied over a full range of operating parameters regarding the heat transfer enhancements at both cold and hot sources.

From these results, it can be concluded that

- Heat pipe-assisted heat enhancement method is approved to be an effective way to improve the TEG performance. Encapsulated heat pipes arrays in the radial direction of the exhaust pipe help take/add heat effectively to an external fluid stream.

- Aligned/staggered heat pipe arrays are effective techniques to enhance heat transfer on the hot side of the TEG to increase the total power output from the device. To date, the heat pipes are most useful thermal components linking sizeable heat sinks with heat sources.
- The features of heat pipes such as temperature flattening, temperature control and thermal diode, may help TEG modules for autonomous, maintenance-free operation under fluctuating heating sources in the future.
- We see the potential research development of TEG system and muffler by such coaxial structure.
- Endurance test versus performance test is needed in future experiment due to the over limitation of conventional copper-water heat pipes. The cascaded configurations of different types of heat pipes could be a solution which will be validated in our experiment.

ACKNOWLEDGEMENT

The work is supported by FAW R&D Centre China and also partially by Ningbo Science and Technology Bureau's Technology Innovation Team Project under Grant No. 2016B10010. The authors further acknowledge the support from China Scholarship Council (CSC) for the second author's scholarship.

REFERENCES

[1] Twaha S, Zhu J, Yan Y, Li B. A comprehensive review of thermoelectric technology: Materials, applications, modelling and performance improvement. Renewable and Sustainable Energy Reviews. 2016;65:698-726.
[2] Karvonen M, Kapoor R, Uusitalo A, Ojanen V. Technology competition in the internal combustion engine waste heat recovery: a patent landscape analysis. J Clean Prod. 2016;112:3735-43.
[3] Singh DV, Pedersen E. A review of waste heat recovery technologies for maritime applications. Energ Convers Manage. 2016;111:315-28.
[4] Liang X, Sun X, Tian H, Shu G, Wang Y, Wang X. Comparison and parameter optimization of a two-stage thermoelectric generator using high temperature exhaust of internal combustion engine. Applied Energy. 2014;130:190-9.
[5] Kleinke H. New bulk Materials for Thermoelectric Power Generation: Clathrates and Complex Antimonides. Chem Mater. 2010;22:604-11.
[6] Vineis CJ, Shakouri A, Majumdar A, Kanatzidis MG. Nanostructured thermoelectrics: big efficiency gains from small features. Adv Mater. 2010;22:3970-80.
[7] Tao C, Chen G, Mu Y, Liu LS, Zhai PC. Simulation and Design of Vehicle Exhaust Power Generation Systems: The Interaction Between the Heat Exchanger and the Thermoelectric Modules. J Electron Mater. 2015;44:1822-33.
[8] Liu X, Deng YD, Li Z, Su CQ. Performance analysis of a waste heat recovery thermoelectric generation system for automotive application. Energ Convers Manage. 2015;90:121-7.
[9] Liu X, Deng YD, Zhang K, Xu M, Xu Y, Su CQ. Experiments and simulations on heat exchangers in thermoelectric generator for automotive application. Appl Therm Eng. 2014;71:364-70.
[10] Lee H. Optimal design of thermoelectric devices with dimensional analysis. Applied Energy. 2013;106:79-88.
[11] Peng Z, Wang T, He Y, Yang X, Lu L. Analysis of environmental and economic benefits of integrated Exhaust Energy Recovery (EER) for vehicles. Applied Energy. 2013;105:238-43.

[12] Wang XZ, Li B, Yan YY, Liu S, Li J. A study on heat transfer enhancement in the radial direction of gas flow for thermoelectric power generation. Appl Therm Eng. 2016;102:176-83.
[13] Li Y, He JB, He HF, Yan YY, Zeng ZX, Li B. Investigation of ultra-thin flattened heat pipes with sintered wick structure. Appl Therm Eng. 2015;86:106-18.
[14] Li Y, Li ZX, Zhou WJ, Zeng ZX, Yan YY, Li B. Experimental investigation of vapor chambers with different wick structures at various parameters. Exp Therm Fluid Sci. 2016;77:132-43.
[15] Diesel Exhaust Gas. https://www.dieselnet.com/tech/diesel_exh.php.

SIMULATION AND CONTROL

Development of an HIL rig for vehicle thermal management systems design and optimisation

M V Morganti, S Longo
Advanced Vehicle Engineering Centre, Cranfield University, UK
R M Shah
Charge Automotive, UK

ABSTRACT

This paper presents an integrated methodology for the design and optimisation of thermal management systems in hybrid electric vehicles. The methodology is based on a hardware-in-the-loop rig where real components of thermal management systems are used to control simulation-driven heating elements. The simulation environment is developed using a multi-physics modelling platform. Thermal management systems optimisation are important because they can effectively improve fuel economy. The rapid penetration of hybrid electric vehicles has increased challenges in thermal management system design as they now need to provide comfort while handling complex thermal loads and endure long period of time without major maintenance.

1. INTRODUCTION

Thermal Management Systems (TMS) design optimisation became crucial over the last decades. TMS have not been subject to significant changes in the last 50 years, however, recent attention to fuel economy and emissions raised several issues [1]. Hybrid Electric Vehicles (HEV) and plug-in hybrid electric vehicles (PHEV) changed the way thermal management systems (TMS) are designed [2]. Leighton outlines how the complexity of Electric Vehicle (EV) TMS has increased throughout the last few years [3].

TMS in modern vehicles needs to provide comfort under highly unstable conditions while handling a relevant thermal load and lasting a long period of time without major maintenance [2]. HEV are multi-component systems featuring different functions. For instance, a PHEV TMS layout features: engine, generator, motor, batteries and a cabin [4]. These vehicles need a separate cooling system for each component, which often implies the presence of additional heat exchangers [3]. Such multi-component systems require sophisticated control strategies. Lack of control leads to over-cooling during warm-up phase and undercooling at high engine loads [5]. This happens especially when cooling is directly powered by the engine, causing high fuel consumption and emissions [5].

Powering components such as coolant pumps with small electric motors for that specific purpose can help reduce energy consumption. Internal Combustion Engines (ICEs) operating conditions, electric machines and battery packs also affect TMS design [4]. The arrangement of TMS devices inside the vehicle affects heat exchange, e.g. in case of constrained space [5]. In addition, batteries and motors are more vulnerable to overheating than ICE, hence heat generation must be efficiently removed to keep good performances and long-lasting life [6]. Heat recovery improves energy efficiency of the system. Quantity and quality of exhaust heat together with the packaging can have a significant impact on heat recovery solutions [7].

When designing a TMS, the effect of the operating conditions on the Internal Combustion Engines (ICEs), electric motors and battery packs needs to be carefully analysed. The external ambient affects performances. High external temperature leads to changes in recirculated air rate and reduced evaporator inlet as the system needs to dissipate solar radiation, heat dissipated by passengers and air routed into the compartment, the first and the latter being highly dependent on driving direction [8]. A high coolant temperature results in a good heat exchange performance [5]. Furthermore preventing all components from running when not needed results in fuel saving, as parasitic loads are avoided; electrically powered components could reduce fuel consumption [5]. Some authors suggest that a partialisation of the air compressor leads to benefits on the system [8]. Others proposed a combined fluid loop to improve EV range [3].

Proposed solutions may be different, but parameters affecting performances of TMS can be always summarised as: selected refrigeration cycle, cooling/heating circuit, passenger compartment, longitudinal dynamics, driving profile (including conditions of road) and climatic conditions [8]. The latter being constant, is an incorrect assumption. It is therefore important to take into account terrain, location and time of the day [8]. As heat exchange involves any sub-system of the vehicle, TMSs affect the performances of vehicles, the energy efficiency, passengers' comfort and polluting emissions. Their design needs to be improved.

1.1. Reasons for the Work

Quality standards in the automotive industry require extensive testing to prove system's robustness. Such systems therefore also need specific tests to be carried out [9]. However, assembling a complete vehicle in order to optimise its TMS is not cost efficient. A model validation approach, allows faults to be detected earlier in the design stage, thanks to specific hardware testing of proposed models [10]. Hardware in the loop (HIL) systems allow precise testing of specific models during development [11]. HIL are built to test and validate design operational strategies, components and their design and to control test-bench components, simulating the real dynamics of the process [10], [12]. HIL are cheaper than a real vehicle simulation, they are available at any time and safe to operate[10].

Building a different test-rig for every single TMS solution under test is not an efficient method. Manually adapting the same testing equipment for every test would also be time consuming. An optimisation-oriented test-rig would therefore be beneficial. The HIL needs to be automatically reconfigured by the optimisation algorithm.

In addition to that, there is a necessity of having a standardised procedure in order to maximise the efficiency of the testing equipment [11]. It is essential to gather all system requirements making sure they do not clash one another, avoiding ambiguities [11]. Then hardware requirements are defined [11]. Final step is the definition of transfer functions for each sensor and actuator and a control unit behaviour model [11].

1.2. State of Art of the Interface between HIL and Multi-Physics Models

Various research groups have tackled the TMS optimisation problem in different ways. HIL systems are usually developed as testing platforms. The majority of HIL environments are developed to individually perform high-fidelity tests on each design option [11][12]. Most test-rigs can therefore be used to test an already fully designed system. This leaves little space for system optimisation. Aachen Technology University researchers, together with FKA mbH (2014) developed a thermo-hydraulic HIL test bench interfaced with a model developed in Dymola [12], which is a multi-physics tool based on Modelica: a unified, object oriented programming language, for physical systems modelling; Modelica is efficient in analysing multi-domain problems, such as vehicle TMS [13]. Baltzer et al. focused on the prevention of side effects and undesired mutual interference of the applications [12]. Physical model and surrounding ambient simulations were hosted by Dymola, while the management of the whole HIL system is granted by LabView. Baltzer et al. also note that, since Dymola is not designed for real time interaction with control elements, it is necessary to integrate its libraries with additional elements, which the authors developed for this project specifically. While running the simulation, the interface is accessible to users allowing their intervention. Aachen Technology University spent some effort in signal synchronisation, this being a severe issue when connecting a virtual model to a physical test-rig. Each HIL component has different timing requirements. An ineffective signal synchronisation would provide poor data exchange or eventually loss of data. The second condition would stop the simulation. Baltzer et al. claim they have achieved a very stable data exchange rate between HIL and the Dymola model between 10 and 20ms , having Dymola a 10-time slower cycle time.

During winter, battery and cabin heating are necessary and therefore the use of a heat recovery system would be beneficial especially in terms of vehicle range improvements. Testing procedures involved the use of a dynamic dynamometer on which the vehicle was placed, simulating different driving cycles. All thermal energy sinks were modelled by the authors in Dymola, including cabin Heating Ventilating and Air Conditioning (HVAC), which uses the measured temperature to determine the heat flow into the cabin. The above described system is not optimisation driven, but it outlines the relevant issues in HIL and multi-physics interface. In our work, we suggest the use of Dymola, as it is flexible and allows complex, integrated multi-physics simulations while having a component representation which resembles the real world [10]. Our methodology performs therefore Multiphysics simulations in a Dymola environment.

1.3. Our Proposal for a Novel Test-Rig

High re-configurability under different scenarios is essential in our methodology. Retrieving results of HIL simulations and use them for control purposes is an effective technique [14]. Results of HIL simulations or results of multi-physics simulations will be used to determine hardware reconfiguration. This means that the system we propose will be composed by different sections featuring a wide range of components. The HIL will include all TMS subsystem, i.e. ICE, lubrication, ICE air intake/exhaust ducts, power electronics, battery system, generator and cabin HVAC. Each sub-system will be designed in order to allow the application of different cooling or heating solutions, i.e. featuring different heat exchange devices assembled in parallel on different branches of the circuit itself. It will therefore be possible to shut down some parts of each sub-system and exclude them from the test-rig. This layout will enable us to perform simulations on a wide range of design solutions.

First virtual simulations will be performed in the multi-physics environment. A discrete optimisation will be carried out to choose the best solution among a range of selected layouts. The latter will be determined on the base of available components on the market. The need for reducing costs leads to standardisation of components. This means it is impossible to randomly select any dimension for the components. Parameters cannot therefore be changed continuously but they rather need to be varied in steps. Once the best solution will be determined, the information will be passed from the multi-physics tool to reconfigure the system. It is possible to feed a HIL system with multi-physics simulation outputs through a control system if the simulation step is minimised [10]. The thermal management controller will then act on the subsystems of the hardware, selecting which branch needs to be run. The simulation will then be carried out on the test-rig. Results will be imported by the real-time computer and compared with those obtained from the virtual simulation. If operating temperature ranges will not be kept, this first solution will be discarded. Results will be re-exported back to the multi-physics environment, the layout will be changed and optimised again. The new solution will be used by the thermal management controller and the HIL layout will be changed accordingly, in order to perform a new simulation. This procedure will be looped until an optimal and robust solution is found. The interaction between virtual simulation and physical hardware can be seen in Figure 1. Information is exchanged via a Controller Area Network (CAN) bus. Data are exchanged by components with each other and with the controller, which communicates with the multi-physics model.

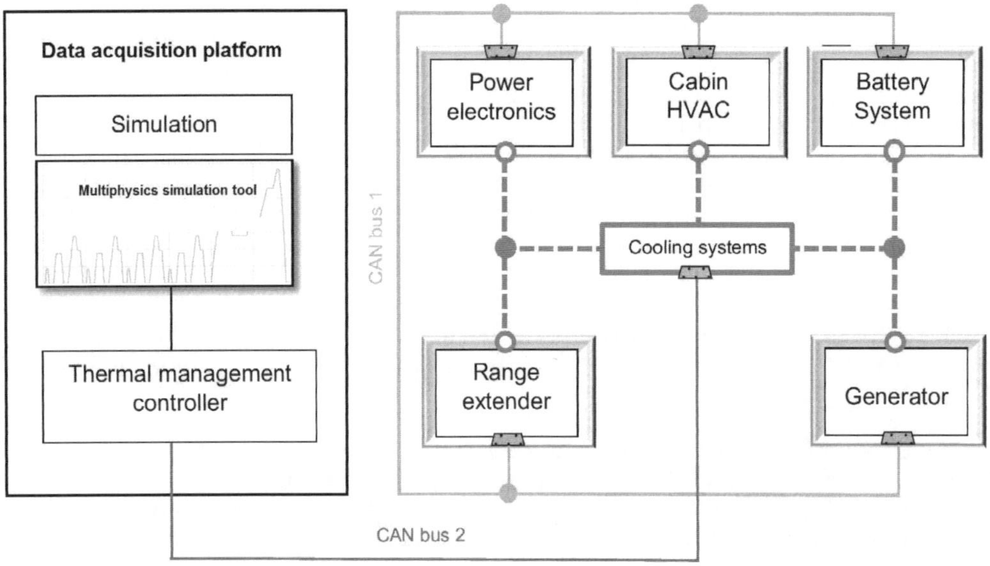

Figure 1: Proposed concept for the test-rig and multi-physics combined simulation.

Authors suggest the combined use of Matlab and Dymola, the former being used for control purposes. When interfacing Dymola with an embedded s-function in Simulink though, using the same time step in both solvers can lead to divergence [15]. The choice of time step in virtual simulation and control system is therefore a critical issue.

Virtual models will be performed in a multi-physics simulation environment. Optimal TMS layouts obtained from simulations, will be tested in the real world by acting on the HIL actuators and reconfiguring the system. Results of HIL simulation will be analysed by the optimisation algorithm. The range of solutions narrowed down.

Other simulations in the multi-physics environment will be then performed adjusting system parameters. Further HIL testing will therefore be carried out, again by letting the results of virtual simulation act on the control system of the actuators. The range of possible TMS layouts will be therefore progressively narrowed down. The method we propose would therefore be much beneficial in reducing the time to market of new vehicles, enhancing their efficiency and performances as well. The proposed rig is a tool that can enable inexpensive, fast and efficient HEV TMS optimisation.

2. NOVEL TEST-RIG LAYOUT DEFINITION

The kind of simulation strategy we use will affect performances and results, as shown in [16]. It is therefore important to refer to a robust and systematic procedure. The definition of a holistic vehicle simulation tool is organised in a four level development structure: (i) vehicle level, (ii) sub-system level (analysing each part of the vehicle), (iii) component level (defining the physical behaviour of each part and each sub-system of the vehicle) and (iv) base level (an investigation on the physical laws at the base of each component) [17]. The vehicle can be divided in the following sections: power supply, air conditioning/heat pump, cooling circuits, passenger cabin and powertrain [17]. It is possible to outline different elements contributing to the whole thermal management system, each one being characterised by a certain temperature and a certain contribution in heat exchanged with other elements and the outer ambient. It is also possible to outline the power flow among the components which are also connected by the TMS. The main elements of the TMS can be identified as shown in Table 1 and Table 2:

Table 1: System components which need to be modelled and their function.

Component	Function
Range extender (ICE)	Providing power to the system
Generator	Converting mechanical energy into current
Generator inverter	Switching DC into AC for the generator
Battery	Chemically storing the energy
Traction motor inverter	Switching DC into AC for the motor
Traction motor	Providing torque to the wheels
Passengers cabin	Internal ambient to be heated/cooled
Cabin HVAC	Device heating/cooling internal ambient
External ambient	All subsystems interact with it

Table 2: Additional models required to run simulation.

Additional model
Driving cycle
Driver's behaviour
Vehicle dynamics model

We developed a model representing the interactions among the various components in terms of power ($\dot{W}$) and heat flow ($\dot{Q}$). We followed two different approaches and compared them [18]:

- **Backward-facing**. This approach (Figure 2) assumes that the vehicle has met the required trace and investigating the performance of each component. This approach does not require a driver model and relies on speed trace at each time step to calculate the force required to accelerate [18]. In this model required torque is determined by the driving cycle. The information is passed to the vehicle dynamics model which then controls the traction motor inverter. The traction motor inverter provides the information about required torque to the traction motor. The power request determines battery discharge cycle and range extender activation. Each component generates heat and exchanges it with the other components and the external ambient. Furthermore, the driving cycle also includes settings for required cabin temperature, which determine HVAC behaviour.

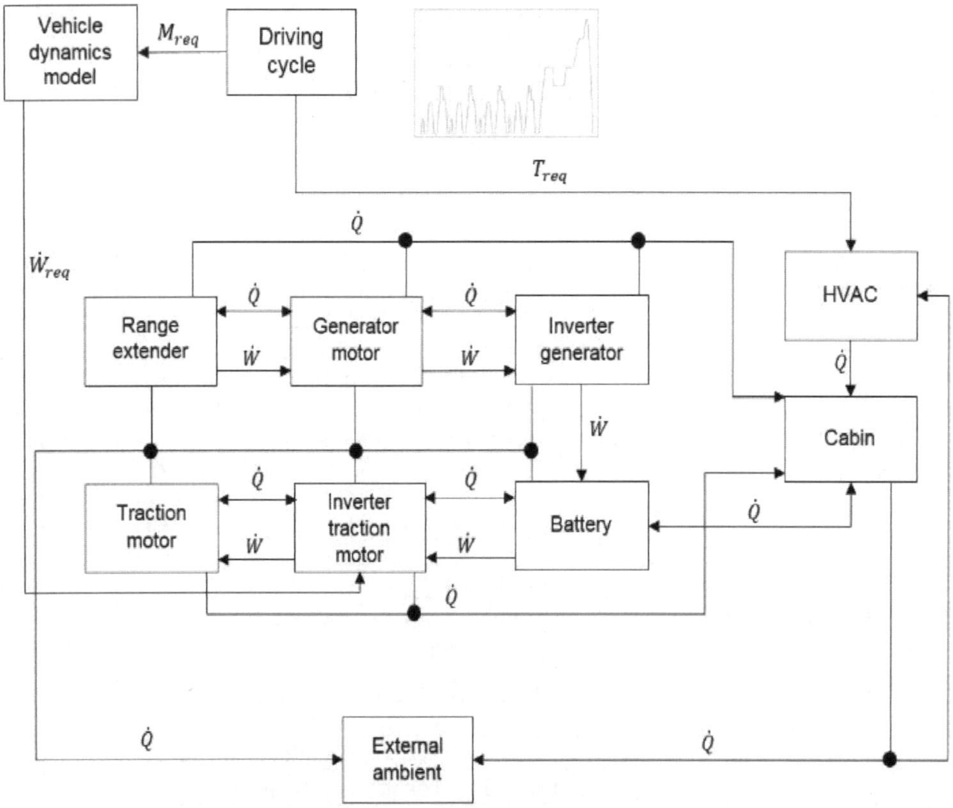

Figure 2: Backward facing modelling approach.

- **Forward-facing**. In this approach (see Figure 3) we rely on a driver's model which has a response to the required speed resulting in a command for the powertrain [18]. The driver acts on the traction inverter through the accelerator pedal. Thanks to the vehicle dynamics model a feedback is provided to the driver, who adjusts his action on the accelerator accordingly. The driver can be a physical person or it can be simulated by an algorithm. In both cases the driver has to perform a specific driving cycle. The driver also selects a required cabin temperature and the HVAC behaves consequently. Heat generation and heat exchange are then influenced by requested torque and requested temperature.

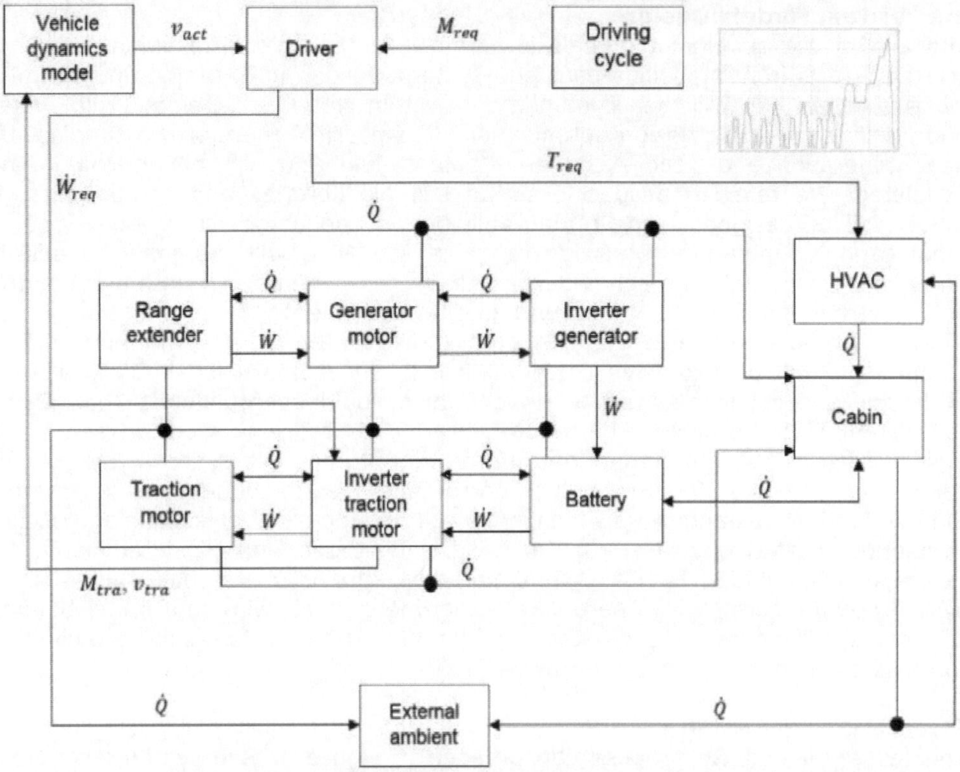

Figure 3: Forward facing modelling approach.

A forward-facing approach, even if more time-consuming than a backward-facing approach, deals with quantities measurable in a physical drivetrain. It is therefore more suitable for hardware development [18].

2.1. Computational Platform Selection for the Test-Rig

The required computational platform needs to be capable of processing both input and output signals in real time. There is a wide number of processing units meeting the aforementioned requirements, prominently the NI CompactRIO and the dSpace range of products. Different research groups adopted different solutions. Some authors suggested the use of CompactRIO [17]. The latter is a smart-structure collateral hardware platform, combining embedded floating point processor (Power PC) with real-time operating system (VxWorks), a high performance field programmable gate array (FPGA) and hot-swappable Input/Output (I/O) modules [19]. A low level of customisation of I/O and timing can be provided by each I/O module connecting it directly to FPGA [19]. A high-speed PCI bus is used to connect the FPGA to the embedded real-time processor [19]. Both processor and FPGA are programmed in LabView language, so that data processing and data logging are available [19]. The use of CompactRIO, however results in a much complicated system. Other researchers proposed of dSpace Automotive Simulation Models, due to its high standards in terms of model robustness [9]. The dSpace platform allows a real-time multi-body model which reflects the physical structure of the system. The I/O blocks can be designed in order to contain real time interface between HIL and dSpace computational platform [9]. Since dSpace products seem to be much more flexible, efficient and effective, our system will be therefore based on dSpace, allowing a direct interface between hardware and model.

2.2. Virtual Models Design

The aim of the developed models is to virtually represent the vehicle TMS. Our strategy starts from modelling each sub-system, validating it, tuning and optimising its parameters. We will then compose the system with the developed sub-systems and finally carry out a design optimisation. It is possible therefore to simulate both heat generation and heat exchange. Coolant fluid and air cooling can be both simulated. We refer to a specific set of Dymola libraries and components. The model will be designed using replaceable blocks and partial sub-models open for modification. Simulation of a wide range of scenarios will therefore be possible. The aim is to simulate the behaviour of system components, information bus, driver and external ambient. It will eventually be possible to place the test-rig in a thermal chamber to simulate the external ambient. Dymola thermal and fluid flow components will be extensively used. Since it is necessary to simulate a driving cycle and collect signals from a driver which could eventually be real, Dymola Vehicle Interface library will be fundamental. This library provides a range of I/O buses and sub-buses to convey the information of each sub-system. Each sub-bus can be connected to the all-inclusive Control Bus, used to connect the sub-systems. Simulating CAN bus and signals in Dymola will provide a more realistic and accurate simulation of the hardware. Models will be evaluated and validated individually before any simulation is run. Actual parts will provide reference values for the system parameters, while hardware experiment data will determine boundary conditions; after that a steady state simulation is run and its results are compared to those of the experiment on hardware [13].

2.3. Hardware Design

The test-rig layout we propose can be seen in Figure 4. Sub-systems composing this layout are: power electronics, ICE range extender, ICE cooling, battery cooling and HVAC, similarly to what was done in [20], [21]. A set of different radiators is available, they are: the high temperature radiator (for the ICE), the medium temperature radiator (for the power electronics), the low temperature radiator (for the battery) and a smaller radiator for the ICE lubricant. The latter can be used as an alternative to a fluid-to-fluid heat exchanger, using ICE coolant fluid. This system is highly flexible since entire portions could be excluded or included in the simulations by simply opening and closing some valves. For instance, it is possible to shut down the entire ICE cooling circuit, activate a supplementary venting for the engine, while excluding fluid-to-fluid oil heat exchanger and activate the lubricant circuit radiator. This option could provide an air-cooled engine layout, which could be used to simulate HEV moped or motorcycles. Battery can be both air or liquid cooled and can interact with the cabin HVAC system if required. It is possible to run battery cooling/heating on the same circuit of the power electronics or on separate circuits. Power-electronics can either be liquid- or air-cooled. Through the use of heat exchangers, it is possible to let the various sub-systems exchange heat with each other. It is therefore possible to see how different layouts could be defined. The methodology needs heat generators which could simulate a different range of heat generation scenarios.

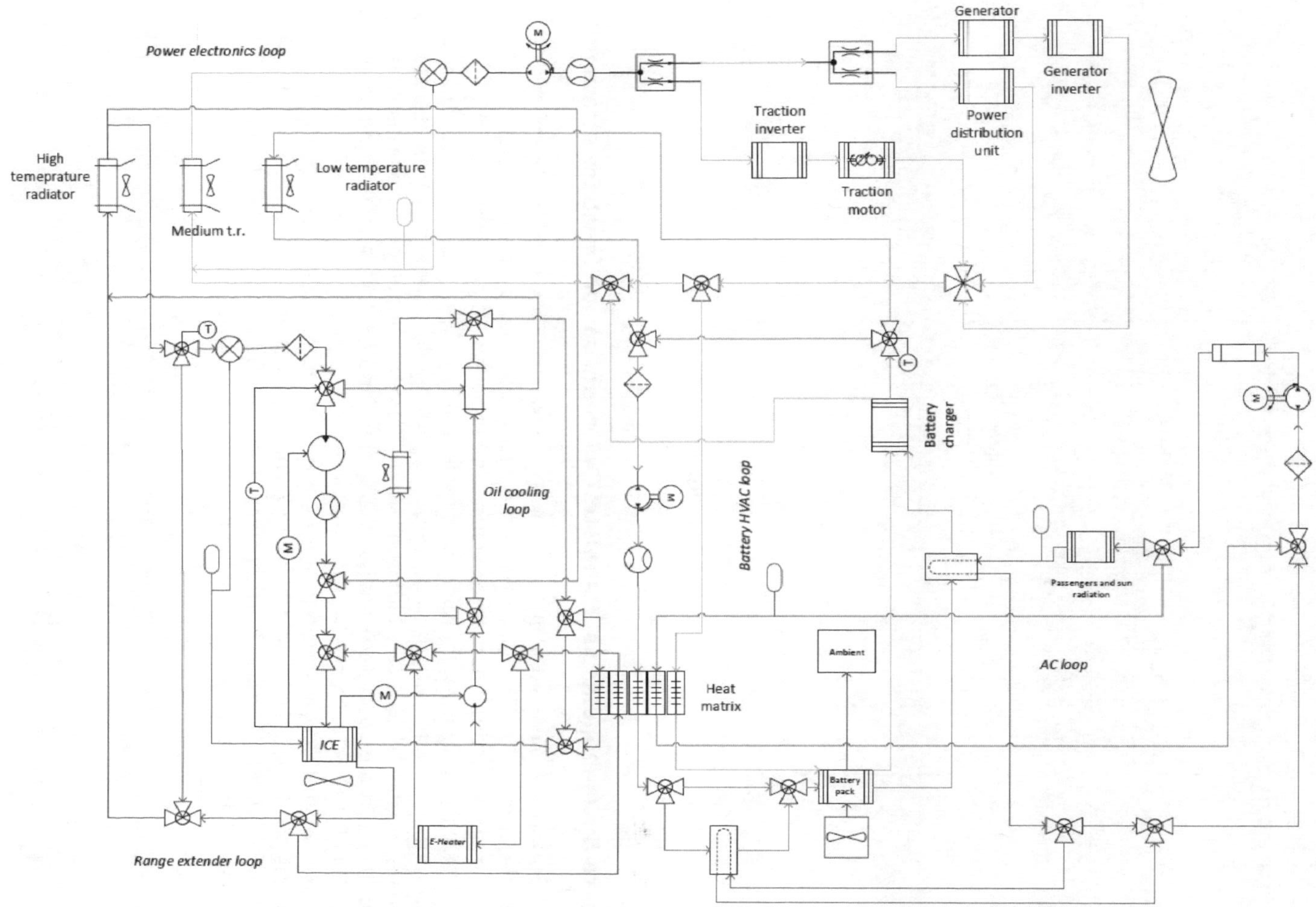

Figure 4: Proposed reconfigurable test-rig, featuring all possible TMS layouts.

2.4. Optimisation

The optimisation strategy for the system we propose, operates both on a component level and on a system level. While a parameter tuning will be performed on the single sub-systems, their interaction will be evaluated as well. The results in terms of complete TMS optimisation act as a feedback on the component tuning. The components are tuned again according to the system optimisation (see Figure 5). The simulation is in a closed loop and progressively adjusts the sub-systems and their interactions. Since automotive parts are standardised, there is a discrete (and often limited) range of choice. Optimisation on components could therefore be performed through an exhaustive search, while more sophisticated strategies (e.g. genetic algorithms) could be used for the interactions among sub-systems.

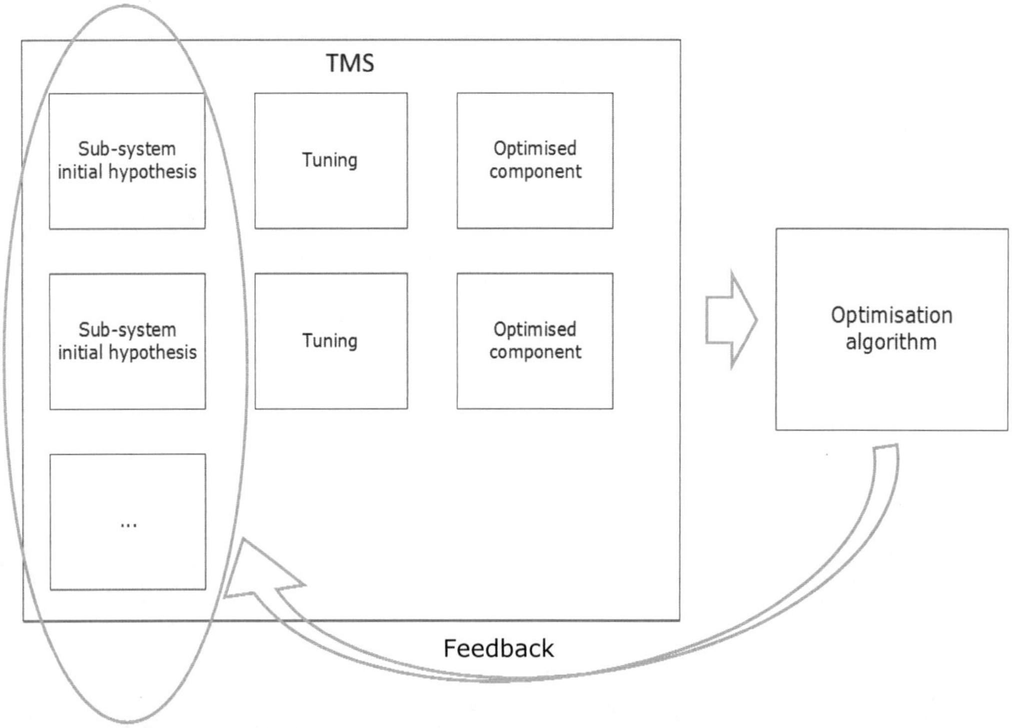

Figure 5: Components vs. complete TMS optimisation feedback strategy.

While performing an optimisation routine, robustness is essential. Solutions must be stable in order to be applicable in the real world. Geng et al. proposed Pareto-based multi-objective particle swarm optimisation algorithm, remarking its robustness in automotive optimisation problems [22]. Being stuck in local optima is a relevant risk for design-optimising algorithm, it is therefore necessary to effectively explore all possible solutions. This methodology uses disturbance operators, in order to mutate analysed solutions and explore further configurations [22]. The latter is a stochastic population-based algorithm in which a swarm of particles subject to a random initialisation move through the search space. Solutions are updated at each iteration using information obtained from their past and from their neighbours. Furthermore, the authors also suggest to incorporate constraints into objective functions as much as possible [22].

3. APPLICATION EXAMPLE

Here we introduce an example on how this methodology can be used. Let us assume that testing vehicle sub-systems are as in Table 3.

Table 3: Vehicle thermal sub-systems.

Front electric motor and power electronics liquid cooling
Rear electric motor and power electronics liquid cooling
Internal combustion engine range extender liquid cooling
Battery pack HVAC
Cabin HVAC

The scheme in Figure 6 refers, more specifically to an all-wheel drive vehicle, i.e. this case study includes all possible power electronics features. This vehicle is an all-wheel-drive and has therefore power electronics on both front and rear. Furthermore, this automobile has an air-cooled battery and a liquid-cooled ICE range extender. It is essential to classify each system according to its operating temperature range, in order to define the different circuits which will be simulated (Figure 6). It is possible then to define the first draft of the test-rig starting from the elements mentioned above. The HIL is designed so that both use of real elements and heat generators are allowed. The system is implemented through steps. The first phase features the definition of pumps and pipes. A first rig is developed to simulate the hydraulic part only. The procedure can be divided into two major sub-steps. The first one is the hydraulic dimensioning which includes the choice of pumps and the choice of pipes and fittings. A second phase is the thermal dimensioning, i.e. dimensioning the circuit on the different thermal ranges to be examined. After the HIL development, it will be possible to vary the TMS and its multi-physics model. The possibility of applying modifications will be essential in reconfiguring while performing the overall TMS optimisation.

It is possible to replace real devices with heating elements dissipating the same heat power of the real components.

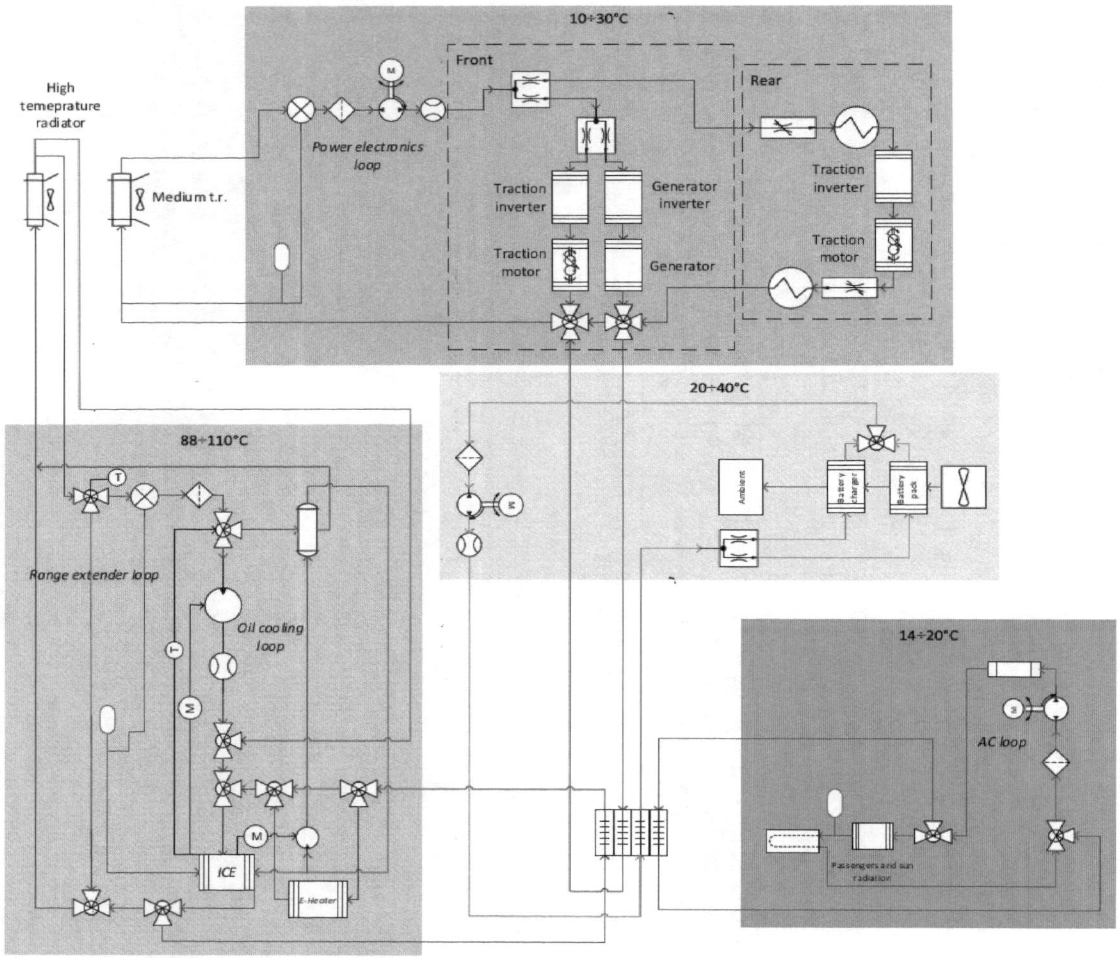

Figure 6: System layout component classification according to temperature range.

3.1. Optimisation Example

An example of the optimisation strategy is the battery HVAC system. Given the required mass flow rate, the operating temperature range and the circuit components, it is necessary to select the correct compressor. Let us assume we are dealing with radial blade centrifugal compressors. Devices delivered by suppliers have different displacement and therefore radius. Each compressor operates within a different range of rotational velocity. It is therefore necessary to select the correct part and the operating rpm. An exhaustive search can be performed throughout the available radius. We measure volume flow rate in different points of the circuit in order to have a feedback. Revolution speed is adjusted accordingly until the optimum value is obtained. Having achieved all the optimum velocities that minimise power consumption and cost, the component is then selected. Feedback is provided to sub-systems from the overall TMS optimisation, in case the correct temperature range is not kept and they are adjusted accordingly. Volume flow rate is therefore adjusted and the sub-system re-configured as shown in Figure 7. It is worthwhile to notice that a correct dimensioning of the battery HVAC could help in keeping the temperature gradient across the battery within 5°C, while keeping the operating temperature between 20°C and 40°C, ideally around 25°C. Our current work is focused on these two objectives and the rig is being developed accordingly.

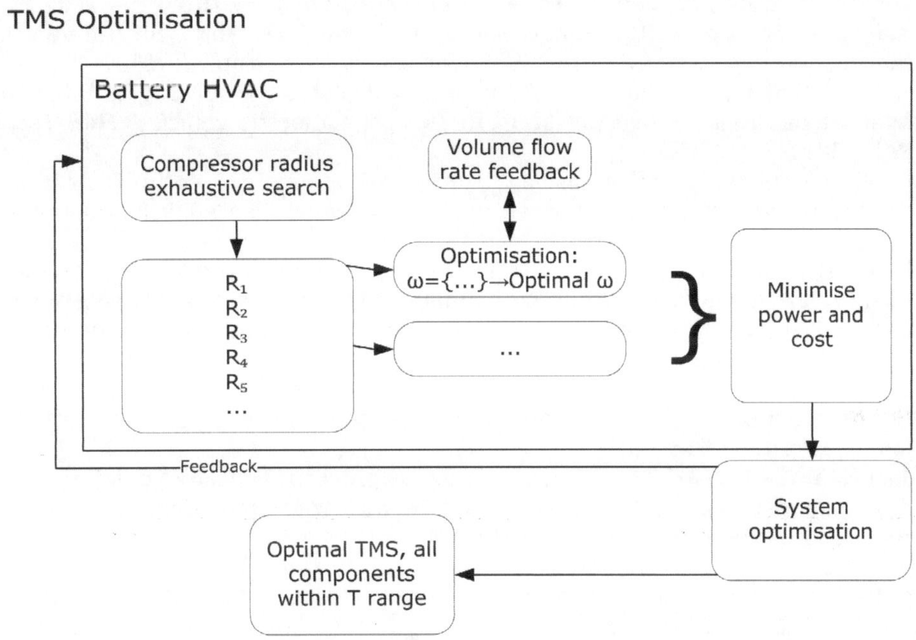

Figure 7: Battery HVAC optimisation example, feedback is provided by the TMS optimisation into the sub-system.

4. CONCLUSION

We have presented a methodology for the development, optimisation and testing of TMS for hybrid vehicles. The methodology is based on a highly reconfigurable (both in terms of software and hardware) HIL set up. In this setup, vehicle components are represented by physical heating elements driven by computer models. The TMS, however, is real but reconfigurable with the use of computer controlled valves and pumps. The proposed methodology allows for the optimisation of the whole TMS before a first prototype is manufactured. The result will be a significant shortening of time-to-market. Furthermore, this system will allow academic research for TMS optimisation. Future work will include the finalisation of the rig and models with the aim of creating an almost completely automatic TMS optimisation suite.

REFERENCE LIST

[1] J. Shutty and R. Pinto, "Advanced thermal managment strategies," *Sae Tech. Pap. Ser.*, 2013.

[2] H. S. Hamut, I. Dincer, and G. F. Naterer, "Analysis and optimization of hybrid electric vehicle thermal management systems," *J. Power Sources*, vol. 247, pp. 643–654, 2014.

[3] D. Leighton, "Combined fluid loop thermal management for electric drive Vehicle Range Improvement," *SAE Int. J. Passeng. Cars - Mech. Syst.*, vol. 8, no. 2, pp. 711–720, 2015.

[4] N. Kim, J. Jeong, A. Rousseau, and H. Lohse-Busch, "Control analysis and thermal model development for plug-in hybrid electric vehicles," *SAE Int. J. Alt. Power.*, vol. 4, pp. 260–268, 2015.

[5] N. Staunton, V. Pickert, and R. Maughan, "Assessment of advanced thermal management systems for micro-hybrid trucks and heavy duty diesel vehicles," *2008 IEEE Veh. Power Propuls. Conf.*, pp. 1–6, 2008.

[6] X. Tao, A. Ivanco, J. R. Wagner, and H. Hofmann, "A hybrid electric vehicle thermal management system - Nonlinear controller design," 2015.

[7] J. Lee, H. Ohn, J. Choi, S. J. Kim, and B. Min, "Development of effective exhaust gas heat recovery system for a hybrid electric vehicle," *SAE Tech. Pap.*, 2011.

[8] R. M. Kossel, M. Loeffler, N. C. Strupp, and W. J. Tegethoff, "Distributed energy system simulation of a vehicle," no. 1. Woodhead Publishing Limited, 2011.

[9] K. Patil, S. K. Molla, and T. Schulze, "Hybrid vehicle model development using ASM- AMESim-Simscape co-simulation for real-time HIL applications," in *SAE 2012 World Congress and Exhibition, Detroit, MI; United States; 24-26 April 2012*, 2012.

[10] S. Thanagasundram, R. Mcmurran, R. P. Jones, and A. Mouzakitis, "Real-time simulation of a vehicle door locking mechanism on a hardware-in-the-loop platform," 2016.

[11] J. (Tata E. L. . Khan, "A standardized process flow for creating and maintaining component level hardware in the loop simulation test bench," *SAE Tech. Pap.*, vol. 2016-1-0, 2016.

[12] A. Palladino, G. Fiengo, and D. Lanzo, "A portable hardware-in-the-loop (HIL) device for automotive diagnostic control systems," *ISA Trans.*, vol. 51, no. 1, pp. 229–236, 2012.

[13] C. Ding, Z. Xu, Y. Zhang, and Q. Tao, "Multi-domain modeling and simulation of vehicle thermal system based on modelica," in *SAE 2014 World Congress and Exhibition; Detroit, MI; United States; 8-10 April 2014*, 2014.

[14] S. Ko, C. Song, and H. Kim, "Cooperative control of the motor and the electric booster brake to improve the stability of an in-wheel electric vehicle," vol. 17, no. 3, pp. 447–456, 2016.

[15] S. J. Liu, D. Faille, M. Fouquet, B. El-hefni, Y. Wang, J. B. Zhang, and Z. F. Wang, "Dynamic simulation of a 1MWe CSP tower plant with two-level thermal storage implemented with control system," *Energy Procedia*, vol. 69, pp. 1335–1343, 2015.

[16] L. I. U. Jingping, L. I. Yingchun, X. U. Zhengxin, X. I. A. Xiaolang, and S. H. "Simulation for series-parallel hybrid electric vehicle powertrain system," in *International Conference on Electric Information and Control Engineering (ICEICE), 2011*, pp. 2–6, 2011.

[17] C. Bouvy, S. Baltzer, P. Jeck, J. Gissing, T. Lichius, and L. Eckstein, "Holistic vehicle simulation using modelica – An application on thermal management and operation strategy for electrified vehicles holistic vehicle simulation using modelica" in *Proceedings of the 9th International Modelica Conference, September 3-5, 2012, Munich, Germany*, 2012, no. November.

[18] K. B. Wipke, M. R. Cuddy, and S. D. Burch, "ADVISOR 2 . 1 : A User-friendly advanced powertrain simulation using a combined backward / forward approach," *IEEE Trans. Veh. Technol. VOL. 48, NO. 6, Novemb. 1999*, vol. 48, no. 6, pp. 1751–1761, 1999.

[19] P. Bilik, L. Koval, and J. Hajduk, "CompactRIO embedded system in power quality analysis," in *Proceedings of the International Multiconference on Computer Science and Information Technology*, 2008, pp. 5–8.

[20] M. Bassett, J. Hall, M. Warth, A. Eilemann, and C. Merkle, "Plug-in hybrid cooling system requirements," *SAE Tech. Pap.*, 2014.

[21] BMW Group, "Technical training: product information. W20 engine, I01 high-voltage components, I01 Heating and A/C Systems." 2015.

[22] B. Geng, J. K. Mills, and D. Sun, "Combined power management / design optimization for a fuel cell / battery plug-in hybrid electric vehicle using multi-objective particle swarm optimization," *Int. J. Automot. Technol.*, vol. 15, no. 4, pp. 645–654, 2014.

Model predictive control for Organic Rankine Cycle applied to hybrid vehicles

A Agurto-Goya
Jaguar Land Rover Ltd, UK

J Kitte, I Trivic, T Tietze
IAV GmbH, Germany

Electric Hybridization and Waste Heat Recovery are technologies with significant potential for fuel saving. As separate applications, both technologies claim a significant reduction of fuel consumption. Deploying a combination of both technologies has synergetic and competitive elements that properly managed, can lead to improved fuel economy.

Waste Heat Recovery technologies like Rankine Systems suffer from poor response and time delay due to large thermal inertia when subject to highly dynamic driving scenarios. This makes it challenging to introduce into light duty vehicle applications. The Model Predictive Control approach represents a possible way to compensate for the existing challenges. The approach allows the definition of a predicted driving horizon for adapting the Rankine System control values such as temperature and pressure, in order to maximize the energy output of expander machine.

This paper investigates the benefits of having a Rankine System with the Model Predictive Control approach on board of a mild hybrid vehicle. The main point the study represents is the embedding of the recovered electrical power into the vehicle drive train, benchmarked against one emission cycle. The gain in power caused by the model predictive control reaches a maximum improvement of nearly 23 %. The correlation between thermal energy input and generated energy output of the plant model leads to a simple rule based control strategy.

1 MOTIVATION

1.1 Rankine cycle in the automotive industry

In the last 15 years the interest in Rankine Cycle technologies has prompted various automotive manufacturers to investigate its potential. Research in light duty OEMs was reported by BMW (Turbosteamer I and II) [1] having both a steam and ethanol based system; and Honda with a steam based system [2]. Both of them claimed to achieve a decrease in fuel consumption of over 10% for their passenger cars. For heavy duty OEMs Chris Nelson reported that Cummins improves fuel consumption by 10% in their WHR program [3]. There are many other research studies on Rankine cycles applied to automotive industry. A summary of those researches are in the table below.

Table 1. Rankine cycle benchmark

Company	System Features	Fuel Savings
Faurecia (2014)	• Ethanol & water • Target cost ~ 450 € • (1.6L -3.0 L)	• 7 % at WLTC (Hybrid) • 6 % at WLTC (1.6L Gasoline) • 4 % at NEDC (1.6L Gasoline)
IAV (2014)	• Ethanol & water • 1.4 TSI Golf VII • Weight 40 kg • Turbine & piston expander	3-4% at NEDC
Feasibility Study of an ORC with applications to Ohio State University EcoCar PHEV (2012) [4]	• R245fa • PHEV	• 7% at peak load • 4.8 % NEDC
Ford Organic Rankine Cycle for Light Duty Passenger Vehicles (2011) [5]	• R245fa • DOE study	• Under EPA highway output close 100% electrical accessory demand • HEV outperform conventional vehicle in highway • Net backpressure can be lowered with a well designed ORC system
BMW Turbosteamer v2 (2012) [1]	• Water • BMW series 5 • Weight 10-15 kg • Impulse turbine expander	• 6 % at cycle ? • Up to 10 % (with predictive thermal management)
EXOES (2014) [6]	• Ethanol & water • Swash plate & e-machine • 1.6L Gasoline turbo engine	• 11% fuel saving (EUDC cycle) • 22 gCO2/km (EUDC cycle)

Although offering significant fuel economy benefits, one of the biggest challenges is the weight and packaging which might restrict the application to large vehicles. Components like the condenser need to be located in front of the existing radiators, forcing those to be oversized or repositioned. The increased usage of space in the engine bay of hybrid vehicles makes packaging even more challenging. Deployment of Rankine systems in light duty vehicles will require manufacturers to find synergies on competing technologies along with the downsizing of components.

1.2 Rankine systems and hybrid vehicles

In recent years the level of maturity and market share of electrified drive trains has continuously increased. Almost every vehicle manufacturer offers at least a hybridized power train. Different range of CO_2 reductions are being reported on vehicles with different degrees of hybridization. Mild hybrid vehicles (MHEV) achieve 5 to 15 % the CO_2-emission reduction in Worldwide Harmonized Light-Duty Vehicles Test Procedure (WLTC) while full hybrid vehicle can achieve a 15 to 20 % CO_2-emission reduction depending on vehicle specification. The recuperation of kinetic energy is the largest adjust lever of fuel saving. Hybrid vehicles with combustion engines also exhibit high thermal energy that properly exploited can boost the fuel savings ever further.

Among the different options of waste energy recovery strategies, the Thermoelectric Generator (TEG) and Rankine Cycle system have been heavily investigated and have stepped into the prototype stage. Figure 1 shows the thermal and mechanical path of recuperation of F-segment vehicle and its exemplary results. The amount of waste heat energy is significantly higher than that of kinetic energy. However it requires more effort to recuperate and convert thermal energy compared to kinetic energy. This can be seen in the difference on efficiency ratio: an integrated Organic Rankine Cycle (ORC) expander machine energy to waste heat energy is 3 % while e-machine energy to overall kinetic energy is 34 %. However the amount of energy recovery is in the same range (1.1MJ vs 1.2MJ).

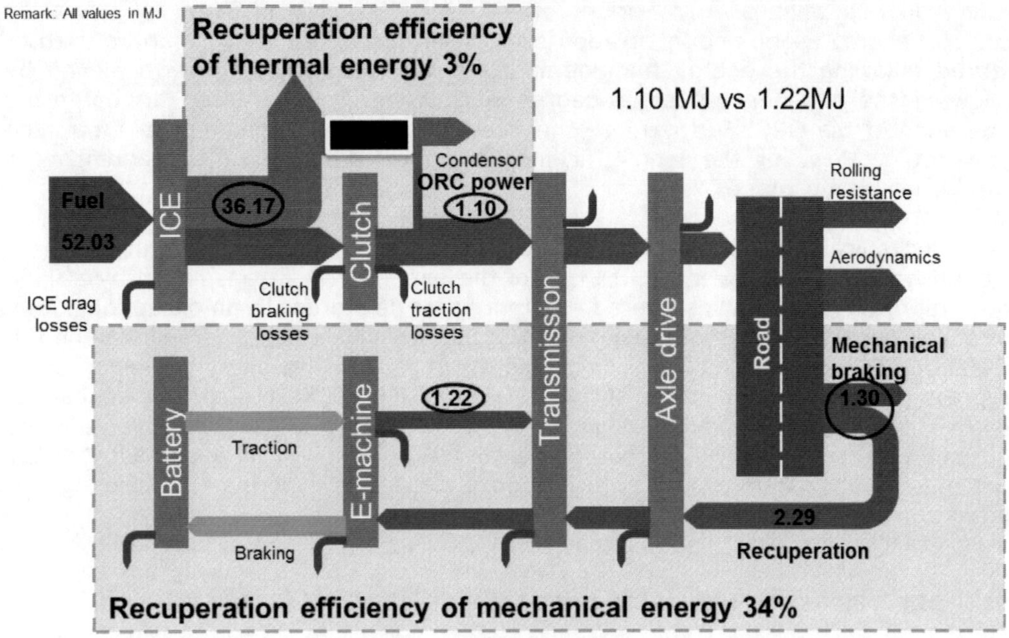

Figure 1. Energetic analysis – thermal and mechanical recovery path (F-segment, WLTC)

The implementation of both recuperation strategies into the drive train of a hybrid vehicle leads to both synergies and conflicts. Therefore net energy recovery is expected to not be fully additive. Synergetic effects can be seen on waste energy conversion of the mechanical coupled expander during braking/driving phases and the damping effect of exhaust gas flow during dynamic driving phases. Both effects increase energy recovery and simplify control of exhaust energy recovery. However there are also potential conflicts and differences. More electric driving diminishes the amount of exhaust energy. The transient behavior is significantly different. While the electrified power train is highly dynamic, the Rankine Cycle System exhibits a high thermal inertia. The Organic Rankine Cycle could be integrated in two different ways (figure 2).

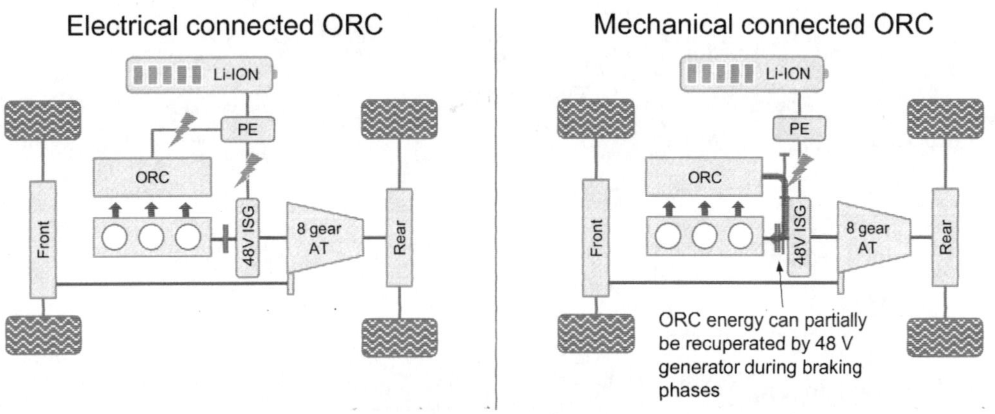

Figure 2. Variants of integration of ORC into a simulated vehicle environment

The electrically connected ORC (e-ORC) is fully decoupled from drive train. The recovered exhaust energy is converted from thermal to electrical energy and utilized when an electrical energy management decides that ORC will have the most effect on fuel savings. This variant offers more flexibility in terms of operation control, energy usage and integration to vehicle. Recovered energy can be used or stored following the energy management strategy. Better control is achieved by allowing the expander speed a degree of freedom to guarantee an optimized operation of the ORC. Furthermore the integration is nearly independent from the powertrain. However the long efficiency chain of electrical paths represents a significant disadvantage.

The mechanical connected ORC (m-ORC) is coupled to the gear box in the drive train. The generated speed and torque of the expander are transmitted by a fixed gear ratio, therefore any recovered energy cannot be stored. In an exceptional case of power generation during braking phases, the drive train clutch is opened and the expander power flows into the e-machine (48V ISG). The short efficiency chain represents the main advantage of the m-ORC. Unlike e-ORC the speed of expander is fixed to speed of transmission input, representing a real disadvantage as optimum operation of ORC is not guaranteed. The position of expander strongly depends on drive train position which impose constrains in terms of packaging and integration into the vehicle.

2 ORC TEST ENVIRONMENT AND MAIN CONTROL VARIABLES

The study was performed at IAV's Ethanol driven ORC test facilities, whose capabilities allowed us to investigate both passenger cars and commercial vehicles. Beyond different heat exchanger generations, expander technologies like a piston expander and a constant pressure turbine were also investigated. The ORC system was integrated into the exhaust gas system after main catalyst and the downstream condenser was integrated into a low temperature coolant circuit. Temperatures and pressures are controlled by PID and PI controllers with integrated pre-control. Figure 3 shows a schematic of the ORC rig for a passenger car application.

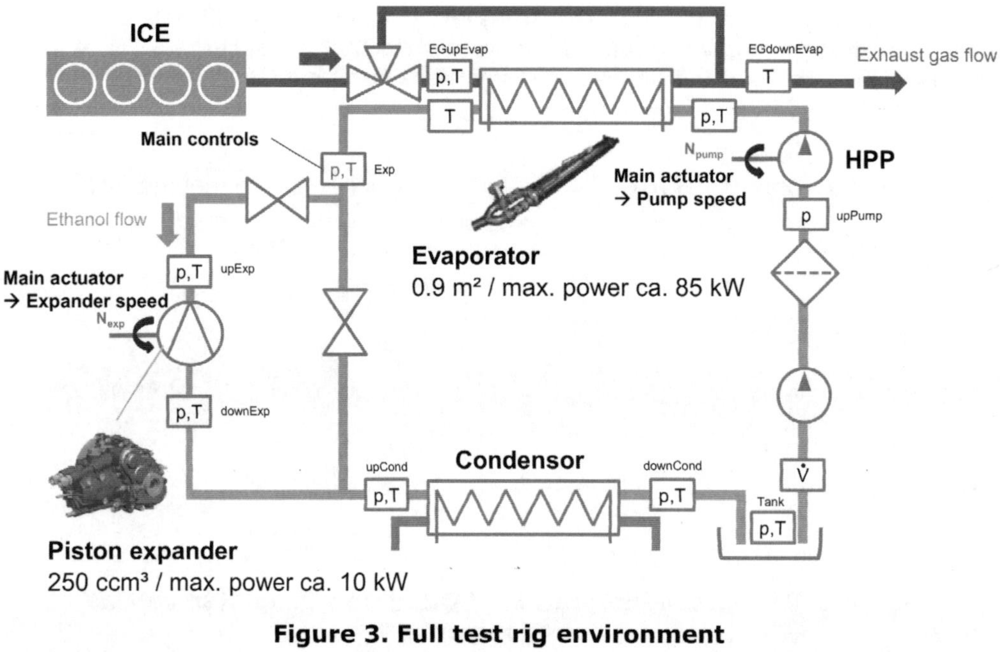

Figure 3. Full test rig environment

The displacement of the single cylinder piston expander is 250 ccm^3 and reaches a maximum power of 10 kW.

Pressure and temperature at the evaporator outlet are the main variables for both control and monitoring. These variables define the efficiency, recovery power and stability of the system. The feed pump and expander speed are the actuators for the control variables (e.g. feed pump speed acts on temperature and expander speed on pressure).

The simulation environment is based on Matlab&Simulink, with real time capability. It is divided into four main domains and able to connect to a test rig. Depending on the task defined, three selectable levels of details are possible. The environment allows us to study both electrified and conventional power trains.

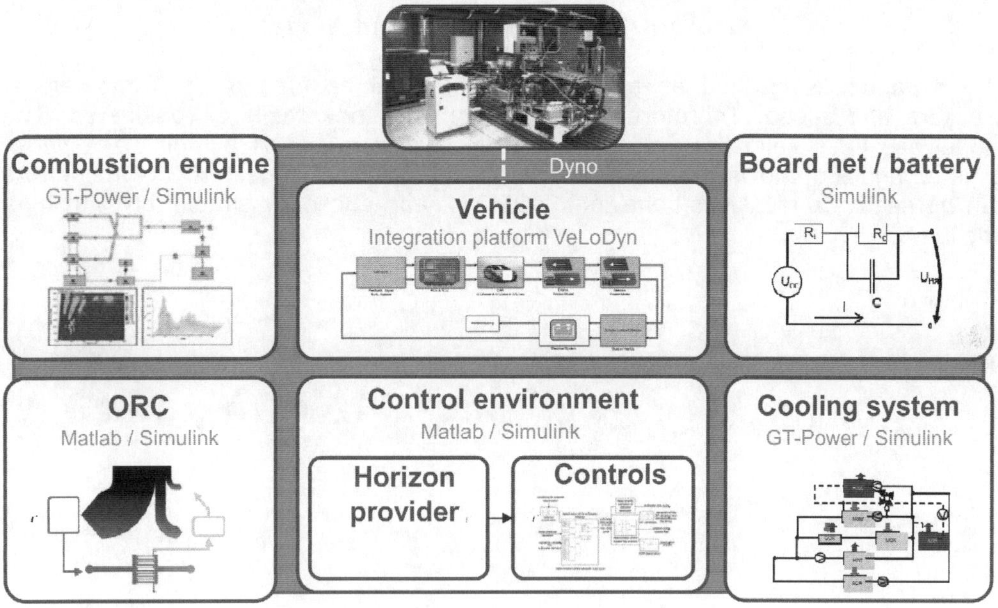

Figure 4. Simulation environment

The control environment is divided into two main parts (figure 5), the horizon provider and control functions. The horizon provider is connected to a digitized map and predicts the route of vehicle which provides the visibility field for the driver. The control functions can work more precisely under real time conditions. The vehicle/powertrain environment feeds the current position (GPS data) of the vehicle back to electronic horizon provider. Another option is the connection to real world, in order to receive information about traffic events (e.g. congestion, traffic lights and weather).

The ORC model is fully dynamic and fast enough for real time for optimization purposes. A transient model needs a variety of controllers. The architecture consists of a plant model and a function model. The ORC plant model is based on a quasi-dimensional model approach for the description of thermal and fluid-mechanical behavior. The heat exchanger model for evaporator and condenser is based on the fixed segment method combined with a switchable moving boundary approach [7].

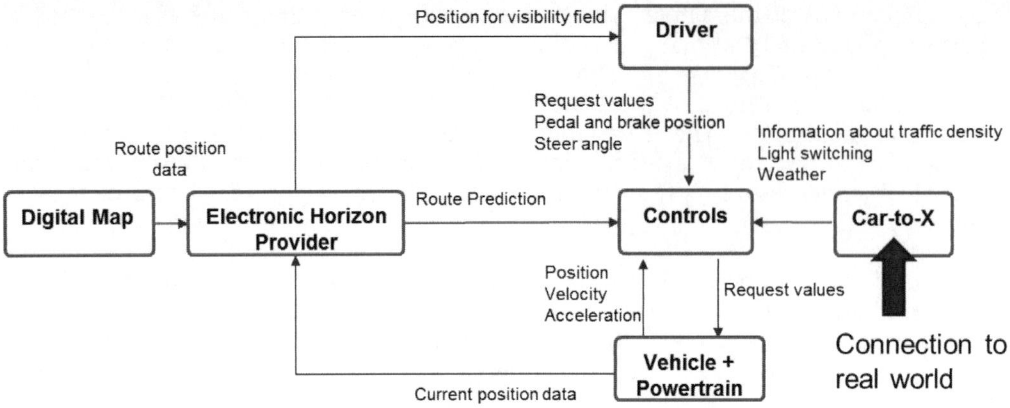

Figure 5. Vehicle environment model

The expander is defined as a quasi-steady model as function of temperature, pressure and speed. Therefore the complete efficiency chain is considered. The model has been validated against transient measurements of a mid- sized petrol engine. Figure 6 shows the validation of the mechanical connected ORC model. As can be seen the transient behavior of ORC is reproduced by simulation with high fidelity.

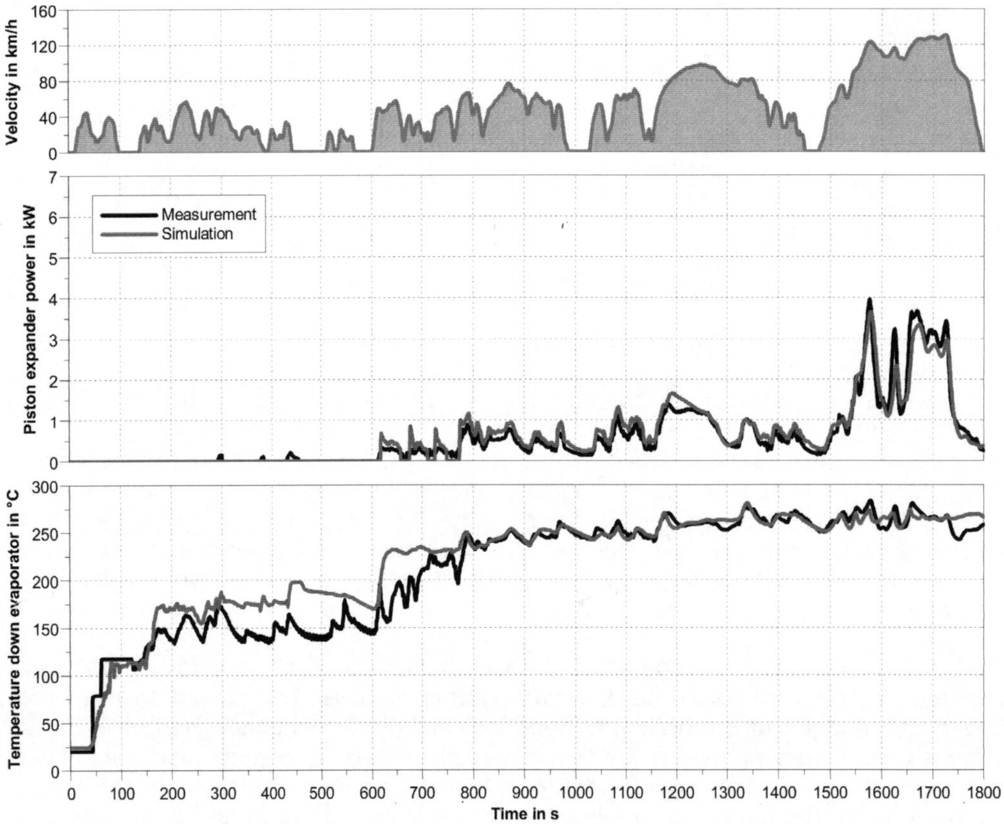

Figure 6. Model correlation based on mid-sized petrol engine: expander power and evaporator temperature output

For the following analysis an electrically connected ORC will be investigated. The main details of the vehicle application are listed in table 2.

Table 2. Vehicle boundary conditions

Parameter	Specification
Vehicle mass	F-segment representative
Aerodynamic drag coefficient c_w	0.35-0.40
Dynamic running radis R_{dyn}	0.31 m
Engine	2.0L I4 gasoline
Transmission	8-gear AT
Cycle and driving conditions	• WLTC, eHorizon • All-wheel drive • Full recuperation power set on both axles
Traction motor	PMM 10 kW / 55 Nm (12 kg) incl. power electronics
Li-ION battery	• 5.1 Ah, Vnom,cell = 3.6 V • SOCrange = 40 – 90% • 1 kWh (48 V / 12 kg)

The electrified power train represents a 48 V mild hybrid system with 10 kW e-machine and 5.1 Ah Li-ion battery. The engine is a 2.0L petrol. The analysis was done by transient simulation based on WLTC.

3 MOTIVATION OF MODEL PREDICTIVE CONTROL

Figure 7 shows ORC recovery power at steady state from the exhaust heat source. The exhaust gas temperature after the main catalyst is 600 °C and the exhaust gas mass flow is 25 g/s for a heat flow of 18 kW. It can be seen that for the same pressure the lower the superheated temperature the higher the expander power. This effect has also published in numerous papers [8]. Furthermore, the sensitivity of temperature rises with gains in pressure. The symptomatic break-in of power at 35 bar can be explained by means of evaporator characteristics.

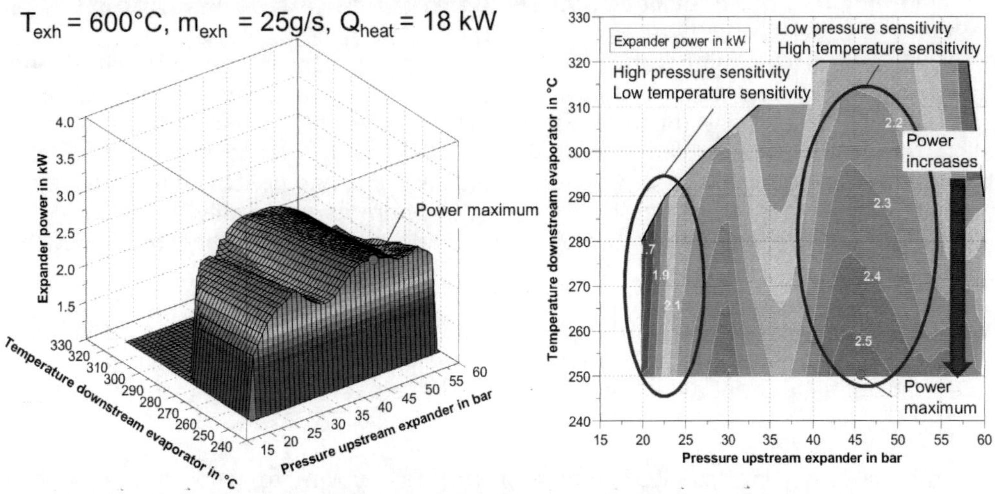

Figure 7. Impact of pressure and temperature on expander power for steady state conditions (typical WLTC point)

Unlike steady state, running the vehicle at transient conditions changes the ORC energy recovery constraints. In figure 8 the temperature at the evaporator outlet and the power of the expander are shown. There are two fixed set points at 250 °C (***a***) and 300 °C (***b***). Both variants reach the same expander energy output at the end of the cycle, which is against the effect found on the steady state investigations. Beyond the overheated temperatures, the evaporating temperature for variant ***b*** is shown in black. This temperature strongly depends upon pressure. The steam quality of the working medium runs between 0 and 1 so it is neither undercooled nor overheated. The expander is only able to work at temperatures higher than the evaporating temperature.

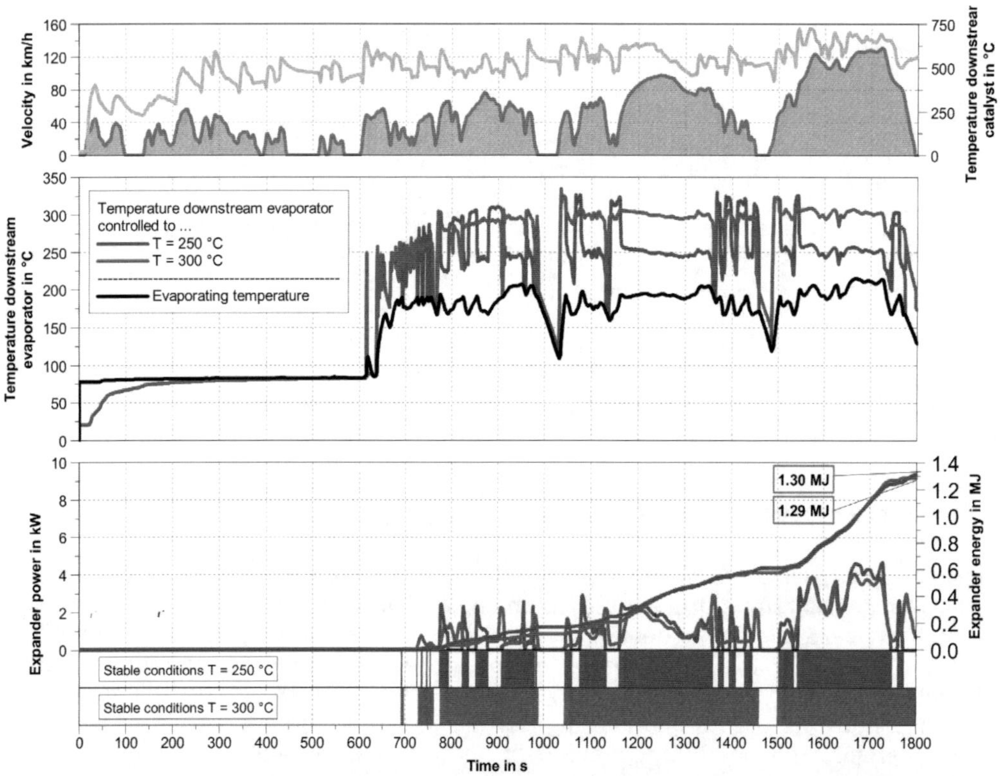

Figure 8. Impact of set point temperature on ORC power output

The ORC controls for temperature and pressure are based on conventional PI and PID concepts with pre-control. After the heat up phase of nearly 600 s the system ***b*** steps into a stable phase much faster than ***a***. Between 600 s and 800 s system ***a*** falls often back into deactivation state. The same behavior is also seen in the break down phase (1300 s – 1450 s). System ***a*** falls back into the deactivation state more often than system ***b***. In the case of the ORC it operates unsteadily at low working conditions, but high set point temperature leads to more stability and a higher power output.

While both variants run unsteady at average and at high ORC working conditions (1100 s - 1300 s) system ***a*** (250 °C) generates more power than system ***b***. In this case transient ORC behavior approximates the steady state behavior shown in figure 7.

In stop phases, superheated temperatures fall down to two phase steam temperature very quickly. Subsequently, a near linear decrease of temperature is detected. In this case, the reduction of thermal energy is only detected by falling steam quality. Changes in temperature result from reductions in pressure.

In order to achieve maximum ORC power output and sustained stable behavior a time dependent set point temperature is recommended. However conventional PID and PI controllers do not allow an optimal set point control because these approaches solely have information about the current system state. In that sense Model predictive control (MPC) represents a more promising approach. The keynote of MPC is the prediction of future incidents by using a fast running ORC model. The computation time dominates the maximum receding horizon. Due to the fact both variants result in the same expander power output, for subsequent investigations variant ***b*** will utilized for comparison as the system runs more stably.

4 MODEL PREDICTIVE CONTROL

Figure 9 shows the MPC approach. Selected control input signals are optimized up to a defined future time horizon. The optimization of control input u has been realized for two independent variables (overheated evaporator outlet temperature $T_{evap,sp}$, pressure $p_{evap,sp}$). In t_1, the optimizer has had to find the optimal control policy that maximizes ORC energy output over predicted horizon A. Between t_1 and t_2, the control policy from Step 1 is used for the ORC operation. Afterwards the MPC plant model is reset to initial conditions in $t_{2.\ Th.}$ For the optimization of the control input u(t) several strategies like genetic algorithms and gradient based algorithms were investigated. The later one showed a lack of convergence and stability at several points. In the end the genetic algorithm proved to be the method with highest reliability and stability for optimization of this kind of technical system.

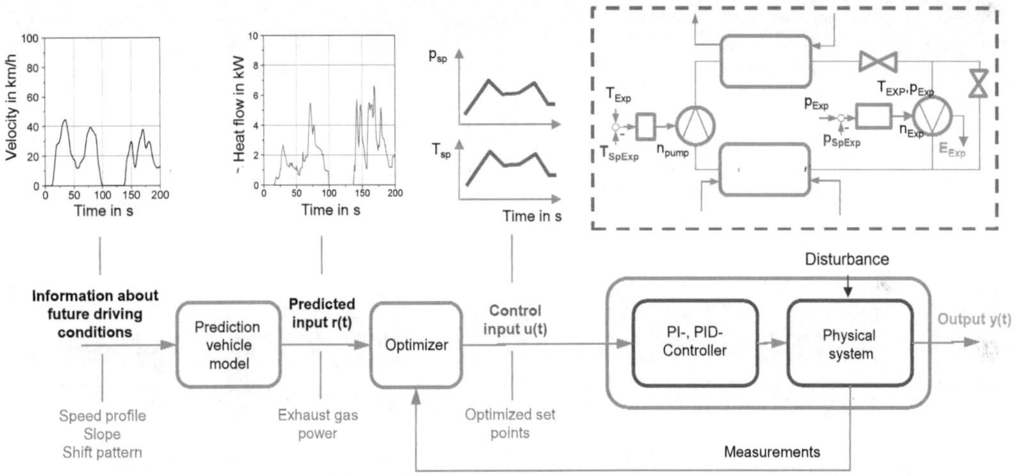

Figure 9. Model predictive control strategy applied to ORC system

A number of virtual and experimental investigations of model predictive control strategies have already been realized. Most of them are focused on maximizing ORC efficiency or the increasing of safety function [9, 10]. In [9] an experimental analysis shows the potential of predictive self-adaptive control for the evaporator outlet temperature. In the study the MPC model is based on a linear approach. When compared to conventional and a switching PI-control algorithm, nearly 20 % more expander energy was generated.

Furthermore in [11] an investigation of linear and nonlinear MPC strategy is presented. There the authors focused on maximizing ORC efficiency for a heavy duty application with two parallel heat exchangers. That study showed significant improvement on waste energy recovery when using nonlinear MPC and linear MPC strategies against the PI controller baseline strategy (25% improvement with nonlinear MPC and 18% with linear MPC). Constrains in computation time discard nonlinear MPC for online optimization.

4.1 Results of optimization

Having discussed the control method impact on energy recovery, the following is a comparison on energy recovery by applying standard control method ***(c)*** versus a predictive control one ***(d).***

The ORC system is continuously working in deactivation (expander off) or activation mode (expander on) when working fluid reaches certain boundary conditions. If evaporator outlet temperature runs 20 K over saturation temperature and pressure larger than 10 bar the system goes into activation mode. In deactivation mode the feed pump runs at minimum volume flow to control the evaporator outlet temperature to 340 °C and bypass the working fluid from the expander. Once the system goes into activation mode the control strategy is switched on again.

For system ***c*** the set point temperature is 300 °C. The set point pressure is linked to current evaporator outlet temperature by an isentropic correlation. This keeps a guarantee of constant distance from the saturation line. If temperature breaks-in abruptly, pressure will be reduced by increasing the speed of expander. If system temperature rises, the set point pressure will be increased by reducing the expander speed.

Figure 10 illustrates the comparison of expander power between ORC system without predictive and with predictive approach. It can be noted that expander of the second system starts 80 s earlier than the one with second system.

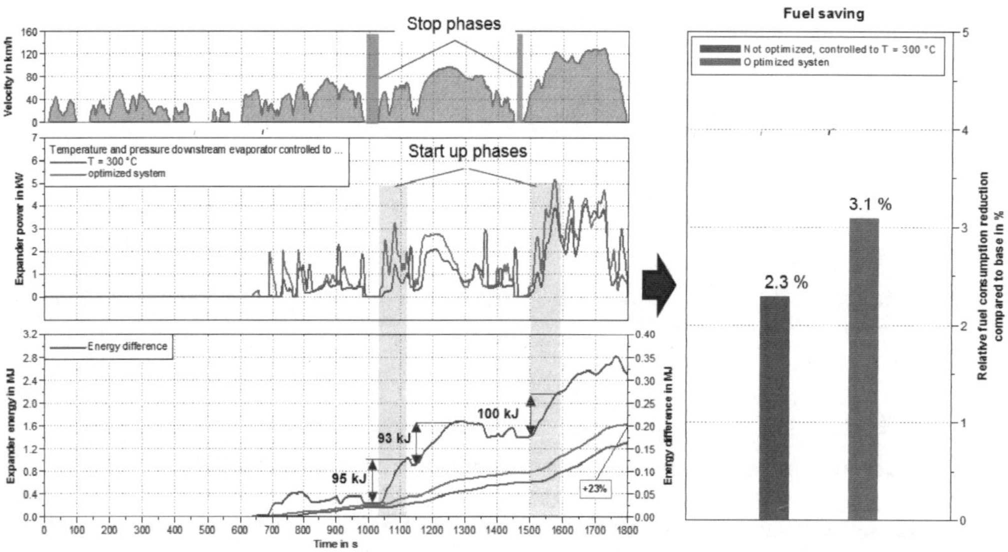

Figure 10. Comparison of expander power for predictive and none predictive system

Both systems run stable at low load working conditions (800 - 1000 s). During the phase right after stop mode (1500 s-1600 s), system ***d*** generates circa 100 kJ more energy than system ***c.*** Furthermore during the averaged operating phase (1150 s - 1300 s) the predictive method generates 93 kJ of surplus energy. Over the whole cycle system ***d*** produces 23 % more energy compared to system ***c***. This leads to an increase in fuel saving of 0.8 %.

Figure 11 shows the comparison of control variables pressure and temperature. The system with predictive approach shows significantly higher pressure level during engine stop (orange zone). Here, the control procedure aims to reduce the expander speed (i.e. dropping power, raising up pressure), in order to guarantee more recovery and high pressure in the acceleration phase (blue zone) after stopping (i.e. system sacrifices energy recovery now by knowing that more energy will be recovered later). Building up of pressure takes time and can be seen clearly on system in system ***c*** where pressure build-up is significantly delayed.

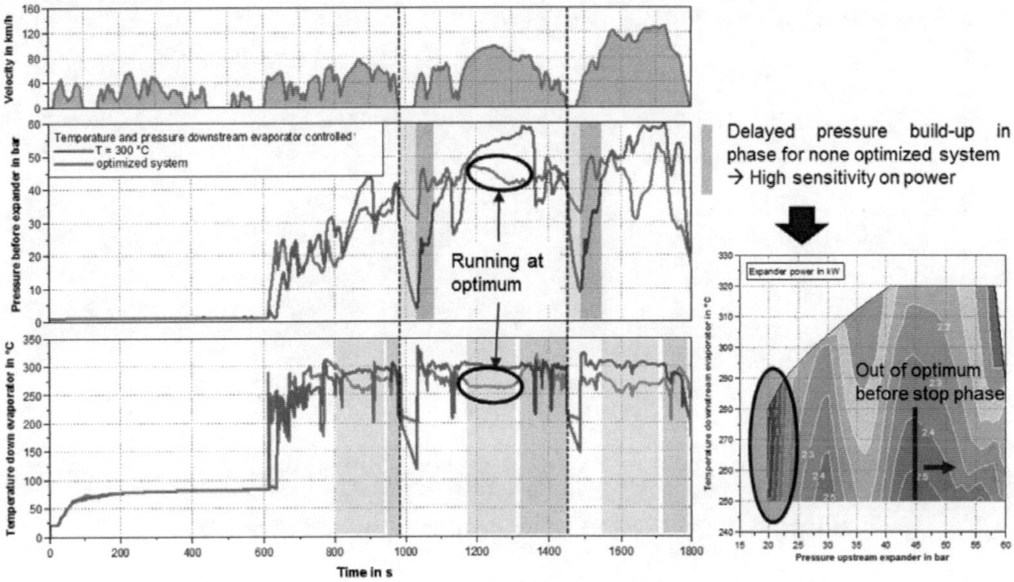

Figure 11. Comparison of pressure and temperature for predictive and none predictive system

In the low load operating area the sensitivity of recovery power to expander pressure is very high (figure 11 - right hand side diagram). Optimization algorithms try to prevent the system falling under 30 bar at low or zero load operating points by raising the pressure before the system steps into the low load operating zone. Raising the pressure loses recovery power, however the pressure can kept over 30 bar in the subsequent stop phase. In the last braking phase, a subsequent start phase does not follow, thereby it is not necessary to keep pressure at a high level.

At averaged operating points with high load working conditions the ORC pressure should be between 40 and 50 bar with low superheated temperature (1200 s - 1350 s). For the predictive control approach this is guaranteed.

Figure 11 shows temperature behavior of system ***c*** and ***d***. The temperature varies between 250°C and 320 °C. For system ***d*** in phases where ORC runs at low load working conditions (e.g. start up and braking mode) the temperature is at 320 °C (green zone). If the ORC goes into high load working conditions, the temperature reduces to near its defined minimum of 250 °C (gray zone).

Control strategy in ***c*** shows insufficient performance in terms of recovery power. It is important that pressure and temperature are decoupled. Control strategy of system ***d*** shows significant increase of expander power output due to decoupling and independent optimization of control variables. This approach represents an essential way to maximize system stability and energy recovery.

4.2 Deduction of simple rules based control strategy

Figure 12 shows the run of optimized evaporator outlet temperature (set point and actual). The studied driving cycle can be divided into three main phases. In the first phase (start/warm-up phase) the ORC algorithm increases the temperature in order to reduce the mass flow and guarantee a smooth start, preventing the system from falling back into the warm-up phase (fast overcome of unstable phase).

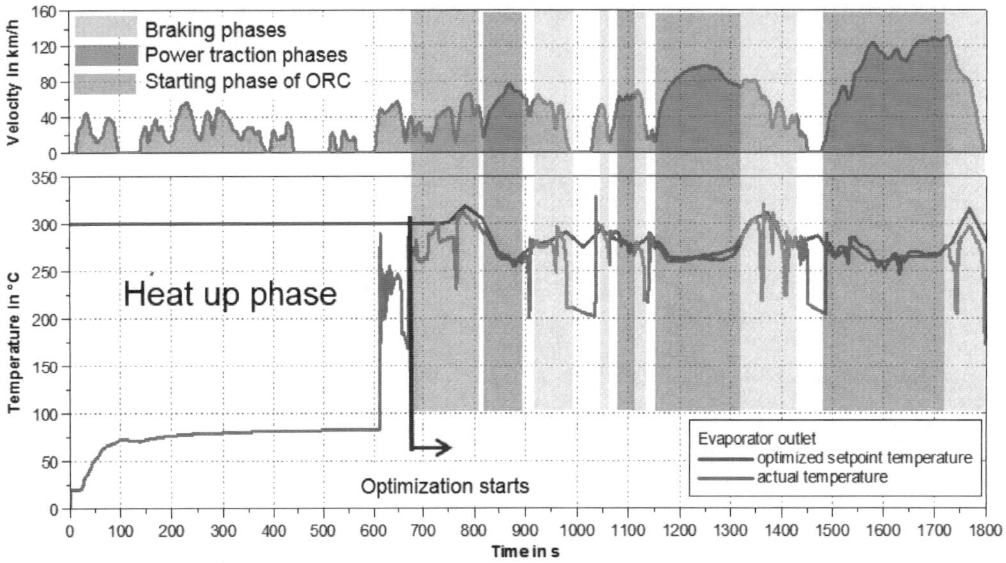

Figure 12. Deduction of simple rule based strategy

In power traction phases, where ORC working conditions are closed to steady conditions, the temperature runs at a low level. The algorithm aims to reduce temperature in order to maximize ORC recovery power.

In braking phases, where ORC works close to its unstable mode, the temperature runs at a high level in order to reduce the volume flow and prevent the cool down effect. This keeps the system in its stable state longer.

5 SUMMARY AND OUTLOOK

Through the paper it has been shown that ORC system behavior at steady state, the variation of pressure and temperature at the outlet of the evaporator leads to significant difference on recovery power from expander. However the behavior at transient conditions of a typical driving cycle shows issues with stability and performance. The study with two fixed constant set points (250 and 300 °C) revealed the same overall energy recovery over the whole cycle. Therefore the expected performance at transient state did not agree with conclusions from the steady state study.

The utilization of the model predictive control approach has shown a gain of 23 %energy recovery over the whole considered driving cycle compared to a system with steady state set points. The control approach applied an online independent optimization of set points variables (pressure and temperature of working fluid at the expander inlet) throughout the driving cycle.

The stability of the ORC system at low load working conditions was kept by an algorithm that was constantly seeking start-up or braking modes and defined the highest possible evaporator outlet temperature set point. The analysis has shown, that optimized control values reflects a periodic pattern as a function of the operating point.

In further steps, the investigation at test bench will allow the utilization of a knowledge based approach that can reduce the runtime of optimization tasks significantly. This approach allows the usage of more complex models and removes the need for linear state space models and its associated disadvantages. Tests at rig will correlate fuel saving figures achieved by the proposed control algorithm.

6 REFERENCES

[1] S. W. O. A. Freymann R, "The turbosteamer: a system introducing the principle of cogeneration in automotive applications," *MTZ,* vol. 69, pp. 20-27, 2008.

[2] http://www.greencarcongress.com/2008/02/honda-researchi.html

[3] Nelson. C, "Waste heat recovery. Directions in engine-efficiency and emissions," in *DEER*, 2008.

[4] http://www.greencarcongress.com/2012/10/orc-20121010.html

[5] http://www.greencarcongress.com/2011/10/ford-study-shows-rankine-waste-heat-recovery-system-on-a-light-duty-vehicle-could-almost-meet-full-v.html

[6] Daccord, R., Darmedru, A., and Melis, J., "Oil-Free Axial Piston Expander for Waste Heat Recovery," SAE Technical Paper 2014-01-0675, 2014, doi:10.4271/2014-01-0675.

[7] M. Esposito M., N. Pompini, A. Gambarotta, M. Canova. A switching moving boundary model for simulation of ORC plants in automotive applications. 15. Stuttgarter Motorensymposium. Stuttgart. 2015.

[8] J. Peralez, P. Tona, O. Lepreux, A. Sciarretta, L. Voise, P. Dufour, M. Nadri. Improving the Control Performance of an Organic Rankine Cycle System for Waste Heat Recovery from a Heavy-Duty Diesel Engine using a Model-Based Approach. IEEE Conference on Decision and Control. Florence. 2013.

[9] A. Hernandez, A. Desideri, C. Ionescu, S. Quoilin, V. Lemort, R. De Keyser. Increasing the efficiency of Organic Rankine Cycle Technology by means of Multivariable Predictive Control. International Federation of Automatic Control. Cape Town. 2014.

[10] A. Hernandez, A. Desideri, C. Ionescu, S. Quoilin, V. Lemort, R. De Keyser. Experimental study of Predictive Control strategies for optimal operation of Organic Rankine Cycle systems. 3rd International Seminar on ORC Power Systems. Brussels. 2015.

[11] E. Feru, F. Willems, B. de Jager, M. Steinbuch. Modeling and Control of a Parallel Waste Heat Recovery System for Euro-VI Heavy-Duty Diesel Engines. Journal Energies. 2014.

Thermoelectric devices for automotive applications to improve efficiency and reduce CO_2 emissions

J Siviter, A Montecucco, P Mullen, M Garland, A Knox
Thermoelectric Conversion Systems Limited, UK
University of Glasgow, School of Engineering, UK

ABSTRACT

Thermoelectric Generators (TEGs) are receiving a great deal of attention from the automotive industry. They are lightweight, reliable, readily available and apparently simple to use. TEGs have been extensively investigated by many vehicle manufacturers as a way of reliably generating electrical power from the waste heat produced by IC engines. Supplemental power generation is regarded as a way to reduce crank torque and therefore fuel load and emissions. In most practical applications, and especially in automotive exhaust gas energy recovery systems, TEGs are subject to limited thermal input thermal energy rather than to a constant temperature difference.

The authors will highlight how recent advances in power converter control algorithms for TEGs simultaneously boost electrical output by 3.7% over conventional MPPT when considering the TEG array as a limited heat system thus simplifying the MPPT converter and reducing implementation costs.

1 INTRODUCTION

1.1 Legislative Motivation

In the European Union (EU) passenger cars account for over 246 million tons of CO2 emissions and forthcoming EU targets require all new cars to achieve 95 grams of CO_2/km by 2021. Every gram of CO_2 exceeding the limit will incur a penalty of €95 payable by each Automotive manufacturer's fleet. In a typical vehicle, only 20% to 25% of the energy produced by burning fuel in the internal combustion (IC) engine is converted to mobility [1]. Approximately one third is rejected to ambient through the exhaust system and gases. Of the remaining energy, the alternator consumes around 5% of which half is dissipated as waste heat [2]. Considering that a car's electrical loads can total over 5kW [3], any supplemental power generation is regarded as a way to reduce crank torque and therefore fuel load and associated emissions. The production of 100W of electrical power leads to a fuel saving of around 1% which reduces CO_2 emissions by around 3gr CO_2/km resulting in a potential cost avoidance of almost €300. Using a thermoelectric device in contact with the heat generated from the exhaust gases of the car it is possible to generate additional electrical power and improve its fuel economy and CO_2 emissions. Clearly the additional cost and weight of the TEG system displaces some of this saving. Thus exhaust gas waste heat recovery (WHR) systems represent an attractive route for car manufacturers to meet the EU emissions target.

1.2 Thermoelectric Devices – Generators and Coolers

A standard Thermoelectric (TE) device consists of a number of *n*- and *p*- type semi-conductor thermos-elements connected electrically in series and thermally in parallel. The elements are usually linked together by small copper strips positioned between two ceramic plates that also serve as both the electrical insulators and thermal conductors. Voltage and current is generated directly from heat flowing across the device and proportional the direction and magnitude of the temperature difference between the two faces of the TE. Further, a thermoelectric device may also be used as a heat pump to force heat flow by applying a voltage and current across the device whereby the polarity of the applied voltage determines the direction of heat flow.

Such TE devices are lightweight and reliable. However, high-volume manufacturing in the automotive market requires non-toxic and cheap semiconducting thermoelectric materials available in great quantities. Common commercial thermoelectric devices use Bi_2Te_3. However, Tellurium is not an abundant material and not suited for the medium-high temperature range (400°C to 700°C) required by automotive applications. Other higher temperature thermoelectric materials, e.g. Silicides and Skutterudites [1] and Oxides [4], are likely to reach higher thermal to electrical efficiency and at lower cost, thus representing interesting candidates for future use in the automotive industry. The manufacturing cost of thermoelectric devices is likely to decrease with high-volume production and improvement in manufacturing techniques.

1.3 Automotive Projects integrating TEGs

Recent automotive R&D projects (VIPER [5], PowerDriver [6]) have demonstrated the technical concept of using a TEG system on a vehicle's exhaust pipe as a WHR system. Academic works have also presented TEG systems applied to diesel engines [7], [8].

Both industrial projects and work published in academic literature [9] [10] [11] have proposed mechanical designs affected by high cost, high mass and increased engine back-pressure. Despite best efforts, simulation models could not predict (or replicate) measured performance. The physical layout of the thermoelectric components has led to high-mass mechanical systems that have proved difficult to properly assemble. In operation, mismatched thermal expansion and uneven mechanical pressure has led to internal component damage. The actual performance of these systems has been below expectation due to an incomplete understanding of the physical behaviour of the devices under operational conditions within the system. Also the choice has usually fallen on high current outputs. As an example [12], Skutterudite cartridges have been proven to work well in real driving conditions, but they produce high currents (almost 50A) at voltages lower than 1V. This leads to clear practical difficulties in the electrical conditioning needed for the energy produced to be absorbed by the vehicle's 12 V system.

In order to tackle these key challenges, this work describes the concept behind an innovative TEG WHR system for automotive exhaust with an extremely low weight and cost penalty. This article analyses distributed maximum power point tracking (MPPT) with the aim of maximising the total electrical power produced and overall system efficiency while minimising the system cost.

2 THERMOELECTRIC WHR SYSTEM

When a TEG device is characterised under constant temperature difference the thermal power is adjusted while the electrical load is modified to compensate for the variation in the effect thermal conductance mainly due to the Peltier effect, but also due to Joule heating [13]. This method effectively masks the complex and subtle device response to variable load current.

However, in most practical applications, and especially in automotive exhaust gas energy recovery systems, TEGs are subject to limited thermal input energy rather than to a constant temperature difference. This is referred to as "constant heat" operation. The available thermal energy may change with time depending on driving conditions, but its rate of variation (in seconds) will be orders of magnitude slower that the TEG's electrical resistance response (in nanoseconds). In considering the constant heat condition, changing the TEG's from open-circuit to at-load operation results in a smaller temperature difference across the device, due to its greater thermal conductivity [14].

A generic TE system for WHR from an automotive exhaust system usually comprises of a hot-side heat exchanger that extracts thermal energy from the exhaust gas, TE devices that convert thermal energy to electrical energy, and a cold-side heat exchanger to remove thermal energy to the circulating coolant.

Previous literature [15] [16] has explained how the effective thermal conductivity of a TEG device changes depending on the electrical current generated and hence increases the thermal flux flowing through the TEG. If the gas were to be able to provide unlimited heat power, it would be convenient to minimise the TEG thermal resistance and hence increase the thermal power flowing through the TEG system leading to increases in the electrical power generated by the TEG. However this is not usually the case: the exhaust gas profile during the Worldwide harmonised Light vehicles Test Procedure (WLTP) often shows a small quantity of thermal power available because, even if the gas temperature is high, the mass flow rate is small. The exhaust gas temperature remains constant while the actual exhaust gas mass flow varies depending on engine speed, load and after-treatment calibration. This will directly affect the amount of energy that can be converted by the TEG array and despite the system A second consideration is that the additional cooling load imposed on the vehicle may negate at least part of the benefit obtained by reduced alternator crank torque.

3 TE ARRAYS AND MECHANICAL COMPLIANCE

As is common with solar photovoltaic applications, it is desirable economically and technically to electrically interconnect in series and/or parallel a number of TE devices into arrays. This is advisable not only to provide the required voltage and/or current magnitudes for the load, but also to minimise the impact of temperature variability on the devices due to mismatches in thermal contact and to the temperature distribution in the exhaust heat exchanger. Ideally, all the TE devices in a single array will be at approximately the same operating temperature and each TEG array is connected to a dedicated MPPT converter. This allows each array to be operated at the electrical operating point at which maximum power can be extracted, as further described in the next section of this article.

However, when in operation TE devices within the same array could be affected by thermal imbalance issues. It is therefore necessary to minimise this risk by preventing the formation of hot-spots: module temperature uniformity on both the hot and cold sides is essential, over the entire surface area of the TEG. In order to achieve this, flatness of the mounting surfaces is critical. Mechanical compliance is also needed for two reasons: a) to accommodate the variability in TEG module thickness due to manufacturing tolerances; and b) to ensure that correct clamping pressure is uniformly maintained over the operating temperature range on all the TE devices in the system. Differential expansion within each TE device due to high temperature differences between the hot and cold sides can result in large lateral shear stresses on pellets, especially in the module corners. This may be induced in a particular module, or may be the accumulation of hot- and code-side heat exchangers' differential expansion: the TE system when assembled is done when all the components are 'cold'.

Although most of the systems presented to date have adopted a multi-planar or "sandwich" structure, these have an inherent mechanical design difficulty due to variable TEG thickness and flatness and the requirement to apply equal pressure across the entire surface of the structure. The clamping beams required result in deflection in the application of excessive pressure on the outer edges of the heat exchanger with reduced pressure in the central region. This has been a cause of device failures due to crushing, and mismatched thermal expansion, especially in the normal direction to the gas flow temperature gradient, leading to high lateral shear stresses causing pellet failures, most often seen in the corners. Insufficient data are available to date to permit a statistically meaningful assessment of systems' fatigue performance and dominant failure mechanism.

It is of fundamental importance to prevent the TE devices from reaching their maximum allowed temperature on the hot-side which could result in permanent failure, usually because of open-circuit due to solder melting. This protection can be achieved mechanically by the installation of a by-pass valve in the TE WHR system, which forces the gas flow to be redirected around the hot-side heat exchanger in a parallel path.

4 MPPT AND TEGS

TEG systems for exhaust gas WHR establish non-uniform temperature differences along the direction of the gas flow. Therefore it is convenient to electrically connect in arrays (series and/or parallel) TEG devices predicted to work at similar temperatures. A dedicated MPPT converter is then required for each TEG array in order to operate the array at its optimal electrical point.

The choice of converter topology depends on the voltage/current rating of the TE devices, on the series/parallel connection of TEG arrays, and the range of operating temperatures over which useful amounts of electrical power can be generated. TEGs with smaller numbers or larger pellets produce smaller voltages and greater currents. This voltage/current rating usually leads to higher system cost (reactive components' cost, size and weight) and is usually more difficult to deal with by the MPPT converters that interface the TEGs to the battery, eventually requiring more complicated converter topologies when high voltage step-up is required. Also, higher currents require larger cable cross-sectional areas and produce considerably higher conducting losses in wiring and electronic components due to greater Joule heating (I^2R), thus decreasing the overall efficiency of the WHR system and increasing the mass carried on the vehicle. Moreover, the results presented in [17] show that parallel-connected TEG arrays are more susceptible to loss of efficiency in the event of inevitable thermal mismatches.

Multiple MPPT converters are used, each controlling the electrical operating point of the TEG array to which it is connected, in what is called a distributed MPPT (DMPPT) architecture. This is beneficial not only because it allows dedicated MPPT for each array, but also because it mitigates the risk of complete failure should one of the TEGs or an electronic component fail. In such cases the output power of only one TEG array would be lost. This is easily identifiable and the vehicle diagnostics and condition-monitoring services can alert the user to the need for repair or replacement. The degree of modularity in the designs of the TEG arrays, MPPT power converters and vehicle interface is largely a matter for the individual manufacturers.

The electronics sub-systems may also have some means of protecting the TE system from excessive temperatures on the hot-side. The TEGs can be electrically short-circuited so that their thermal conductivity greatly increases thus reducing the differential temperature across the TEG but also the absolute temperature on the hot side. This operation can be performed using the power converters with a closed-loop temperature control. It may also be appropriate to cascade the TEGs by placing a high-temperature TEG thermally in series with a lower-temperature TEG.

5 PROPOSED TE SYSTEM FOR EXHAUST WHR

This section proposes an innovative TE system for automotive exhaust WHR. This solution is compact, low-cost, light-weight and has a low assembly cost. It also addresses the mechanical clamping issues discussed in the previous section and it integrates the parallel-path bypass valve. The hot-side extruded Aluminium Nitride (AlN) heat exchangers ensure low thermal gradients. Tension bands around the outer periphery of the cold-side heat exchanger ensure correct clamping of the TEs and apply uniform mechanical pressure with a low weight penalty. Only two coolant connections are necessary and the concentric butterfly valve fitted at the exit of the system can easily control the exhaust gas route. As the bypass is located internally to the TEG system, when the valve is closed, the exhaust gas is forced towards the fins of the hot-side heat exchangers, improving the extraction of thermal energy.

This system fits eight 25mm x 25mm TEs radially on the octagonal circumference of the exhaust gas heat exchanger (and five per each face), therefore it is convenient to connect in 8 arrays of TE devices progressively from the gas inlet towards the gas outlet. As the operating conditions of the TE devices in each array are likely to be similar (and also different from the other arrays), it is advisable to connect each array to a dedicated MPPT converter, whose peak power rating needs to be 50W (assuming a target power production density of $1W/cm^2$). The total number of converters depends on the size of the TE system; in this case five MPPT converters are needed, for a total system rating of 250W, achieved with forty 25mm x 25mm TEs. For the purposes of the comparison and validation of the power converter algorithm, a single TE device was characterised and the three conversion algorithm techniques were implemented.

6 EXPERIMENTAL SETUP

6.1 Test system description

In order to demonstrate the principal of TEG in a 'limited heat' application the following test apparatus was built (Fig. 1). The system operates on the following principle: the heat source for the thermoelectric modules is a Copper block containing a single Silicon Nitride igniter-heating element. The igniter is supplied from a DC power supply that is controlled to a desired input power level. The Copper block temperature rises through the heat exchanger to contact one face of the thermoelectric device.

At the cold side, a chiller unit with a water-glycol mix is connected to an Aluminium block and circulates to remove heat from the block. The temperature of the water entering the Aluminium cold block is controlled at a desired temperature by a chiller unit capable of removing up to 1kW of heat.

In order to ensure mechanical compliance the test setup with hot block, cold block and TEG are held under compression to compensate for any thermal expansion.

6.2 Electrical and Control

A Keysight 34970A data acquisition and control (DAQ) unit was interfaced with a custom electronic controller and used to control and monitor the system. The unit controls the DC power supply and interfaces to IV-trace and MPP modes. Moreover, the unit monitors the power ratings of the TEG, the DC load on the heater block, the output voltage and current from the TEG and various temperatures throughout the test system. A 12V 7Ah lead acid battery was used as a sink for the Thermoelectric Conversion Systems KM2 MPPT and BM4 MPPT converter [18], thus allowing for constant loading on the TE device and energy collected to be analysed.

There is a total of six thermocouples places around the test system as follows: one next to the test fixture to measure ambient temperature; two in the hot Copper block; one in contact with the Hot side of the Thermoelectric device; one in contact with the cold-side of the Thermoelectric device; two for the inlet and outlet of the coolant water. These are connected to the DAQ unit.

Figure 1: Thermo-mechanical Test System and MPPT electronics

The DAQ is then connected to a PC through GPIB. An Agilent VEE program was designed to automatically send the control signals according to the reference parameters set by the user (i.e. input power to the heaters) and to log the measurements into Excel spreadsheets.

6.3 Test configuration

For the testing presented in this article a single TEG is characterised in three different operating modes to demonstrate the concept of limited heat. Here, a single TEG is tested in a thermomechanical fixture with the TEG sandwiched between the hot Copper block and the cold Aluminium block. The cold block is set to a fixed temperature at the chiller unit and the hot block then supplied with power at various fixed levels.

6.4 Electronic Power Converters with MPPT

Three maximum power point tracking algorithms were selected for direct comparison. For the purposes of this experiment it was important to ensure the same thermal conditions remained throughout all the experiments. The first MPPT algorithm is known as "Fractional Open-Circuit" (FOC) method. This method follows directly the 'maximum power point theorem' stating that maximum power from a power source can be obtained when the internal resistance of the source matches that of the load. Therefore in terms of a thermoelectric device the internal resistance of the TEG is set to exactly 50% of the Voc as the TEG is assumed to have the same performance of an electrical resistor.

The second method is that which is typically found in Solar-PV MPPT and known as Perturb and Observe (P&O). Usually both a voltage and current are measured. In this method the power output is directly measured and at a fast rate (~500Hz) the algorithm adjusts to find the maximum power point.

The final method uses a modification to the FOC method whereby the thermal condition of the thermoelectric device is accounted for in the algorithm. This differs from following the maximum power transfer theorem but it is adapted to account for the thermal gradients across the thermoelectric device. In the following section the detailed results of each algorithm will be compared and contrasted.

7 CONSTANT HEAT EXPERIMENT

The DC heater controlling the Copper hot block of the thermomechanical test system was configured to apply power in a series of steps of fixed time interval to the hot side of the thermoelectric device. For the purposes of comparing each different algorithm, the same conditions were applied across the same time interval.

The heater steps follow: 100W, 120W, 80W each step applied for 10 minutes (Fig. 2). The chiller unit was set to maintain 25C at the cold side of the TE and temperature, voltage, current and power from the thermoelectric device monitored.

Following the fractional open-circuit voltage test, the output power rapidly rises following the heater block temperature increase. The peak power at the maximum temperature difference across the TE is 4W or a cumulative energy of 5.11kJ over the course of the test. The Perturb and Observe test resulted in a slightly lower cumulative energy result of 5.03kJ and a small drop in peak output power. However, the Modified FOC test resulted in both a higher output power and a higher peak power of 5.3kJ and 4.3W respectively. These are shown in Fig. 3. A summary table of results is presented in Fig 4.

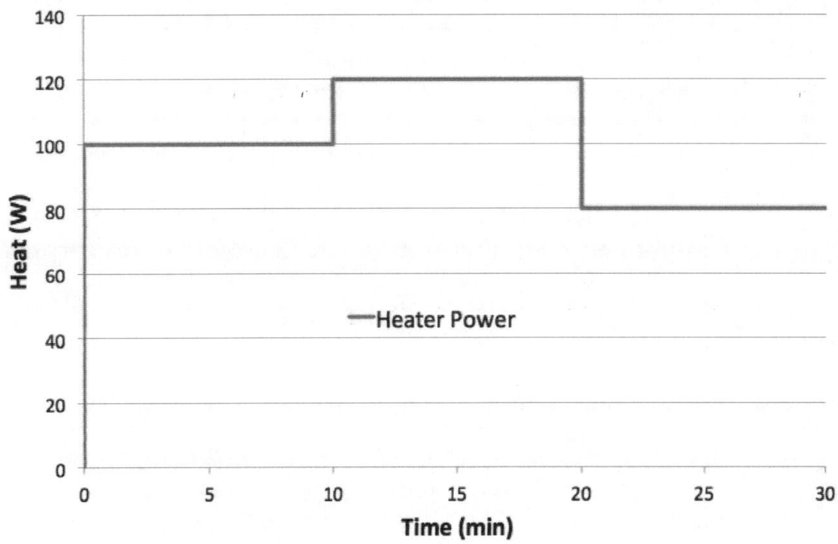

Figure 2: Constant Power test regime

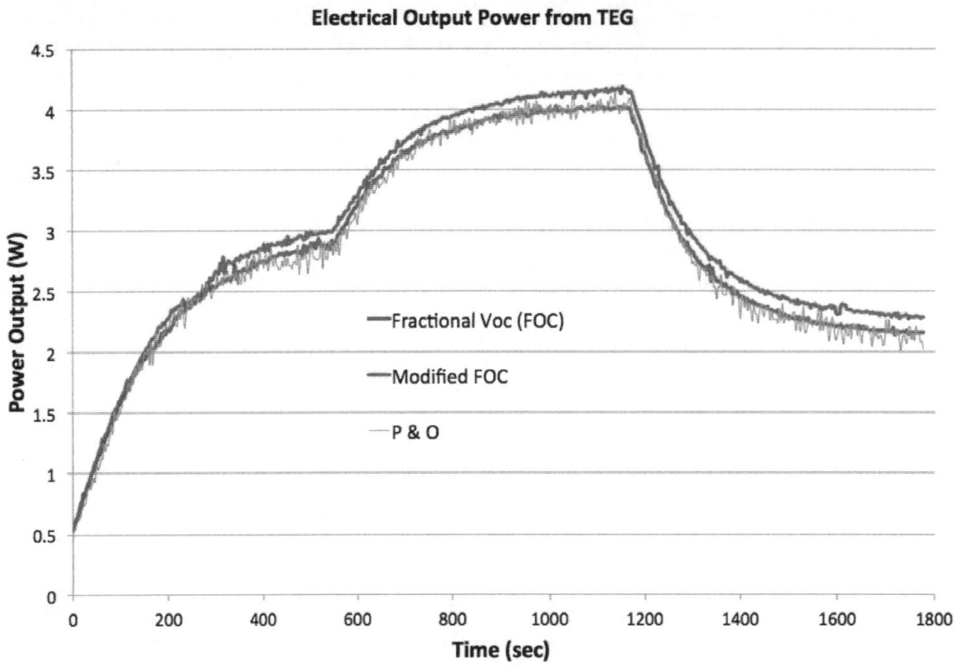

Figure 3: Summary of Results

	FOC (50% V_{OC})	P & O	Modified FOC
Total Energy Obtained (kJ)	5.11	5.07	5.30
Percentage increase from FOC (50% V_{OC})	/	-0.7%	+3.7%
MPPT Efficiency (100% when $V_L=V_{OC}/2$)	99.9%	99.3%	**103.7%**

Figure 4: Results and comparison of power converter algorithms

8 CONCLUSION

From the test results presented it is clear that the modified FOC algorithm outperforms both the traditional Maximum Power Point theorem and the more commonly used Perturb and Observe. Given that the Perturb and Observe method requires additional computational power and current sensing (which is not required in either of the other two methods) the benefit of the Modified FOC is clear.

Thermoelectric generators for automotive waste exhaust gas recovery has been the subject of numerous research projects however, the device inefficiency and complex heat exchanger requirements has limited its further integration in large scale integration and uptake in the marketplace. The results presented in this paper form part of a series of steps that should be adopted by automotive manufacturers and suppliers to ensure the TEG produces maximum power while minimising the cooling load implications.

9 REFERENCES

[1] M. Brignone and A. Ziggiotti, "Impact of novel thermoelectric materials on automotive applications," *9th European Conference on Thermoelectrics,* pp. 493-496, 2011.

[2] J. Martins, L. M. Goncalves, J. Antunes and F. P. Brito, "Thermoelectric Exhaust Energy Recovery with Temperature Control through Heat Pipes," *SAE Tech. Pap. ,* Vols. 2011-01-0315, 2011.

[3] K. Rhodes, D. Kok, P. Sohoni and E. Perry, "Estimation of the Effects of Auxiliary Electrical Loads on Hybrid Electric Vehicle Fuel Economy," *SAE Technical Paper,* no. 2017-01-1155, 2017.

[4] J. W. Fergus, "Oxide materials for high temperature thermoelectric energy conversion," *J. Eur. Ceram. Soc.,* vol. 32, no. 3, p. 525–540, 2012.

[5] VIPER Consortium, "TSB Project VIPER (Vehicle Integrated Powertrain Energy Recovery)," 2011. [Online]. Available:
http://gershon.ucoz.com/landrover/bob-gilchrist-jlr.pdf.
[Accessed 06 Feb 2017].

[6] B. Stirrup, "Final Report Summary - POWER DRIVER," European Union, 2015.

[7] D. Crane, J. LaGrandeur, V. Jovovic, M. Ranalli, M. Adldinger, E. Poliquin, J. Dean, D. Kossakovski, B. Mazar and C. Maranville, "TEG On-Vehicle Performance and Model Validation and What It Means for Further TEG Development," *J. Electron. Mater.,* vol. 42.

[8] D. Crane, J. LaGrandeur, V. Jovovic, M. Ranalli, M. Adldinger, E. Poliquin, J. Dean, D. Fang and D. Kossakovski, "Cylindrical Cartridge as a Building Block for Thermoelectric Generator Applications," in *International Conference on Thermoelectrics*, 2013.

[9] J. G. Haidar and J. I. Ghojel, "Waste heat recovery from the exhaust of low-power diesel engine using thermoelectric generators," *International Conference on Thermoelectrics (ICT),* p. 413–418, 2001.

[10] R. K. Stobart, A. Wijewardane and C. Allen, "The Potential for Thermo-Electric Devices in Passenger Vehicle Applications," *SAE Tech. Pap,* Vols. 2010-01-0833, 2010.

[11] S. Risse and H. Zellbeck, "Close-Coupled Exhaust Gas Energy Recovery in a Gasoline Engine," *Res. Therm. Manag.,* vol. 74, pp. 54-61, 2013.

[12] E. F. Thacher, B. T. Helenbrook, M. a. Karri and C. J. Richter, "Testing of an automobile exhaust thermoelectric generator in a light truck," *Proc. Inst. Mech. Eng. Part D J. Automob. Eng,* vol. 221, no. 1, pp. 95-107, 2007.

[13] A. Montecucco, J. R. Buckle and A. R. Knox, "Solution to the 1-D unsteady heat conduction equation with internal Joule heat generation for thermoelectric devices," *Appl. Therm. Eng.,* vol. 35, pp. 117-184, 2012.

[14] A.Montecucco, J.Siviter and A.Knox, "Constant heat characterisation and geometrical optimisation of thermoelectric generators," *Applied Energy,* vol. 149, pp. 248-258, 2015.

[15] G. Min and N. M. Yatim, "Variable thermal resistor based on self-powered Peltier effect," *J. Phys. D. Appl. Phys.,* vol. 41, no. 22, p. 222, 2008.

[16] A. Montecucco, J. Buckle, J. Siviter and A. R. Knox, "A New Test Rig for Accurate Nonparametric Measurement and Characterization of Thermoelectric Generators," *J. Electron. Mater.,* vol. 42, no. 7, pp. 1966-1973, 2013.

[17] A. Montecucco, J. Siviter and A. R. Knox, "The effect of temperature mismatch on thermoelectric generators electrically connected in series and parallel," *Appl. Energy,* vol. 123, pp. 47-54, 2014.

[18] Thermoelectric Conversion Systems Ltd, "Thermoelectric Conversion Systems Ltd," [Online]. Available: http://www.teconversion.com. [Accessed 06 02 2017].

[19] Y. Wang, C. Dai and S. Wang, "Theoretical analysis of a thermoelectric generator using exhaust gas of vehicles as heat source," *Appl. Energy,* vol. 112, pp. 1171-1180, 2013.

Matlab/Simulink based in-vehicle predictive thermal environment model

B Stegmann, N Dutta, I Hughes, W Jansen
Jaguar Land Rover, UK

ABSTRACT

A Matlab/Simulink model capable of generating temperature profiles at specific positions in a vehicle is introduced. The calculated temperatures depend on user input such as geographical location, operating status (i.e. driving or parking) and package zone within the vehicle. Based on measurement data of warm-up and cool-down behaviour at various positions in the car, the tool is generating temperature profiles accounting for the influence of local geographical ambient temperature, solar radiation and time of the year. The model can be used to quickly calculate temperature boundary conditions for component lifetime analyses.

1 INTRODUCTION

The focus of thermal simulation work can be differentiated with regards to the different time scales of the simulation duration. A common test case for thermal robustness analysis for example is the momentarily temperature distribution under a high load cycle and the subsequent check against long term or short term allowable temperatures. The results of such simulations only portray one moment in time usually under steady-state conditions.

Longer term analyses deal with maximum and average temperatures during drive cycles and e.g. the corresponding energy consumption. These cycles may vary in time between minutes and maybe hours. For both mentioned scenarios the ambient temperature is considered constant, which, considering the comparably small time scales, is a fair assumption.

The durability of any product is predominantly but not only influenced by its usage, other effects being temperature, humidity, pressure, etc. Supplying local ambient temperature data that can be used in lifetime analyses is the aim of the presented simulation model. In order to do so, temperature profiles of locations within the passenger cabin are calculated. The time scale dealt with is one year, covering the most important changes between summer and winter.

The afore-mentioned differentiation with respect to time scales of the simulations almost goes hand-in-hand with the tools used to do so. A general assumption can be made that the longer the time to be simulated is, the more likely it is to step away from 3D solutions and move towards 1D or 0D models. A good comparison of the pros and cons of each simulation method is given in (1). The most apparent trade-off is clearly between simulation time on the one hand and resulting spatial resolution as well as accuracy on the other. Figure 1 shows a qualitative graph comparing simulation time and accuracy with the length of the test case. In terms of lifetime analysis, requirements may be slightly changed compared to a steady-state worst case scenario. Main effects that need to be accounted for are the maximum and minimum temperatures seen over a year. The quick generation of a set of boundary conditions is a predominant request.

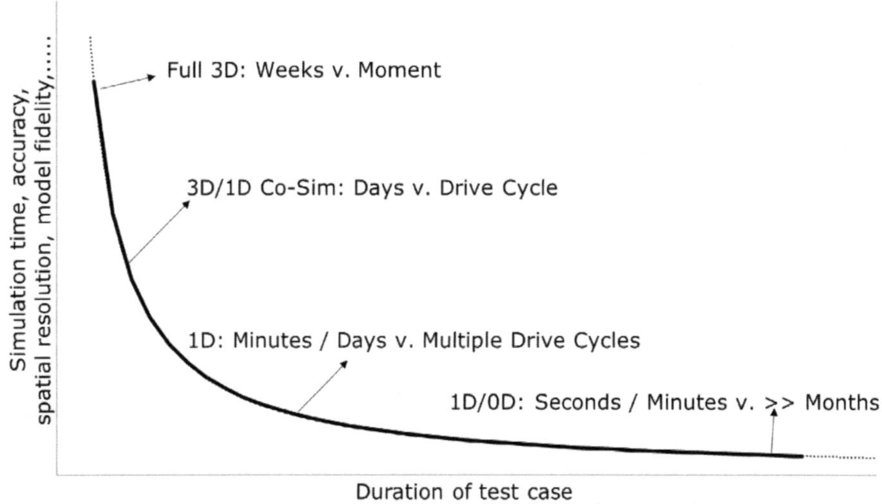

Figure 1: Qualitative comparison of real time against simulation time, accuracy and simulation methods used

Looking at the temperature distribution inside the vehicle, it is worth mentioning that local temperatures vary significantly depending on the HVAC system settings. Furthermore, the form of the cabin and its volume play an important role in the stratification of the temperature profile within the cabin, mainly due to the greenhouse effect after a heat soak. All these aspects drive the requirement for a quick model, which is able to predict temperatures at discrete locations within the vehicle cabin. While (2) and (3) are good examples of physical models of the temperature in a vehicle cabin, (4) includes an overview of both physical modelling and black-box modelling approaches.

During the lifetime analysis of any component it is important to realistically account for the ambient conditions under which this component will operate. Otherwise a component or system can either be over-engineered, and therefore unnecessarily increase development cost, or on the other hand it can be undersized and therefore not tolerate the loads that it is exposed to. Either case is undesirable and should be avoided. In order to account for ambient temperatures over the lifetime of a component the model presented in this paper has been set up.

2 MODEL SETUP

Before the model is set up, an interface definition needs to be done specifying the target outputs of the model. By doing so, the required input information can be defined. In the presented case the definition of boundary conditions looked like shown in figure 2. The thermal environment of a component is influenced by various factors such as the load the component is exposed to, causing the component to heat up and affect the temperature around it. This influence is shown on the lower left hand side of the boundary diagram. If a component is actively cooled the effect of heat rejection to a cooling system is shown in the top left area of the diagram. Additionally, the local environment of the component within the vehicle cabin is affected by rather global conditions such as the global ambient temperature and solar radiation (top right in figure 2). The method discussed in this paper deals with the input derived from these factors.

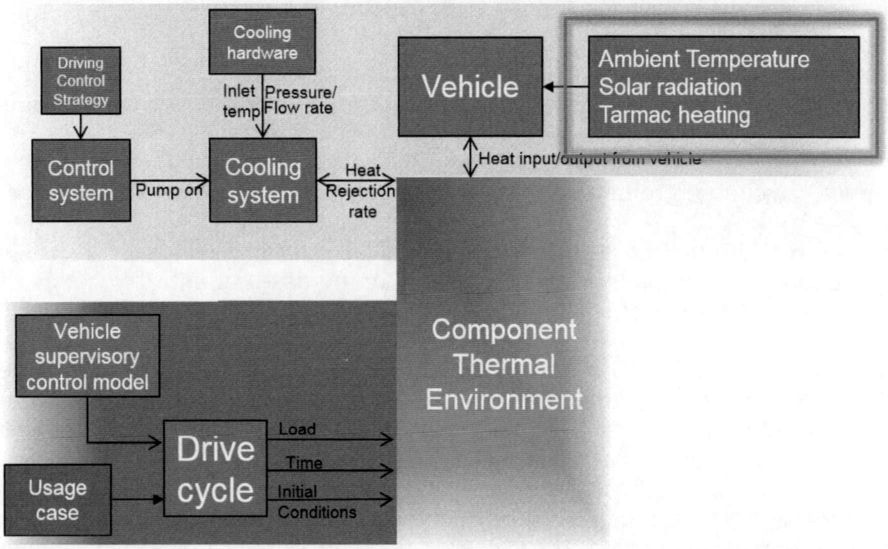

Figure 2: Boundary diagram for simulation tool requirements

Based on the boundary diagram, the deliverables were defined as shown in figure 3. In order to define the global ambient temperature and the radiative heat input, the location in the world needs to be specified. Since these parameters change during a year, a complete temperature and radiation profile for any location needs to be gathered. Other variables needed are position of a component within the cabin, cabin size, and operating status of the car.

Once the inputs and outputs have been defined a categorisation of the required information needs to be carried out in order to decide which inputs will be pre-set and which ones will be to be defined by the user before the simulation. In order to generate local ambient temperature profiles for a certain position in a vehicle cabin, several inputs need to be at hand.

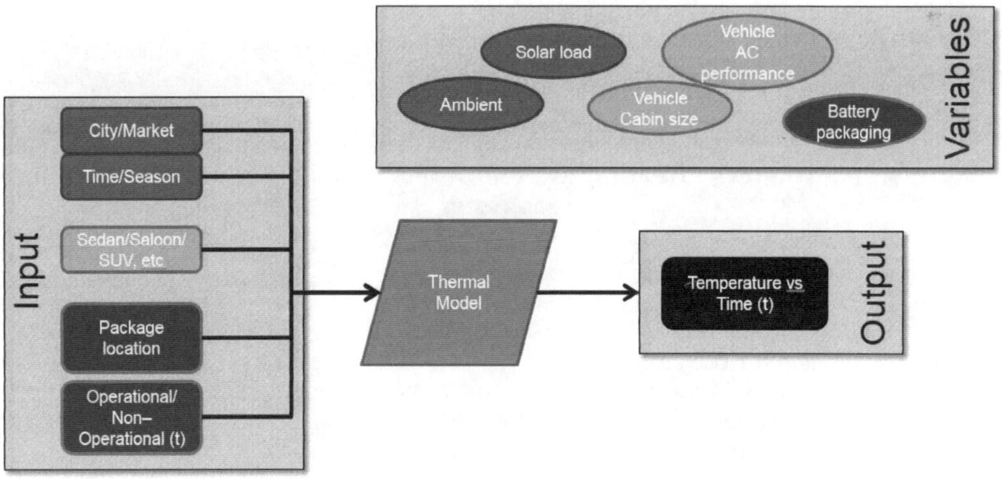

Figure 3: Deliverables of simulation tool

The so-called hard inputs can be modelled prior to usage and will not need to be changed by the user. An example of a hard input is data of solar radiation for a set of global locations. Other inputs are case-specific and therefore depend on the choice of the user. Hard inputs in the model at the current state are:

- Global ambient temperature data (NREL) (5)
- Solar Radiation data (NREL) (5)
- Driving cycle pattern
- Local thermal response behaviour from parking condition to driving and vice versa

Soft inputs, which need to be selected by the user during model execution, are:

- Location of the test case (i.e. city/"market")
- Position of the component that required lifetime assessment
- If required: position of cooling air flow

The positions in the cabin and its vicinity which can be described by the presented method are limited by their source of heat input. Inside the passenger compartment the air volume is predominantly affected by the global ambient and the solar radiation during warm-up while the AC performance is defining the cool-down behaviour. All of these parameters are mostly independent of the driving cycle and cabin heat exchange with environment – especially if the HVAC operation is set to automatic operation. The closer a position is to a drive cycle dependant heat source, the less likely it is to be described correctly with the presented approach. Mainly positions close to exhaust pipes and especially the areas like the engine bay sees mainly heat from the engine and radiation from components of the exhaust system. Hence the output of this model is limited to prediction within the vehicle cabin.

Figure 4 gives and some examples of positions and their predominant heat input sources. The slightly greyed-out "Underbonnet" position is an example of a position that cannot be characterised by the tool.

	Underbonnet	Underseat	Underfloor	Rear Side Trim
External Ambient	+	+++	+++	+++
Solar radiation	+	+	0	++
Exhaust *(Drive cycle dep.)*	+++	0	+	0
Cabin Cooling *(Drive cycle dep.)*	0	++	0	+

Figure 4: Parameters affecting the temperature behavior in various cabin positions

3 METHOD

Mathematical basis for the simulation model is a formula that takes into account the global ambient temperature around the vehicle as well as the solar radiation. Both of this input information are derived from NREL databases and have been further refined by measurement data acquired during several years of real world testing. Main reason for adjusting ambient temperature data is the fact that this data gets measured mostly on airport fields or meteorological weather stations outside bigger conglomerations where temperatures are lower than at the centre of the city.

The global ambient temperature profile for each location is condensed from an hourly profile for the complete year (i.e. 8544 hours per year) into an hourly profile that covers the length of the driving cycle pattern. In the example presented later on the temperature trace of each month was averaged in order to account for a week-long cycle that was to be repeated four times each month. While it seems obvious to average temperature values at the same time of each day, the question of which days to average is up for debate and might be answered differently in different departments (i.e. take the 1st 4 days, the 2nd 4 days, etc. or take each Monday, each Tuesday, etc. or use the hottest 4 days, etc.). As always when averaging, the extreme values will most likely be lost and it needs to be decided which approach of averaging delivers the closest result for the task to be investigated.

Solar radiation has been accounted for using a freely-available Matlab function called "suncycle" (6). Since this function calculates the absolute solar irradiance that is being emitted by the sun, it needs to be scaled to what is being experienced inside the earth's atmosphere. In this case, a linear scaling from a maximum of 1350 W/m² to 1050 W/m² was conducted. Using the Matlab function, the radiation profile of any location in the world can be calculated for a complete year. With this information it is possible to describe the temperature at a certain position within the vehicle cabin as:

$$T_{local}(t) = T_{global}(t) + \frac{Q_{solar}(t)}{c_i} \qquad 3.1$$

where the coefficient c_i is constant and only depending on the position i of the car. This pragmatic approach also correctly states that when soaking a car at a certain global ambient overnight, the local temperature will eventually be equal to the global ambient since solar radiation is equal to zero. The coefficients c_i can be determined during measurement procedures that include a soaking period without solar load followed by a soak with a specified solar load at a predefined ambient temperature. Once the soak is finished, the temperatures are not time-dependant anymore and eq. 2.1 can be rewritten t

$$c_i = \frac{Q_{solar}}{T_{local} - T_{global}} \qquad 3.2$$

The same procedure needs to be repeated in order to define the coefficients c_i when cooling the car using the HVAC system. Depending on the position i in the cabin the steady-state cool-down temperature with activated HVAC system differs from the set point on the HVAC interface. Generally speaking, the further away from the outlet ducts in the cabin and the higher up in the cabin, the hotter will the steady-state turn out to be. These temperatures are not only a function of when the car is being operational (i.e. being driven), but also depending on how big the cabin vehicle is and more detailed factors like the power and COP of the HVAC system. However, as these factors do not change for a car line, their effect is covered by the above test procedure when starting any drive cycle after the soak period. An example of a temperature trace during such a test scenario is shown in figure 5.

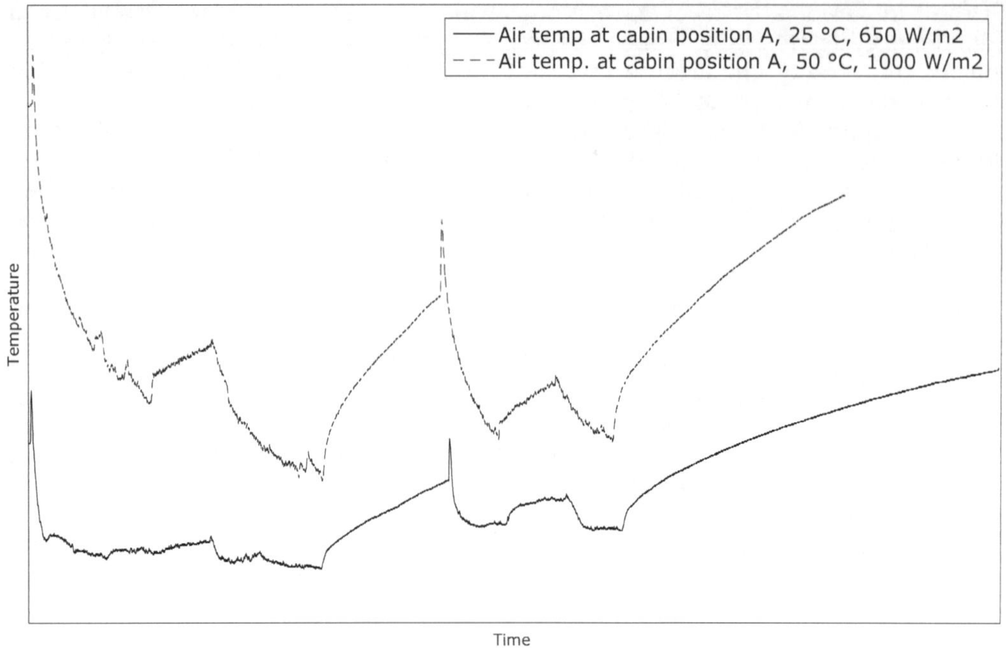

Figure 5: Temperature profiles in the vehicle cabin for two different ambient conditions

Other than the quasi-steady-state temperatures, it also needs to be accounted for the time taken to reach these temperatures during either warm-up or cool-down. As the cool-down process is mainly driven by the HVAC performance it takes a lot less time to cool the cabin down. The warm-up and cool-down behaviour can be approximated using Newton's law of cooling, so that the temperature trace at any position I can be written as:

$$T_{t+1} = T_t + \Delta T * a_i \qquad 3.3$$

For each position i in the vehicle cabin the coefficient a_i describes how quickly a local ambient temperature changes to the temperature calculated with formula 3.1. The temperature difference ΔT in formula 3.3 therefore refers to the difference between actual temperature and target temperature. This concludes that there is a set of four parameters for every position i in the cabin that describe the steady-state warm-up and cool-down temperatures as well as the respective "time constants".

The drive cycle scheme used to uniformly gather data for each position is shown in figure 6. It succeeds an 8h long soak period and consist of four driving cycles separated by phases of idle or engine off. After the drive cycles a soak period is added. This was mainly done to compare against the thermal behaviour during the "engine off"-phase between cycle 2 and 3. The four drive cycles are of very different speed profile types. This was chosen on purpose in order to see the effect of the HVAC system during different load scenarios. The test vehicle was equipped with more than 50 thermocouples in relevant cabin positions measuring temperatures at breath level, foot level and various vent outlets for first and second row, passenger side and driver side.

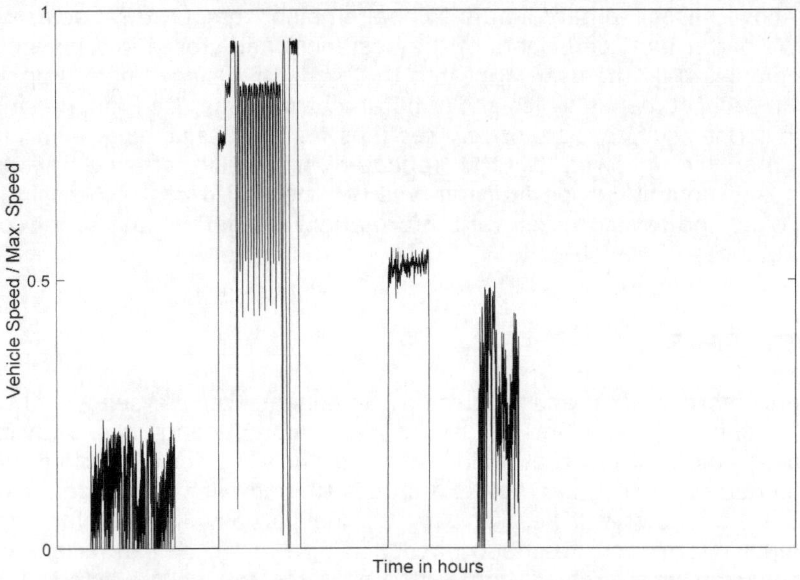

Figure 6: Drive cycle profile used for data acquisition

The test procedure was carried out at a selected range of constant global ambient temperatures and different rates of radiative heat input. This way it is possible to consider the resulting temperature profiles as a step response either to solar radiation or the HVAC operation. Table 1 shows the boundary conditions at which the measurements were conducted.

Table 1: Applied boundary conditions during testing

	-20 °C	25 °C	35 °C	50 °C
1000 W/m²			X	X
650 W/m²		X	X	X
0 W/m²	X	X		

After gathering measurement data, the results need to be analysed and a curve-fitting process is applied to the temperature profiles from the respective locations. By doing so, the parameters a_i can be specified for both warm-up and cool-down. Since formula 3.3 is an approximation the curve fitting not always matches the temperature profile from the measurement in which case the focus was rather on reaching the correct temperature than matching the gradient of the curves. The rationale behind this is that for most of the time of the year the car will be stationary. It is therefore more important to predict the quasi steady-state temperature then the correct time trace.

The final input that needs to be provided is the driving cycle pattern stating when the car is driven or not. It is important to mention that the actual power request or speed profile is not needed for the model to run. It will only be distinguished between a "running" car and a stationary car. This approach can be further refined, especially in the case when the driving cycle has significant influence on the operating temperature of the component. This in consequence leads to an increase in the local ambient temperature of the component itself. Before a driving patter has been defined, it is possible to set the operation mode manually in order to account for various driving scenarios. Once the usage profile has been agreed on in order to account for a realistic power demand for the analysed component this most likely does not change any more and can be set as a hard coded input.

All of the above described hard inputs (global ambient temperature, solar radiation, driving cycle patter and coefficients of the positions) and stored in a parameter file. They are completed by the user input into the model by using simple graphical user interfaces in order to define which set of parameters will be used for the simulation. The Matlab script itself consists of few sections for the user dialogs and loading of the parameter file followed by the repeated simulation of a Simulink model generating the monthly temperature curves. At the end a post-processing section generates plots and saves the relevant information. Altogether, the script comprises less than 200 lines of Matlab-code.

4 MODEL USAGE

As more and more electric and electronic components can be seen in the vehicle cabin, it gets more important to assure these components can withstand the thermal loads coming from the cabin warm-up and cool-down overlaid with their own heat generation. Most common example is the infotainment system that has a considerable amount of self-heat as well as the partially direct solar irradiation. With the input information described in section 3, a test case can be analysed and the corresponding local ambient temperature profile can be predicted. The result generated by the model is shown in the following figure 7. Next to the actual temperature profile for each month the corresponding driving pattern is shown (Driving / Parking) and the respective solar radiation for each month at the selected location. Additionally, a histogram of relative occurrences depending on the local ambient temperature of the component is generated. This can be used to assess maximum and minimum temperatures and statistical information such as median temperature, etc. in order to quickly compare different configurations.

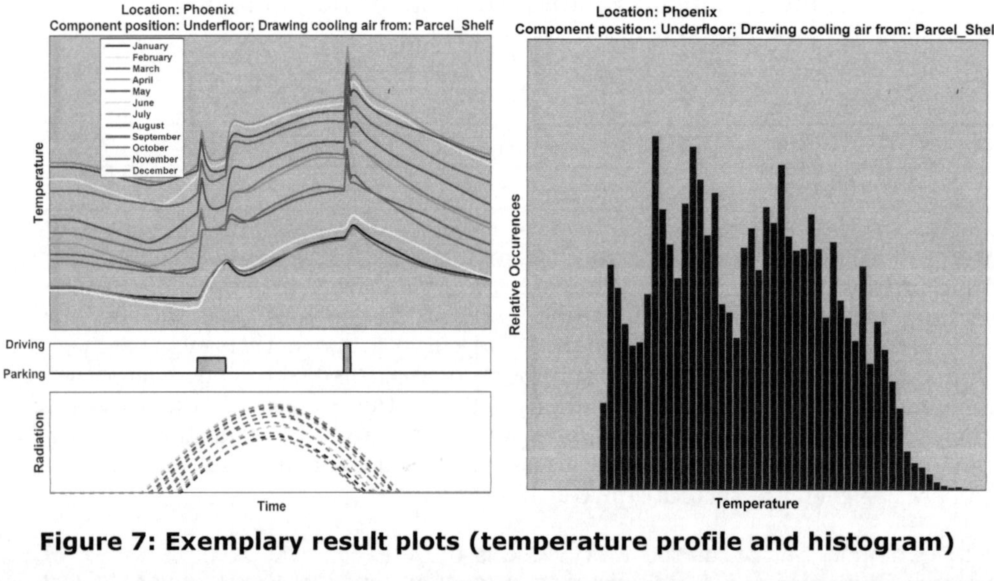

Figure 7: Exemplary result plots (temperature profile and histogram)

In order to use the model to feed into a lifetime analysis, usually temperature profiles need to be generated. An exemplary application for the simulation tool is to choose air from a certain position within the cabin to cool a component at another position in the cabin. In this case, two sets of temperature profiles need to be regarded at the same time in order to assess if the air at the position of the cooling air is cooler than the local ambient of the component. Especially after a soaking period under heavy solar load this information can be valuable when it comes to choosing the right cooling air position and cooling strategy.

In the case of two positions being compared in parallel, an additional coefficient needs to be defined which accounts for the transitional behaviour between the two positions once the cooling starts or ends. Spatial distance of the two selected positions and existence of fans e.g. obviously affect the delay time between the two temperature profiles. As these effects may vary for each analysis, a fixed delay coefficient has been set.

Main benefit of the model is its short runtime. The results shown in figure 7 were calculated in less than 10 seconds, making the model not only easy to use but a helpful tool in the generation of boundary conditions. The applicability of the model is diverse. From a thermal point of view both 1D- and 3D-simulations require either steady state ambient temperatures or transient temperature behaviour as boundary condition. The presented model is able to deliver this input information with very short running times. Especially with regards to lifetime predictions the temperature in the vicinity of a certain component can be generated easily for different markets around the world. Questions of packaging can also be assessed using temperature profiles at different positions within the vehicle cabin in order to carry out sensitivity analyses.

5 CONCLUSIONS AND FURTHER IMPROVEMENTS

An empirical method has been introduced in order to characterise the thermal behaviour of positions inside a vehicle cabin. This information can be used in order to generate temperature profiles at these discrete positions for various ambient temperatures and solar radiation magnitudes. The user input is kept at a minimum as is the runtime of the model. However, as fast as the model may be in generating local ambient temperature profiles, it comes with a set of limitations that shall not be left unconsidered. The fact that the actual speed profile of each driving cycle is not required as an input does not only facilitate pre-processing but also helps in achieving the short computational time. Biggest drawback of this approach is that the thermal behaviour of all locations around a car which are dominated by the heat generation of the engine cannot be considered by the model. This includes the whole engine bay compartment as well as the area around the exhaust pipes and mufflers. Areas within the cabin that may be influenced by e.g. the exhaust heat flow need to be analysed with regards to how much they are affected by the power request of a drive cycle.

Furthermore, only air temperatures inside the cabin in hot ambient environments have been analysed in the process. Due to the temperature dependency of material properties, surface temperatures have not yet been accounted for. For now, the focus has been laid on high ambient temperatures rather than sub-zero environments.

At the current phase of the model-build, the presented data is only valid for rather large SUVs. Since cabin size affects the thermal stratification in the cabin the temperature profile at e.g. the parcel shelf position will look differently for a sedan not only for the maximum temperature but also the reaction time in cool-down and warm-up phases. In order to make the tool more versatile it is planned to acquire data for the most common vehicle segments at Jaguar Land Rover This will include complete datasets for sedans and cross-over products. A uniformly applied scheme of positions that should be characterised at during test procedure will help in post-processing the data and generating a database which can be used company-wide.

Additionally, a lot of effort is put into the optimisation of the curve fitting process. The described approach makes use of Newton's law of cooling and is therefore only an approximation as the law should only be applied to bodies with dominating conductive and nearly constant heat transfer characteristics. Therefore, the way the thermal behaviour of the individual locations is described will change towards non-physical model approaches of system identification such as transfer functions or state-space models.

In the future, temperature measurement data as well as 3D simulations of validated cabin models will help in setting up a database of thermal behaviour at strategically interesting locations of a vehicle. 3D simulations can furthermore help in the verification of the model. While the model currently only generates input for subsequent simulations, the integration into other component-specific models is possible.

REFERENCE LIST

(1) Paulke, S., Ellinger, M,. Wagner, S. (2007): Air Conditioning Cabin Simulation with Local Comfort Rating of Passengers. 2nd European Workshop on Mobile Air Conditioning and auxiliary systems - ATA/CRF (p. 6, 16). Turino: P+Z Engineering.

(2) Marcos, D., Pino, F., Bordons, C., Guerra, J. (2014): The development and validation of a thermal model for the cabin of a vehicle. Applied Thermal Engineering Volume 66, p. 646 – 656.

(3) Torregrosa-Jaime, B., Bjurling, F., Corberan, J., Di Sciullo, F., Paya, J. (2015): Transient thermal model of a vehicle's cabin validated under variable ambient conditions Applied Thermal Engineering Volume 75, p. 45 - 53.

(4) Björklund, M. (2004): A Model of the Air Temperature in a Truck Cabin. Master's Degree Project, KTH Royal Institute of Technology – Sensors and Systems, Stockholm, Sweden.

(5) National Renewable Energy Laboratory (NREL), Weather Data - NREL System Advisor Model. Retrieved from https://sam.nrel.gov/weather (Last accessed on 26 January 2017).

(6) http://mooring.ucsd.edu/software/matlab/doc/toolbox/geo/suncycle.html Last accessed on 13 April 2017.

Combined engine-ORC thermodynamic analysis based on a second law and thermo-economic approach

S Lion [1,2], M G Momesso [1,2], I Vlaskos [1], C Rouaud [3], R Taccani [2]
[1] Ricardo Deutschland GmbH, Germany
[2] University of Trieste, Italy
[3] Ricardo UK Ltd, UK

ABSTRACT

In this work an example of a second law and thermo-economic based analysis approach is presented, through the study of a new combined engine-ORC (Organic Rankine Cycle) system architecture, in order to be able to improve the fuel economy of medium to heavy duty Diesel engines. The Ricardo proprietary 1D engine simulation code, WAVE, has been used in order to perform a detailed engine process simulation. A post-processing routine has been developed, with the aim of studying in detail the performance of the engine concept via the First and Second Law of Thermodynamics. The boundary conditions from the topping engine side have been used in order to perform an analysis of a bottoming ORC system, recovering engine wasted heat both from an energy and exergy point of view, whilst considering different working fluid possibilities. The first two analysis steps can be considered as a necessary starting point for a subsequent thermo-economic analysis based on new exergy costing techniques. The approach can be useful both for the study of new combined engine-waste heat recovery system concepts in a preliminary feasibility analysis stage of a project, as well as for the assessment of the thermal management of an engine, based on any possible operating point and architecture that can be simulated in WAVE.

NOMENCLATURE

a	specific exergy or availability (J/kg)
A	exergy or availability (J)
$\dot{A}$	availability (or exergy) rate (W)
b	specific flow availability (J/kg)
$\dot{C}_i$	exergy flow cost ($/s)
c_i	unit cost of exergy stream ($/J)
h	specific enthalpy (J/kg)
h_0	specific enthalpy at dead state (J/kg)
I	irreversibility (J)
$\dot{I}$	irreversibility rate (W)
m	mass (kg)
$\dot{m}_{wf}$	ORC working fluid mass flow (kg/s)
ORC	Organic Rankine Cycle
q	heat exchanged (specific, J/kg)
s	specific entropy (J/kg*K)
s_0	specific entropy at dead state (J/kg)

T	temperature (K)
T_0	temperature of dead state (298.15 K)
$\dot{W}$	power (W)
$\dot{Z}_k$	sum of capital investment, operation and management cost rates ($/s)
φ	crank-angle (°c.a., degree crank-angle)

1. INTRODUCTION

Heavy Duty Diesel Engines (HDDE) are among the main contributors to ambient pollution, in particular CO_2, NO_x and PM, since they are the most used technology for commercial vehicle and ship propulsion, as well as for small-medium size distributed stationary power generation applications. Moreover, emissions standards, such Euro, Tier and Stage (1), are pushing engine manufacturers and fleet owners to further decrease emissions whilst, at the same time, improve engine and powertrain efficiency.

In the last decades, Organic Rankine Cycles (ORC) have been among the most considered and developed technologies to retrofit HDDE with the aim of improving powertrain efficiency and, at the same time, reducing fuel consumption and environmental impact. Many publications can be found in literature about HDDE waste heat recovery using ORC systems for different applications, such as commercial vehicles, stationary engines or ships. Several heat sources are usually considered, such as exhaust gas, Exhaust Gas Recirculation (EGR), engine cooling, Charge Air Cooler (CAC) or lubricant heat, some showing more potential than others, depending on the engine configuration and on the temperature levels. Some examples can be found in (2–10), whilst a rather good introduction about ORC technology in general can be found in (11). Most of the literature refers to thermodynamic feasibility studies and, in particular, to the choice of the most effective working fluid, to the optimization of the ORC architecture and the operating parameters. A limited amount of studies considers the evaluation of the transient behaviour of these systems over a typical application duty cycle, in order to assess the possible fuel economy benefit. An even smaller number of studies is related to experimental developments, in particular about commercial vehicles and ships, mostly due to the challenging application and to the remaining high development, investment and operational costs.

Generally, ORC systems are applied in the stationary power generation sector, whilst in the marine and in the commercial vehicle sectors, prototypes are under tests and development, being expected to enter the market in the next decade. However, one of the actual drawbacks is that most of the systems are currently only a retrofit of existing engines, there is still potential for further performance and costs improvements, especially when considering the engine and the ORC as a combined system, rather than two separate entities. This would allow an optimization of the overall powertrain, consisting of engine, ORC, thermal management and power management system. It is, indeed, for this purpose that a Second Law of Thermodynamics (exergy) analysis is becoming more popular, in close synergy with thermo-economics and optimizations techniques.

2. FIRST AND SECOND LAW ANALYSIS APPROACH

In the first phase of a project considering waste heat recovery systems, the attention is usually focused on which heat sources are available to be recovered from a HDDE, and which cycle layout can be used in order to obtain the most favourable configuration, both regarding performance and techno-economic feasibility. However, generally ORC heat sources boundary conditions from an engine are supplied as standalone data, not considering the overall system as a unique entity, which should behave in the most optimized way possible. The scope of this paper introduces a possible method that can be used in order to assess combined engine-ORC both from a First and from a Second law of Thermodynamics perspective. The approach can be also useful in order to estimate the various heat rejection contributions of the engine when considering thermal management studies in a preliminary project stage or when considering possible improvements of an existing engine. The application example reported in this paper is a medium size recreational boat powered by a medium-duty 200 kW brake power HDDE, but the method can be applied to any kind of application in principle for example a vehicle, once the boundary conditions are known or assumed.

2.1. Engine side modelling

Industry standard 1D engine modelling software, such as WAVE (12), is ubiquitous tool in every phase of the engine-system development. However, the approach in these tools considers all the energy fluxes with the same potential of producing work (First Law approach). This is actually not true, which is why exergy (or often called availability, a) can be used to understand where inefficiencies in the system components concentrate, in order to propose possible improvements. Exergy or availability, can be defined as the potential of doing work or, referring to a system, it is the maximum useful work that can be extracted during a process towards equilibrium with environment, following reversible transformations.

The First Law of Thermodynamics is a law of energy conservation, which states that energy cannot be destroyed but only transformed from one form to another (e.g. work to heat). Whilst in a Carnot closed cycle heat engine, heat cannot be converted completely and continuously into work (Carnot law) (13), an engine, which operates on an open cycle, is theoretically not limited by Carnot efficiency. However, a 100% thermal efficiency cannot be reached due to internal inefficiencies (irreversibilities), and, in particular, friction and mechanical losses, fluid flow losses in the pipes, throttling or valve losses, pumping work losses, combustion losses (usually the highest), heat losses, expansion and compression losses in the turbocharger and heat transfer to the cooling system and to the environment. It is in this context that a Second Law of Thermodynamics and, in particular, exergy analysis can help. Indeed, unlike energy, exergy (or availability) can be destroyed, increasing the entropy of the system and generating irreversibilities, whilst, like energy, exergy can be transferred from system to system or to the surroundings. It is specifically in this last case when waste heat recovery systems, such as ORC, can help to improve the overall powertrain efficiency, recovering transferred exergy, which otherwise would be a loss for the engine system.

Regarding the engine side, a model developed in WAVE of a medium to heavy duty Diesel engine, with single stage turbocharger, six cylinders in-line and 200 kW brake power, has been used as a baseline in order to develop a combined MATLAB post-processing routine able to calculate both energy and exergy balances for every operating point and configuration that can be simulated in WAVE. The results of the post-processing have then been used as boundary conditions for the ORC system and, in this particular example, to recover exhaust gas heat after the turbine. Moreover, a complete energy and exergy assessment of the engine model can be easily done, through the post-processing routine, for every possible operating point, this being useful in order to assess the complete engine heat balance in a quite accurate way.

The engine model has been simulated over a typical possible operational map at different load and speed points. For every operational point the First and Second Law balances can be calculated as soon as the post-processing routine has been developed for the engine configuration simulated.

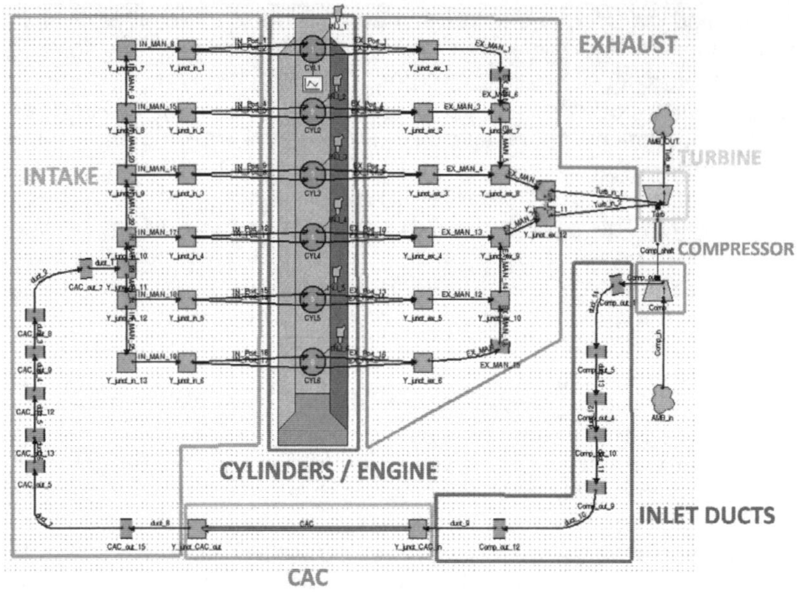

Figure 1. Engine WAVE model divided in control volumes

In Figure 1, the engine WAVE model is shown, divided in several control volumes, one for each main sub-component (or sub-system) of the engine. In each control volume, a conservation of energy and an availability balance, based on crank-angle resolution [J/°c.a.] have been applied. For example, for the control volume related to the cylinders/engine, as done by Rakopoulos *et al.* (14), the availability balance can be written as:

$$\frac{dA_{cyl}}{d\varphi} = \frac{m_{in}b_{in} - m_{out}b_{out}}{6N} - \frac{dA_{work,brake}}{d\varphi} - \frac{dA_{frict,loss}}{d\varphi} - \frac{dA_{HT,loss}}{d\varphi} + \frac{dA_{fuel,burned,ch}}{d\varphi} - \frac{dI}{d\varphi} \quad (1).$$

With:

- A_{cyl} is the cylinder availability [J]. In particular, this term is 0 over a cycle in steady state, $\int_0^{720} \frac{dA_{cyl}}{d\varphi} d\varphi = 0$
- m_{in} and m_{out} are the incoming and outgoing mass of gas [kg]
- N is the engine rotational speed [rev/min]
- $A_{work,brake}$ is the availability term related to the brake power [J]
- $A_{frict,loss}$ is the availability term related to the friction losses [J]
- $A_{HT,loss}$ is the availability term related to the cylinders walls heat transfer, in particular to the coolant and oil circuits [J]
- $A_{fuel,burned,ch}$ is the chemical availability of the burned fuel (in this case diesel, with LHV=42.8 MJ/kg), as reported in (14)
- I is the term related to irreversibilities [J], which is the only unknown of the equation
- b_{in} and b_{out} are the specific flow availability terms [J/kg] at the inlet and outlet of the cylinders, and have been calculated using the flow exergy formulation, considering only thermo-mechanical exergy terms, as (for the inlet term):

$$b_{in} = (h_{in} - h_{0,\text{in}}) - T_0(s_{in} - s_{0,\text{in}}) \quad (2)$$

in which, the terms with 0 are related to the so-called dead state (298.15 K and 1 bar), but are calculated considering the 11 gas species composition used in the WAVE simulation following the ideal gas formulations.

It is out of the scope of this paper to consider in depth the overall theory behind a Second Law analysis applied to internal combustion engines. In particular, as good references, the publications by Rakopoulos *et al.* (14) and Edwards *et al.* (15) can be considered for what concerns the theory and approach used. The formulations have then been adapted using the thermodynamic outputs provided from WAVE.

Generally, the same type of availability balance has been applied to every control volume, in order to calculate the irreversibilities for each system sub-component. All terms have been integrated over an engine cycle in order to obtain cycle-averaged values for availability flows and irreversibilities.

2.2. ORC side modelling

An Organic Rankine Cycle (ORC) system has been proposed in order to recover heat from the exhaust gas at the turbine outlet of the engine. In this preliminary work, the boiler backpressure effect on the engine side has not been evaluated: in forthcoming work the increased backpressure penalty could be assessed.

The ORC system in this example is simple, recovering exhaust gas from the engine tailpipe line and rejecting it to a sea water stream, since this case study is a diesel engine for small-medium size recreational boats. However, other applications can be considered, such as a road vehicle.

The ORC scheme and the relative processes are reported in Figure 2.

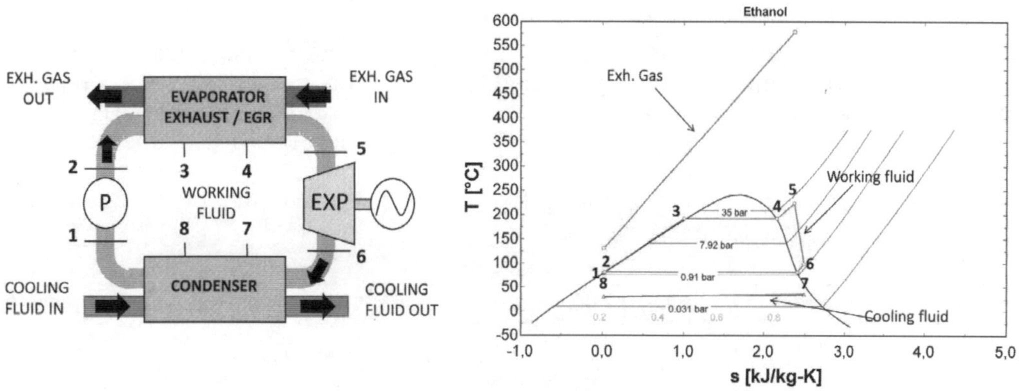

(a) **(b)**

Figure 2. ORC scheme and T-s diagram example for ethanol as the working fluid

The ORC system works with the following processes. The working fluid is pumped from the condensing pressure to the evaporation pressure (1-2), then it is preheated by the heat source in the preheater, still being in a liquid phase (2-3). The fluid is then evaporated in a counterflow evaporator (3-4) and superheated in a superheater (4-5). Then expansion is carried out, from the evaporation pressure back to the condensing pressure, in the expander (5-6). De-superheating happens from superheated vapour to saturated vapour (6-7). In the counterflow condenser, the condensation process happens from saturated vapour to saturated liquid (7-8). Finally, the fluid is sub-cooled in the sub-cooler (8-1) and starts the process again.

The heat exchangers have been modelled divided in three parts, as counterflow, and applying the fixed boundaries method. For the expander and the pump, respectively, an isentropic efficiency of 75% and 60% have been imposed. Electricity production has been considered, with a generator efficiency of 95%. The model has been developed using the principles of conservation of mass and energy, regarding the First Law analysis.

Concerning the Second Law analysis, each of the four main components has been considered as a control volume, and the following formula (16) has been applied to calculate the irreversibility rate for steady-state conditions:

$$\dot{I} = T_0 \dot{m}_{wf} \left[\sum s_{out} - \sum s_{in} + \sum_j \left(\frac{q_j}{T_j} \right) \right] \qquad (3).$$

Five different working fluids have each been simulated for the ORC system: ethanol, methanol, water-steam, toluene and R245fa. An optimization procedure has been applied in order to maximize the net power output of the ORC system, whilst both the First Law and the Second Law performance of the system have been evaluated. The optimization procedure applies a genetic algorithm, in order to find the right combination of the five independent variables (working fluid mass flow, cooling fluid mass flow, pump pressure ratio, degree of superheating and condensing pressure) that maximizes the net power output of the system. The model and the optimization procedure have been implemented in Engineering Equation Solver (EES, (17)).

3. RESULTS

It is out of the purpose of this paper to report a complete overview of the results obtained from the combined engine-ORC analysis. For this reason, just an overview of the information that can be extracted with the approach considered will be reported. As an example, only one engine operating point has been considered in the details given, regarding both engine and ORC side results. An overview of the performance of all the fluids will also be reported, considering other operating points covering most of the engine map.

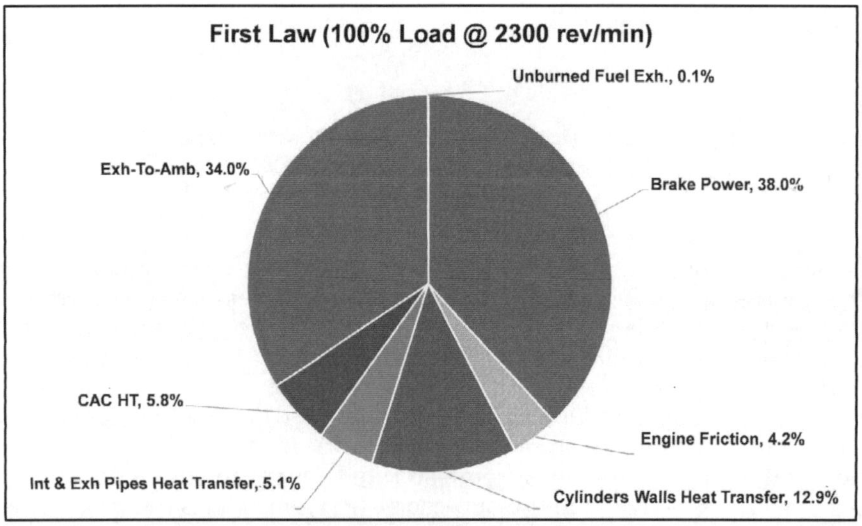

(a)

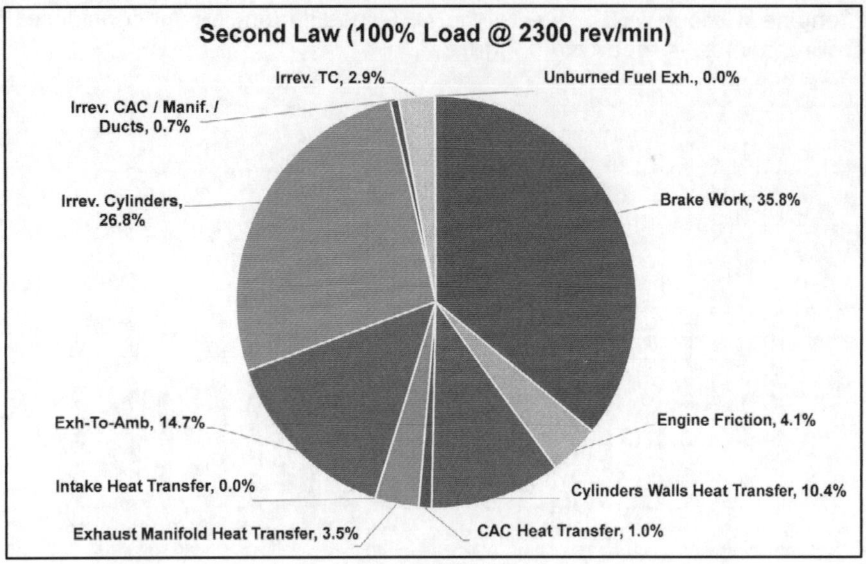

(b)

Figure 3. First (a) and Second Law analysis (b) for the engine at 100% Load and 2300 rev/min

In Figure 3, a graph for the First Law analysis (a) and a graph for the Second Law analysis (b) of the engine side, operating at 100% Load and 2300 rev/min (around 200 kW brake power output) are given.

As it can be seen from the reported charts, in the case of the First Law, only energy fluxes can be observed, in particular regarding brake power output (38% of the injected fuel energy), friction lost energy (4.2%), heat transfer to the cylinder walls (12.9%), heat lost through intake and exhaust pipes (5.1%), heat exchanged in the Charge Air Cooler (CAC, 5.8%) and thermal energy available in the exhaust gas (34%). In the case of the Second Law analysis, contributions such as brake work availability (35.8% of the injected fuel availability), engine friction availability loss (4.1%), cylinder walls heat transfer availability exchange (10.4%), CAC exchanged availability (1.0%), exhaust manifold heat transfer-related availability (3.5%) and availability of the exhaust gas (14.7%) can be observed. In this case, the irreversibilities, introduced during the engine working processes, can be observed as well, demonstrating, in particular, how the cylinder's irreversibilities, most of them being combustion-related, waste 26.8% of the injected fuel availability. Other irreversibilities are related to the heat transfer processes in the CAC, manifolds and ducts (0.7%) and in the turbocharger (compressor and turbine, 2.9%).

In the chart in Figure 3(b), it can be seen how improvements in the combustion processes could lead to advantages in terms of overall engine performance. Indeed, it is during combustion that most of the injected fuel availability is wasted, not producing effective work. Moreover, heat transfer related availability is also a contribution which should be considered. Indeed, the higher the temperature of the availability flux, the higher is the potential of producing additional work, for example, through the use of a bottoming cycle like an ORC system. In particular, the exhaust gas shows a high potential for further work production, yet contributions from the cylinder walls and CAC heat transfer could also be considered in order to improve overall engine efficiency.

Concerning the ORC side and imposing a 130°C exhaust gas cooling limit because of possible sulphur deposition problems during condensation, the Brake Specific Fuel Consumption improvement (BSFC, reported in percentage compared to the baseline engine without ORC), due to the ORC installation, for all considered fluids and operating points, is reported in Figure 4.

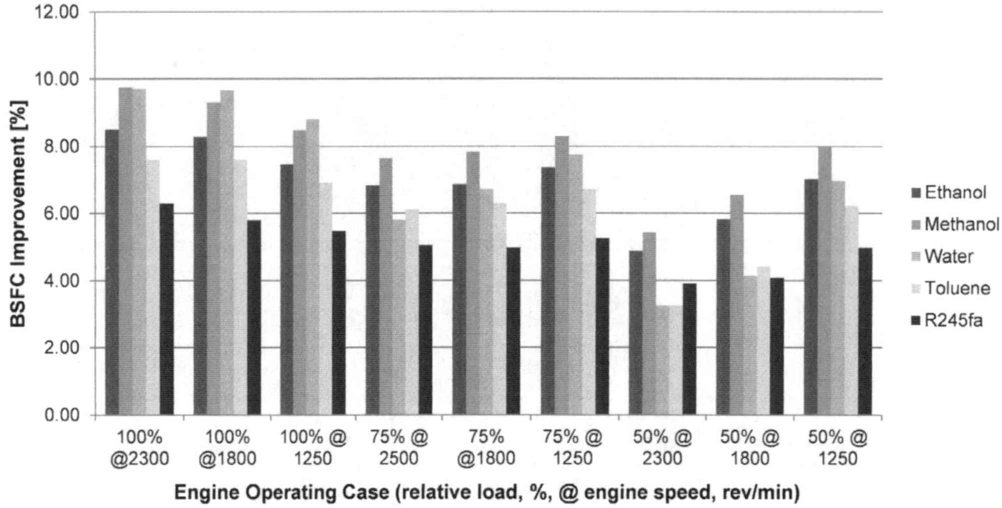

Figure 4. Combined engine-ORC system BSFC improvement [%] compared to the baseline engine without ORC

As can be observed, the best results, in almost all operating points, have been obtained using as fluids methanol, ethanol, water-steam or toluene. In particular, at 100% Load and 2300 rev/min, the best fluid is methanol, with a BSFC improvement of 9.8% compared to the baseline engine. This is mostly due to the fact that these fluids are more suitable for the typical exhaust gas after-turbine temperature. In the 50% Load and 2300 rev/min case, the R245fa shows better performance compared to, for example, water or toluene, because of the lower temperature level of the exhaust gas (274°C from simulations). However, for exhaust gas waste heat recovery, usually R245fa can be considered as not the best choice, due to the expected lower thermodynamic performance. However, other considerations, such as for example availability and flammability, could still lead to the choice of this fluid, or its future replacement R1233zd(E), which will be possibly used in future developments because more environmentally friendly.

Considering the 100% Load and 2300 rev/min case, the First (a) and Second Law (b) analysis charts are reported for the ORC system in Figure 5.

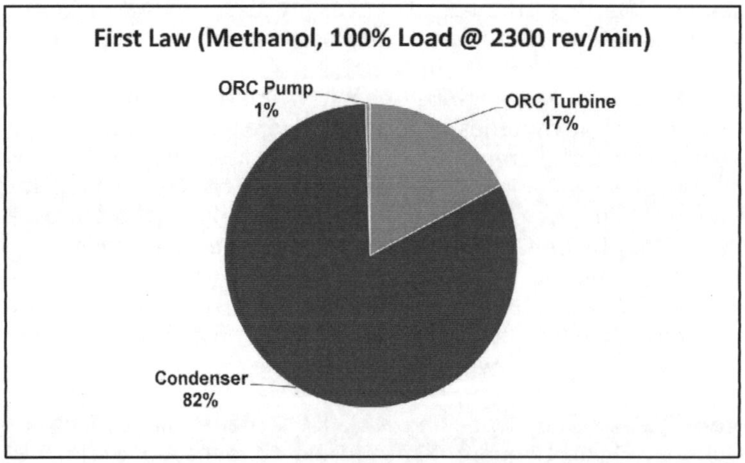

(a)

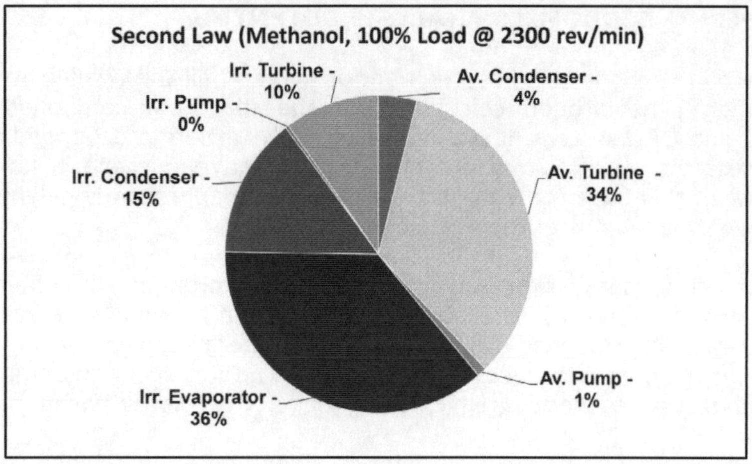

(b)

Figure 5. First and Second Law analysis for the ORC at 100% Load and 2300 rev/min

As it can be observed from the above charts, the major part of the ORC system irreversibilities occurs in the evaporator. This is mostly due to the finite temperature difference heat transfer between the exhaust gas (heat source) and the ORC working fluid. Indeed, as stated in exergy theory, the heat exchange between two fluids can be considered as a reversible process only if it happens under infinitesimal temperature difference.

In Figure 5(a), the recovered energy in the evaporator, split between the ORC component contributions, is reported, showing how most of the energy entering the ORC is rejected in the condenser without producing useful work. However, in Figure 5(b) it is possible to observe the distribution of the recovered availability, introducing four new ways in which the exergy is wasted. Two of them are related to the irreversibilites that occur in the turbine and in the pump (10% and 1% respectively), the others refer to the heat exchange processes in the evaporator (36%) and in the condenser (15%). Concerning the heat exchange irreversibilities, which are the main contributions, the First Law states that around 82% of the heat recovered by the ORC evaporator is rejected (and wasted) in the condenser. On the other side, through the Second Law, it is shown that the availability lost in the condenser is only the 4% of the total recovered in the evaporator, and the irreversibilities are estimated as the 15% (still lower than those in the evaporator). Actually, this can be explained with the consideration that the temperature of the fluid in the condenser is lower than in the evaporator, this means that the ideal work potentially obtainable from this flow is lower (definition of availability). For this reason, an improvement of the performance of the condenser is not as effective as an improvement in the performance of the evaporator would be.

A way to improve the global performance of the ORC system is to reduce the irreversibilities, especially those in the evaporator. A solution could be the adoption of a supercritical system. This system would allow a better match of the temperature curves of the heat source(s) and working fluid but, particularly in transport applications where on-board safety is very important, a very high cycle pressure could be disregarded as an option. Another possibility could be using a zeotropic mixture, which has the property of evaporating and condensing at variable pressure, thus improving the thermal match with the heat source. Further, a more efficient expander could lead to cycle performance enhancement. For this reason, a more accurate expander model will be considered in the forthcoming work.

4. THE THERMO-ECONOMICS ANALYSIS POTENTIAL

Once the Second Law analysis has been performed and all the availability rates (or availability flows) have been calculated for the main sub-components of the combined engine-ORC system, it is possible to apply a thermo-economic method in order to assess the system from a combined thermodynamic and costs feasibility point of view. In this paper only an introduction to the approach is given, since the implementation will be part of the forthcoming work.

Thermo-economics, sometimes called exergo-economics, is the method that combines together exergy (or availability) analysis with conventional cost analysis, in order to assess and optimize the performance of energy conversion systems both from a thermodynamic and a costs perspective. Thermo-economics must not be confused with techno-economics, which does not refer to exergy-based methods.

Exergy-based (Second Law) analysis helps to understand where inefficiencies are concentrated in the system, calculating exergy destruction (irreversibilities) and exergy losses. However, a Second Law analysis alone does not help to understand what is the monetary cost of system inefficiencies. Additionally, through the use of a thermo-economics approach, it is possible to assess or estimate what is the cost of production of the system's products (for example, electricity and steam for a cogeneration system or mechanical energy to be applied to the crankshaft for an engine-ORC system for vehicles applications) through cost allocation techniques, also evaluating what is the estimated cost of the system irreversibilities.

As reported by Bejan *et al.* (18), the objectives of a thermo-economic analysis might be: calculate the costs of each product stream of a system (e.g. electricity, steam, mechanical power and so on), understand the cost formation process and the flow of costs in the system, optimize specific variables of a system or of a single component, design cost-efficient systems in the first phase of a project, diagnose system inefficiencies, defects or faults, and propose solutions to improve the performance, as well as quantify the inefficiencies in terms of costs.

For an energy conversion system, operating at steady-state, there are many materials entering and exiting (e.g. fuel), heat and work streams that interact with each other and the surroundings. The main step of a thermo-economic analysis is addressing all these streams and calculating the relative exergy rates ($\dot{A}_i$), which can transfer exergy inside or outside the system, as well as those related to exergy destruction terms (irreversibility rates, $\dot{I}_i$). This is what has been done, for example, for the reported engine-ORC system. As a second step, costs have to be associated or assigned to every exergy (or availability) stream, since exergy but not energy (which cannot be destroyed) is the only rational base of cost assignment. As reported by Bejan *et al.* (18), this procedure is often called Exergy Costing.

According to Exergy Costing techniques, the cost stream [$/s] associated with an exergy stream can be written as:

$$\dot{C}_i = c_i \cdot \dot{A}_i \qquad (4).$$

With c_i the unit cost of the stream [$/J] and $\dot{A}_i$ the availability flow rate [W] associated to the considered stream. Following this approach, for every exergy stream entering or exiting the considered system, component or sub-component analysed, there will be an associated cost stream.

For example, for a k-th component (or control volume) it is possible to write, for the entering and exiting streams, the associated cost stream. An example of a possible component receiving heat and producing work (Figure 6) has been reported below, together with the exergy-cost balance.

Inlet: $$\dot{C}_{in} = c_{in} \cdot \dot{A}_{in} = c_{in} \cdot (\dot{m}_{in} \cdot a_{in}) \quad (5).$$

Outlet: $$\dot{C}_{out} = c_{out} \cdot \dot{A}_{out} = c_{out} \cdot (\dot{m}_{out} \cdot a_{out}) \quad (6).$$

Work: $$\dot{C}_W = c_W \cdot \dot{W} \quad (7).$$

Heat: $$\dot{C}_q = c_q \cdot \dot{A}_q \quad (8).$$

Cost balance: $$\sum_{out} \dot{C}_{out,k} + \dot{C}_{W,k} = \dot{C}_{q,k} + \sum_{in} \dot{C}_{in,k} + \dot{Z}_k \quad (9).$$

As can be seen in the cost balance in eq. (9), the sum of the exiting cost streams equals the sum of the entering cost streams plus the change in cost due to the capital investment cost $(\dot{Z}_k^{CI})$ and the operating and maintenance expenses $(\dot{Z}_k^{OM})$, which sum to give the term reported as $\dot{Z}_k$.

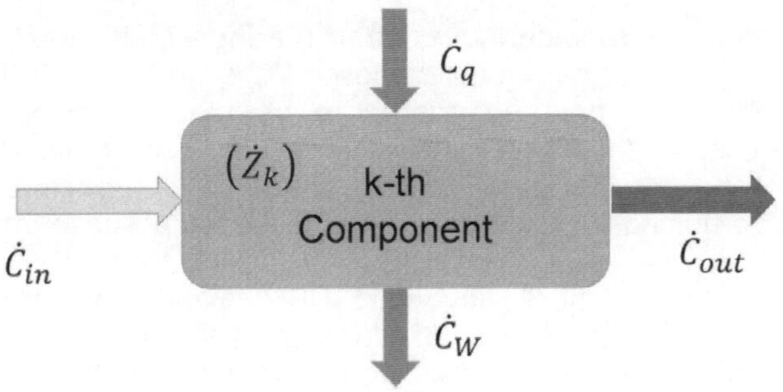

Figure 6. Scheme of the exergy/cost balance for a hypothetical component receiving heat and producing work

Once exergy streams have been calculated, one of the main issues in a thermo-economic analysis is the estimation of the term $\dot{Z}_k$ and, particularly, of the cost of the system components (Purchased Equipment Costs, PEC). The best way to obtain this information is through supplier quotations, old purchase costs information or company databases, these last one usually covered by industrial secret. Some estimation approaches also come from literature and are often used is ORC studies. They usually refer to some well-known texts in the sector, such as Turton (19) or Peters (20), but the values always refer to equipment costs applied to the chemical industry, since valuable databases for ORC component costs are not available, as reported by Lemmens (21), because experience in the field is limited and data not available for high production numbers, especially considering ORC systems for vehicles. Moreover, some authors, e.g. (22, 23), propose some components cost correlations based on typical size parameters developed through experience or available literature data.

The scope of the forthcoming work will be to introduce a thermo-economic analysis combined with the study of more advanced engine-ORC concepts, considering new cycle architectures and the exploitation of different heat sources, as well as optimizing the system in order to decrease irreversibilities and improve overall efficiency, which has the goal of decreasing concept costs.

5. CONCLUSIONS AND FUTURE WORK

The purpose of the paper is to introduce and describe a method developed to consider the synergy between state-of-the-art 1D gas dynamics simulation software and a thermo-exergo-economic modelling approach, with the final objective of investigating, in a preliminary phase of a development project, more efficient and cost effective combined engine-waste heat recovery systems.

Through the developed post-processing routine, the overall engine heat balance, as well as the Second Law balance, can be evaluated in detail, with the possibility to assess all energy fluxes of the system, as well as to determine where inefficiencies (irreversibilities) concentrate. The approach can be used both for combined engine-waste heat recovery systems development as well as for the study of the engine thermal management, having a complete overview of the heat rejection of the engine for every operating strategy which can be simulated.

The method proposed can be applied to any kind of engine model (CI, SI, 2-stroke, 4-stroke, dual fuel, EGR, multiple stages turbocharging etc) and any kind of engine concept or architecture which can be simulated, as well as for any possible waste heat recovery system that can be thermodynamically modelled.

In the example reported considering a small marine application, an ORC system recovering exhaust gas heat has been proposed. The maximum BSFC improvement when using the ORC has been estimated to be around 9.8%. In addition, all combined system heat fluxes, exergy fluxes and irreversibilities have been investigated, showing the potential of the proposed approach to enable better understanding of combined systems performance and to propose improvements.

In the forthcoming work, more advanced ORC layouts will be proposed, the recovery of heat from a combination of different heat sources (e.g. coolant, CAC and exhaust gas) will be assessed, and the engine exhaust backpressure effect of fitting the ORC boiler on the engine tailpipe could also be considered. More advanced expander models will be proposed, or performance correlations considered, in order to investigate the system at off-design conditions, evaluating also different combined engine-ORC operating strategies and concepts (e.g. low heat rejection engines with coupled ORC).

As a final step, a thermo-economic approach will be implemented for the studied concepts in order to assess the systems from a cost-effectiveness point of view and to analyse the trade-off between costs and overall powertrain performance.

ACKNOWLEDGMENTS

The authors would like to thank the Directors of Ricardo for permission to publish this paper. The support of colleagues, in particular Dr Simon Edwards, in the preparation of this paper is recognized.

The research leading to these results has received funding from the People Programme (Marie Curie Actions) of the European Union's Seventh Framework Programme FP7/2007-2013/ under REA grant agreement n°607214.

REFERENCE LIST

1. Ricardo. EMLEG - Emissions Legislation Database [cited 2015 Sep 1]. Available from: http://www.emleg.com/
2. Sprouse C and Depcik C. Review of organic Rankine cycles for internal combustion engine exhaust waste heat recovery. Appl. Therm. Eng. Elsevier Ltd; 2013;51(1–2):711–22.
3. Lion S, Vlaskos I, Rouaud C and Taccani R. A review of waste heat recovery and Organic Rankine Cycles (ORC) in on-off highway vehicle heavy duty diesel engine applications (Under Review). Renew. Sustain. Energy Rev. 2017.
4. Lion S, Michos CN, Vlaskos I and Taccani R. A thermodynamic feasibility study of an Organic Rankine Cycle (ORC) for Heavy Duty Diesel Engine (HDDE) waste heat recovery in off- highway applications. ECOS 2016. Portoroz, Slovenia; 2016.
5. Howell T, Gibble J and Tun C. Development of an ORC system to improve HD truck fuel efficiency. Ricardo. DEER Conference. 2011. p. 1–21.
6. Maybin B, Hanna A, Rouaud C, Porteous S, Baxter J, Douglas R et al. The Energy and Thermal Management of A Hybrid Double Deck Bus. 23rd Aachen Colloquium Automobile and Engine Technology. 2014.
7. Nolan C, O'Shaughnessy R, Douglas R, Rouaud C, Hanna A and Seaman R. Waste heat recovery to improve fuel economy on a hybrid bus. MTZ Heavy Duty On and Off Highway Motoren. 2013.
8. Shu G, Liang Y, Wei H, Tian H, Zhao J and Liu L. A review of waste heat recovery on two-stroke IC engine aboard ships. Renew. Sustain. Energy Rev. Elsevier; 2013;19:385–401.
9. Singh DV and Pedersen E. A review of waste heat recovery technologies for maritime applications. Energy Convers. Manag. Elsevier Ltd; 2016;111(X):315–28.
10. Michos CN, Lion S, Vlaskos I and Taccani R. Analysis of the backpressure effect of an Organic Rankine Cycle (ORC) evaporator on the exhaust line of a turbocharged heavy duty diesel power generator for marine applications. Energy Convers. Manag. Elsevier Ltd; 2017;132:347–60.
11. Quoilin S, Van Den Broek M, Declaye S, Dewallef P and Lemort V. Techno-economic survey of organic rankine cycle (ORC) systems. Renew. Sustain. Energy Rev. 2013;22:168–86.
12. Ricardo Software. Ricardo WAVE. [cited 2016 Feb 4]. Available from: http://www.ricardo.com/en-GB/What-we-do/Software/Products/WAVE/
13. Klein SA and Nellis GF. Thermodynamics. Cambridge Press Editor. 2012.
14. Rakopoulos CD and Giakoumis EG. Second-law analyses applied to internal combustion engines operation. Prog. Energy Combust. Sci. 2006;32(1):2–47.
15. Edwards KD, Wagner RM and Graves RL. Identification of Potential Efficiency Opportunities in Internal Combustion Engines Using a Detailed Thermodynamic Analysis of Engine Simulation Results. SAE 2008 World Congress. Detroit; 2008.
16. Somayaji C, Mago PJ and Chamra LM. Second Law Analysis and Optimization of Organic Rankine Cycle. ASME Conf. Proc. 2006;2006(42053):591–6.
17. Klein SA. EES (Engineering Equation Solver) - Professional Version V9.710-3D. F-Chart Software [Internet]. [cited 2016 Jan 20]. Available from: http://www.fchart.com/
18. Bejan A, Tsatsaronis G and Moran MJ. Thermal Design and Optimization. John Wiley & Sons, Inc.; 1996. 542 p.
19. Turton R, Bailie RC, Whiting WB and Shaeiwitz JA. Analysis, Synthesis, and Design of Chemical Processes. 3rd ed. Pearson Education, Inc.; 2009.
20. Peters MS and Timmerhaus KD. Plant Design and Economics for Chemical Engineers. 4th Ed. Plant Design and Economics for Chemical Engineers. McGraw-Hill Chemical Engineering series; 1991. 478-570 p.

21. Lemmens S. Cost engineering techniques and their applicability for cost estimation of organic rankine cycle systems. Energies. 2016;9(7).
22. Quoilin S, Declaye S, Tchanche BF and Lemort V. Thermo-economic optimization of waste heat recovery Organic Rankine Cycles. Appl. Therm. Eng. Elsevier Ltd; 2011;31(14–15):2885–93.
23. Panesar AS. An innovative organic Rankine cycle approach for high temperature applications. Energy. Elsevier Ltd; 2016.

ENERGY MANAGEMENT

Oil thermal management during engine transients from a cold state

D Di Battista, R Cipollone
Department of Industrial and Information Engineering and Economics,
University of L'Aquila, Italy

ABSTRACT

The paper presents an experimental activity aimed at assessing the influence of a faster warm-up of the lubricant oil in a F1C (3 L) turbodiesel engine used both for light and heavy duty applications. The engine is ran on a dynamometer test bench and transient homologation cycles have been implemented considering different initial oil conditions. Lubricant and cooling thermal dynamics have been evaluated and their mutual thermal interactions have been discussed. In particular, an initial hotter oil showed fuel consumption and CO_2 emissions saving. Harmful substances (CO, HC, NOx, particulate matter) have been measured too, displaying significant reduction mainly due to the modified engine temperature dynamics.

1. INTRODUCTION

CO_2 levels in the atmosphere is one of the most important concerns in the future and emission reduction represents the most important driver of the technological evolution of the energy related transformations. Transportation sector accounts for 30-35 % of the overall CO_2 emissions: a wide portfolio of new technologies are under consideration as short-mid term response to the CO_2 concerns. International governments, therefore, have set limits on the carbon dioxide emission from vehicles and imposed fines for manufactures that will not respects the targets. In Europe, for instance, the limit is 95 gCO_2/km by 2021 for passenger vehicle and 170 g CO_2/km for light duty vehicles (1). In addition, pollutant emissions (HC, CO, NOx and PM) have to be reduced in the path from Euro 5 and Euro 6 regulation. In particular, for diesel engine this restriction is very stringent (2) at a level that, in a mean future, the specific engine market could have production repercussions.

Literature studies have shown that a large portion of engine-out emissions (both CO_2 and harmful pollutants) are largely induced during cold start-up (3) and in the first 75% of engine homologation cycles (4). This is mainly due to higher engine internal frictional losses (5), ineffective combustion and low efficiency of catalytic converters (6).

Focusing on frictional losses, relative surfaces are lubricated with engine lubricant oils (7). The viscosity of engine oil is designed to operate efficiently at warmed-up temperatures (8); therefore, the high viscosity at cold start-up sets a heavy burden on fuel economy (9). It is demonstrated that low lubricant temperature (as in cold start-up conditions) increases the friction mean effective pressure (FMEP) of the engine by 400% (10). Moreover, the majority of vehicles operate at a temperature lower than the designed one because of the short lengths of journeys (11). Therefore, most part of vehicles generate excessive quantity of engine-out emissions and consume more fuel than it is expected. Several studies (12) (13) (14) have indicated that the interaction between ring pack and cylinder liner is the major source of mechanical energy loss of IC engine (15), but also crankshaft main bearings account for a substantial portion of friction losses during cold engine phases (16)(17). Consequently, the investigation of engine transient operation has become an important objective to engine manufacturers, intensified by the fact that significant deviations are experienced when comparing instantaneous transient emissions with their quasi-steady counterparts (18) (19) (20) (21). This gives great importance to the reduction of the engine warm up phase (22) (23) and this rebounds on interventions on the cooling and lubricating circuits (24) (25) as well as in the combustion progression. Under misfiring conditions, in fact, combustion cannot supply sufficient power to drive the engine and to overcome the increased friction losses. It is interesting that, unlike acceleration or load acceptance, during cold starting, naturally aspirated engines suffer equally to their turbocharged counterparts (26). For example, it has been found that a diesel engine may emit up to seven times more particulate matter during cold operation than under warm conditions (27) coupled also to a prolonged period of unacceptable smoke emissions (28).

The paper presents an experimental activity aimed at evaluating the benefit related to the reduction of the oil warm up time in a Iveco F1C (3 L) turbodiesel engine used both for light duty and heavy duty applications. The engine is ran on a dynamometer test bench and transient homologation cycles have been implemented. Lubricant warm up time has been shortened by heating up the oil, thanks to an external heater, and the behavior of the lubrication circuit has been reconstructed and compared to the cold engine reference case (29). Also the interactions with the engine cooling circuit have been investigated. The thermal power needed to heat up the oil, however, can be easily taken from the exhaust gases, which represent an almost free internal thermal source in the engine. Exhaust gases, in fact, reach almost immediately suitable and useful temperatures (30), but the additional gases/oil heat exchanger introduces pressure drops on the engine exhaust line that must be kept under control (31) (32). The fuel consumption reduction obtained with a warmer oil is measured in the range of 2-3% (and, therefore, CO_2 emissions). Harmful substances (CO, HC, NOx and particulate matter) have been measured too, displaying significant reduction mainly due to the modified temperature engine dynamics.

2. EXPERIMENTAL SETUP

The engine used as reference has been the IVECO F1C 3.0L. The complex behavior of its lubrication has been extensively studied thanks to an experimental test bench properly realized. In particular, it is a turbocharged diesel engine, on which the operating conditions of a NEDC homologation cycle was reproduced on a dynamic test bench.

In particular, as showed in Figure 1, the lubricant oil circuit was characterized in order to have wide information about the hydraulic and thermal behavior of the circuit, its control strategy and the intersection with the coolant circuit:
- the pump is of a gear type, with a maximum flow rate of about 55 L/min, almost linear with its revolution speed. It is mechanically linked to the engine via chain;
- oil amount in the oil sump is about 7 L;
- in order to control the pressure of the oil inside the engine, a relief valve is placed inside the pump case. A control pressure signal is taken from the main gallery to open this valve and recirculate part of the oil flow rate in the sump: the maximum pressure allowed is about 4.5 bar;

- the heat exchanger (HX) between oil and coolant is placed just after the pump and before the oil filter. In particular, it is a plate heat exchanger completely immersed in the coolant that comes from the radiator. Therefore, there is coolant flow only when the thermostat is open, otherwise the coolant inside the HX is stagnant and heated up by the oil and engine walls during warm up.
- the reference vehicle used to perform driving cycle is IVECO Daily, on which this engine is mounted.

In Figure 1, also, it is shown the positioning of all the thermocouples and pressure transducers placed in the experimental test bench. The engine test bench has been equipped, moreover, with all the instrumentation necessary to measure fuel consumption and emissions. Table 1 summarizes instruments used and relative uncertainties. In particular, two channels of CO_2 measurement have been installed: one on the exhaust and the second one on the intake manifold, in order to evaluate the EGR rate.

Table 1: Uncertainties of sensors used

Quantity	Instrument	Uncertainty
Temperature	K-Thermocouple	2.2°C
Pressure	Gems 2200 series	0.1 bar
Fuel consumption	AVL 733s fuel balance	0.05%
Engine charge air flow	Bosch HFM 5	< 3%
HC	FID	< 2%
CO	NDIR	< 2%
CO_2	NDIR	< 2%
PM	MicroSoot Sensor	5%
NOx	CLD	< 2%

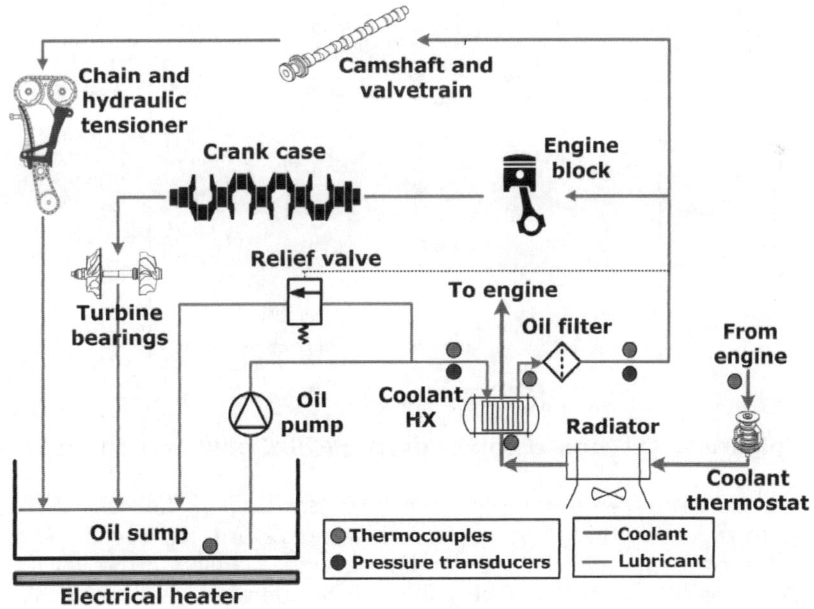

Figure 1: Oil circuit of the IVECO F1C engine and sensors positioning: blue lines represents coolant circuit, grey lines the oil one

In order to assess the benefits related to a faster oil warm up time, an electrical heater of 1500W was placed around the oil sump. The oil, so, was pre-heated before the NEDC starting and then, continuously heated during the warm up. The results coming from this so-called "heated oil NEDC" was compared to the result of a NEDC cold engine test. This would represent the best conditions in which the lubricant can be; hence, the results expected can be considered as the best one reachable intervening on the oil circuit. After the start up having the oil hot, in fact, oil flows through cold engine walls and in about 50 s, it reaches a much lower thermal value. This in parts reduces the benefits of having an initial hot oil. A much effective contribution can be reached if oil is continuously heated and a way to sustain its temperature should be found.

3. THERMAL BEHAVIOUR

Figure 2 shows the oil temperatures dynamic during the reference NEDC cold engine test. The thermocouples positioning has been represented in Figure 1, following the lubricant path. In particular, the temperature difference across the oil/coolant heat exchanger has been measured; it is always negative and it is close to zero during warm up time. This means that the oil that enters in the engine (black line) is always slightly colder than the oil coming from the pump. When thermostat opens (about 900 s), this delta temperature increases till to about 4 °C. This demonstrates that the oil/water heat exchanger is fed by the coolant coming from the radiator. In the figure is clear, moreover, how the oil does not reach a steady state during a NEDC cycle.

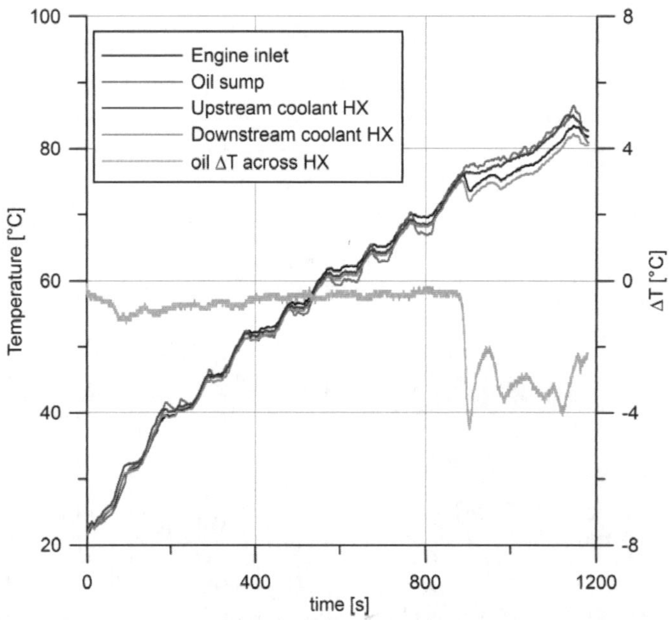

Figure 2: Oil temperatures during NEDC cold engine test

In Figure 3, the same temperatures have been showed in the case of heated oil NEDC test. In this test, the oil starts from a temperature close to 80°C and then, because of the cold engine walls and components, it is cooled in the very beginning of the NEDC cycle (about 100 s). The minimum temperature reached is about 40°C and, then, the oil dynamics follows the temperature in Figure 1. However, thanks to the heat given by the electrical heater, the oil temperature is quite high than the one in the cold engine mode, and oil warm up time is lower. In particular, in Figure 3, the oil temperature instantaneous difference across the coolant heat exchanger is higher in the first 100 s and, then, it is asymptotic to zero. After 750s, it reaches the value of -4°C (thermostat opening).

This behavior invites to consider technologies able to sustain oil temperature during the first passages inside the engine: even though oil is heated before engine start, when this happens very rapidly it is cooled down. When it is remixed inside the oil sump, the hot oil present sustains the temperature but after a short period of time, it is definitively cooled and the benefit of the initial hot state is vanished.

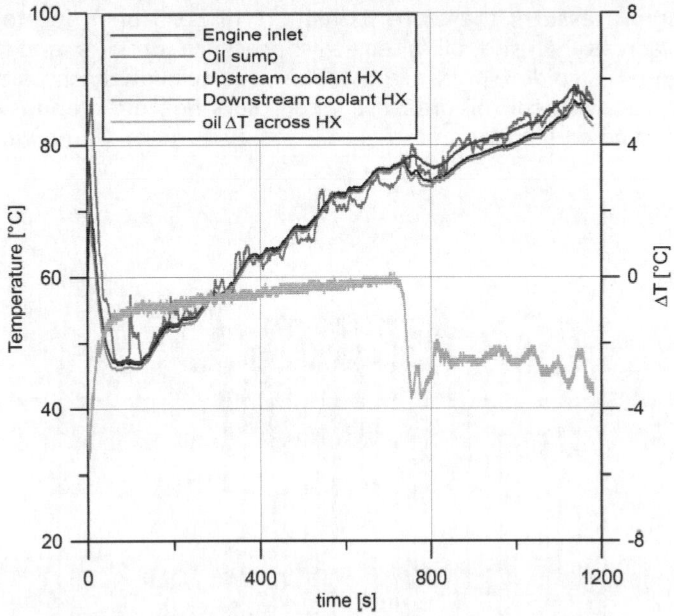

Figure 3: Oil temperatures during NEDC heated oil engine test

The warm-up time reduction is demonstrated by the coolant temperatures showed in Figure 4: coolant temperature in the engine (before the thermostat) is always higher in the case of heated oil NEDC and, so, thermostat opens about 150s before. This can be noticed, also, observing the coolant temperature in the HX: it is lower than the engine one and increasing during warm up, but when thermostat opens, it suddenly falls to 20°C, due to the radiator cooling.

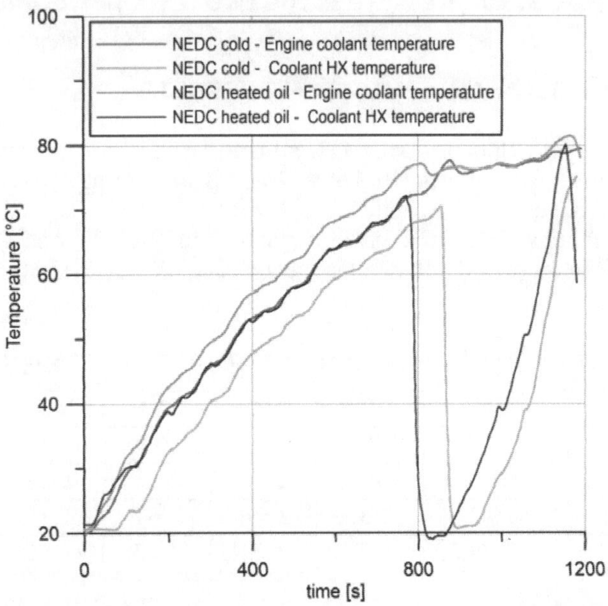

Figure 4: Coolant temperatures in the two NEDC tests performed

It is just the point to observe that, in this configuration, the coolant is quiescent in the water/oil heat exchanger and it flows only when the thermostat is open (and water is cooled by the radiator). This is the OEM arrangement of the engine tested, and, surely, a different cooling layout could be used in order to promote oil warming up. In figure 5, the behavior of the pressure is represented. The oil pressure in the engine is controlled by the relief valve and it does not overcome 4.5 bar. The oil pump pressure, therefore, is higher (till to 5 bar) due to the pressure drop occurred across HX and oil filter: this pressure drop is about 0.6 bar. In particular, when oil is colder, pressure is higher due to higher viscosity. In the cold phase of the drive cycle, the oil pressure when oil is heated is about 0.3 bar lower than the case of cold oil starting and this reduces also the oil pump work.

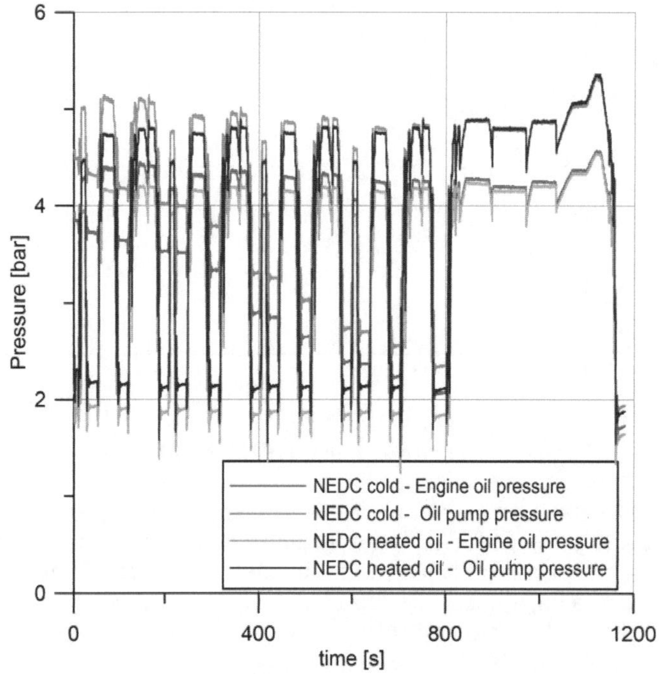

Figure 5: Oil pressure in the two tests performed

4. FUEL CONSUMPTION AND EMISSIONS RESULTS

The warm-up time reduction of about 150 s brings to a significant fuel consumption reduction and emissions savings. In Table 2 are summarized the results obtained. Fuel consumption is reduced of about 2-3% and similarly CO_2. Great reduction of unburned hydrocarbons (HC) and carbon monoxide (-4.2% and -3%) have been achieved. Also particulate matter (PM) is reduced of about 3.5% and NOx of about 7.6%.

Table 2: Fuel consumption and emissions results

	fuel	**CO_2**	**HC**	**CO**	**NOx**	**PM**
	L/100 km	*g/km*	*g/km*	*g/km*	*g/km*	*mg/km*
Cold engine	8.56	199.5	1.80	2.55	1.21	77.5
Heated oil	8.37	193.1	1.74	2.47	1.12	74.8
Delta	- 2.2%	-3.2%	-4.2%	-3.1%	-7.6%	-3.5%

Thanks to the lower friction when the oil is heated up, the same brake torque can be achieved with a lower fuel and, therefore, with higher air/fuel ratio (AF, Figure 6). At the very beginning of the NEDC, the AF ratio is about 20% higher in the heated oil case: this means a more complete combustion and justifies the reduction of HC, CO and PM.

In Figure 6, Air to Fuel (AF) ratio is represented, comparing the two situations with oil cold and hot, when engine is run from a cold state. The very low engine load justifies very high AF ratio, reaching mean values close to 80, with instantaneous peaks at 120-180. Figure 6 reports the differences between cold engine-cold oil with respect to cold engine-hot oil state. The circulation of hot oil produces a sensible friction reduction which requires a higher AF ratio, so a lower fuel consumption. The differences tend to decrease as the homologation cycle proceeds, bringing the engine to a thermal stabilized state.

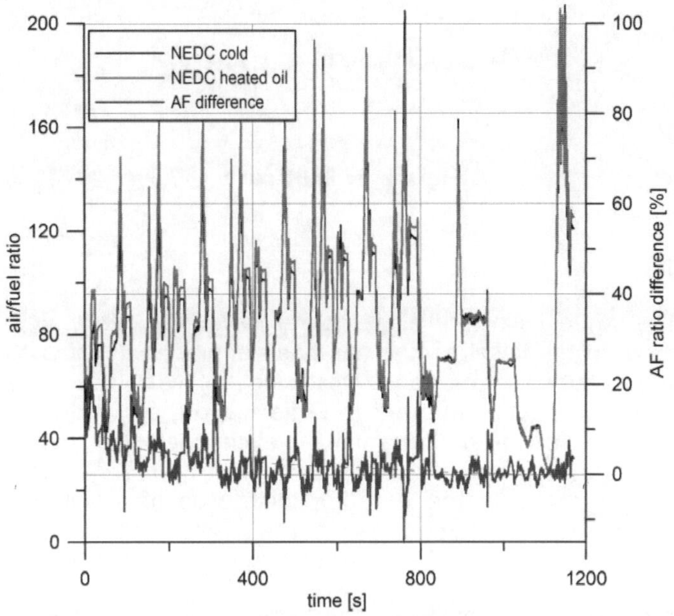

Figure 6: Air/fuel ratio

However, higher AF ratio, would bring higher NOx formation (or although the same formation rate, if the combustion temperature is not so high). The decrease in the NOx emissions resulted, therefore, can be related to the different EGR rate imposed by the ECU and measured thanks to the CO_2 difference between intake air and exhaust (eq. 1). EGR rate results from the following definition:

$$EGR_{\%} = \frac{x_{in}}{x_{exh}} \frac{[x_{exh} M_{CO_2} + (1 - x_{exh}) M_{air}]}{[x_{in} M_{CO_2} + (1 - x_{in}) M_{air}]} \qquad \text{(eq.1)}$$

where x is the concentration of the CO_2 (ppm) and M is the molecular weight of the specie. In fact, the lower fuel rate changes the operating point of the engine: the same instantaneous torque is achieved with lower fuel and ECU reacts modifying the EGR control valve opening degree. Figure 7 shows the EGR rate evaluated as in eq.1 from the CO_2 measured in the exhaust and in the intake manifold. It is clear, how in the case of heated oil EGR is always higher during warm-up time till to reach similar values in the extra urban part of the drive cycle. Higher EGR rate is responsible of the NOx reduction.

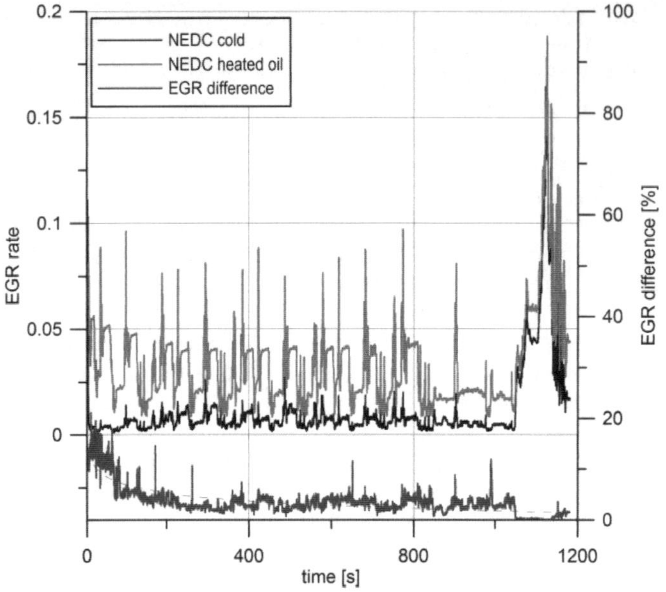

Figure 7: EGR rate

5. CONCLUSION

In this work, the influence of the oil temperature during NEDC was assessed. An IVECO F1C turbocharged diesel engine has been equipped in order to measure fuel consumption and pollutants emissions. Most relevant variables were measured on the oil and cooling fluid circuits in order to understand the two thermal dynamics. A slightly negative effect of the oil/coolant heat exchanger was demonstrated in the cold phase of the NEDC cycle: the temperature of the oil at the engine inlet is always lower than the one at the oil pump discharge of about 0.5-1 °C. In this specific engine, the heat exchanger is on the branch after the thermostat, so cooling fluid is flowing only when the radiator is crossed by the fluid itself.

In order to observe the effect of the oil temperature during engine warm up the oil was pre-heated and then continuously heated, with a 1500 W electrical pad, during a NEDC cycle. It is clear that this power should come from the engine itself: when exhaust gases are used, for instance, to heat up the lubricant oil, the thermal energy can be considered free, but backpressure related to the presence of a gas/oil heat exchanger on the engine exhaust line must be kept under control. The higher oil temperature in the early phase of the drive cycle produces lower engine friction losses and lower pressure drop across the lubricant circuit (up to 0.3 bar). This leads to a lower fuel consumption of about 2.2% and similar CO_2 emission saving (-6 gCO_2/km).

Moreover, a warm up time reduction of about 150 s have been demonstrated: higher temperatures brings also better combustion conditions with benefits in terms of HC (-4%), CO (-3%) and particulate (-3%) emissions. NOx emissions deserve particular attention: in fact, the new value of fuel consumption leads to different engine working points and a higher value of EGR rate which has the effect to a significant NOx reduction (-7%).

ACKNOWLEDGMENTS

This work has been realized in the framework of the Italian project STEVE "Thermal Systems for the efficiency of Light Commercial Vehicles".

REFERENCE LIST

(1) Regulation (EU) No 253/2014 of the European Parliament and of the Council of 26 February 2014 amending Regulation (EU) No 510/2011 to define the modalities for reaching the 2020 target to reduce CO 2 emissions from new light commercial.

(2) Regulation (Ec) No 715/2007 Of The European Parliament and of the Council of 20 June 2007 on type approval of motor vehicles with respect to emissions from light passenger and commercial vehicles (Euro 5 and Euro 6) and on access to vehicle repair and maintenance information.

(3) Liu, Z., Li, L., and Deng, B., "Cold Start Characteristics at Low Temperatures Based on the First Firing Cycle in an Lpg Engine," Energy Conversion and Management, Vol. 48, No. 2, pp. 395-404, 2007.

(4) Gumus, M., "Reducing Cold-Start Emission from Internal Combustion Engines by Means of Thermal Energy Storage System," Applied Thermal Engineering, Vol. 29, No. 4, pp. 652- 660, 2009.

(5) Daniels, C. C. and Braun, M. J., "The Friction Behavior of Individual Components of a Spark-Ignition Engine during Warm- Up," Tribology Transactions, Vol. 49, No. 2, pp. 166-173, 2006.

(6) Tronci, S., Baratti, R., and Gavriilidis, A., "Catalytic Converter Design for Minimisation of Cold-Start Emissions," Chemical Engineering Communications, Vol. 173, No. 1, pp. 53-77, 1999.

(7) Ahmadi, H., Rashidi, A., and Mohtasebi, S.-S., "Investigation of the Anti-Wear Properties of Nano Additives on Sliding Bearings of Internal Combustion engines," Int. J. Precis. Eng. Manuf., Vol. 14, No. 5, pp. 805-809, 2013.

(8) Taylor, O., Pearson, R., and Stone, R., "Reduction of CO2 Emissions through Lubricant Thermal Management During the Warm Up of Passenger Car Engines," SAE Technical Paper 2016-01-0892, 2016, doi:10.4271/2016-01-0892.

(9) Roberts, A., Brooks, R., and Shipway, P., "Internal Combustion Engine Cold-Start Efficiency: A Review of the Problem, Causes and Potential Solutions," Energy Conversion and Management, Vol. 82, pp. 327-350, 2014.

(10) Shayler, P., Allen, A., Leong, D., Pegg, I., Brown, A., et al., "Characterising Lubricating Oil Viscosity to Describe Effects on Engine Friction," SAE Technical Paper, DOI No. 10.4271/2007-01-1984, 2007.

(11) Datoo, A. N. and Fox, M. F., "An Initial Investigation of the Lubricant Condition in the Automotive Ring Zone under Cold Start Conditions," Tribology Series, Vol. 41, pp. 517-522, 2003.

(12) Becker, E. P., "Trends in Tribological Materials and Engine Technology," Tribology International, Vol. 37, No. 7, pp. 569-575, 2004.

(13) Yong, Z., Guohua, C., and Bing, L., "Two Dimensional Numerical Analysis of Piston Ring Lubrication of an Internal Combustion Engine," SAE Technical Paper, DOI No. 10.4271/1999-01-1222, 1999.

(14) Bolander, N. W. and Sadeghi, F., "Deterministic Modeling of Honed Cylinder Liner Friction," Tribology Transactions, Vol. 50, No. 2, pp. 248-256, 2007.

(15) Ali Usman, Cheol Woo Park - Transient Lubrication of Piston Compression Ring during Cold Start-Up of SI Engine - International Journal of Precision Engineering and Manufacturing-Green Technology January 2016, Volume 3, Issue 1, pp 81-90.

(16) Zammit, J., Shayler, P., Gardiner, R. and Pegg, I., "Investigating the Potential to Reduce Crankshaft Main Bearing Friction During Engine Warm-up by Raising Oil Feed Temperature," *SAE Int. J. Engines* 5(3):2012, doi: 10.4271/2012-01-1216.

(17) Shayler, P., Cheng, L., Li, Q., and Wahab, E., "A Modified Oil Lubrication System with Flow Control to Reduce Crankshaft Bearing Friction in a Litre 4 Cylinder Diesel Engine," SAE Technical Paper 2016-01-1045, 2016, doi:10.4271/2016-01-1045.

(18) Hagena, J. R., Filipi, Z. S., and Assanis, D. N. Transient diesel emissions: analysis of engine operation during a tip-in. SAE paper 2006-01-1151, 2006.
(19) Gullett, B. K., Touati, A., Oudejans, L., and Ryan, S. P. Real-time emissions characterisation of organic air toxic pollutants during steady-state and transient operation of a medium duty diesel engine. Atmos. Environ., 2006, 40, 4037–4047.
(20) Eastwood, P. G., Tufail, K., Winstanley, T., Darlington, A., Karagiorgis, S., Hardalupas, Y., and Taylor, A. M. K. P. Estimation of deviations in NO and soot emissions between steady-state and EUDC transient operation of a common-rail diesel engine. SAE paper 2009-24-0147, 2009.
(21) Rakopoulos, C. D., Dimaratos, A. M., and Giakoumis, E. G. Experimental study of transient nitric oxide, smoke and combustion noise emissions during acceleration of an automotive turbocharged diesel engine. Proc. IMechE, Part D: J. Automobile Engineering, 2011, 225(2), 260–279.
(22) Cipollone, R., Di Battista, D., Gualtieri, A., and Massimi, M., "Development of Thermal Modeling in Support of Engine Cooling Design," SAE Technical Paper 2013-24-0090, 2013, doi:10.4271/2013-24-0090.
(23) Osman, A., Muhammad Yusof, M., and Rafi, M., "Vehicle Testing and Development Involving a Simplified Split Cooling with Integrated Exhaust Heat Recovery and Reuse," SAE Technical Paper 2016-01-0647, 2016.
(24) Bent E., Shayler P. J., Rocca A. L. and Rouaud C., "The effectiveness of stop-start and thermal management measures to improve fuel economy," Proc. of IMechE Int Conf C1365, VTMS 11, C1365, paper C027, Coventry, 2013.
(25) R. Cipollone, D. Di Battista, A. Gualtieri, A novel engine cooling system with two circuits operating at different temperatures, Energy Conversion and Management, Volume 75, November 2013, Pages 581-592, ISSN 0196-8904.
(26) Rakopoulos, C. D., Dimaratos, A. M., Giakoumis, E. G., and Peckham, M. S. Experimental assessment of turbocharged diesel engine transient emissions during acceleration, load change and starting. SAE paper 2010-01-1287, 2010.
(27) C D Rakopoulos, A M Dimaratos, and E G Giakoumis - Investigation of turbocharged diesel engine operation, exhaust emissions, and combustion noise radiation during starting under cold, warm, and hot conditions - Proceedings of the Institution of Mechanical Engineers, Part D: Journal of Automobile Engineering September 2011 vol. 225 no. 9 1118-1133.
(28) Arcoumanis, C. and Yao, X. G. Transient smoke and unburnt hydrocarbon emissions during coldstart in a turbo-charged DI diesel engine. In Proceedings of the IMechE Seminar on Transient performance of engines, 1994, pp. 43–60, 1994 (Mechanical Engineering Publications Ltd, London).
(29) D. Di Battista, R. Cipollone, Experimental and numerical assessment of methods to reduce warm up time of engine lubricant oil, Applied Energy, Volume 162, 15 January 2016, Pages 570-580, ISSN 0306-2619.
(30) Roberto Cipollone, Davide Di Battista, Marco Mauriello, Effects of Oil Warm up Acceleration on the Fuel Consumption of Reciprocating Internal Combustion Engines, Energy Procedia, Volume 82, 2015, Pages 1-8, ISSN 1876-6102.
(31) D. Di Battista, M. Mauriello, R. Cipollone, Waste heat recovery of an ORC-based power unit in a turbocharged diesel engine propelling a light duty vehicle, Applied Energy, Volume 152, 15 August 2015, Pages 109-120, ISSN 0306-2619, http://dx.doi.org/10.1016/j.apenergy.2015.04.088.
(32) Di Battista D, Mauriello M, Cipollone R. Effects of an ORC based heat recovery system on the performances of a diesel engine. SAE technical paper 2015-01-1608; 2015 http://dx.doi.org/10.4271/2015-01-1608.

HEATING AND AIR COOLING

Predicting thermal comfort and draught discomfort for an automobile cabin cooldown

M Hepokoski[1], S Peck[1], C Gingras[1], A Curran[1], S Gullman[2], D Jacobsson[2]
[1] ThermoAnalytics, USA
[2] Volvo Car Corporation, Sweden

ABSTRACT

Discomfort due to draught is a significant concern in the design of automotive environments since climate control systems typically rely on the movement of air to modulate the ambient temperature within the passenger cabin. In this study, the percentage dissatisfied (PD) due to draught (Fanger et al. 1989) was used to evaluate an automobile cabin environment. Draught discomfort was predicted as a function of the measurements obtained from a test manikin. The comfort predictions from the draught discomfort model and those of a thermal comfort model were used to evaluate a vehicle cabin undergoing a transient cooldown.

1. INTRODUCTION

Human thermal models have often been used to predict the comfort of occupants within vehicle passenger compartments (1). Thermal sensation is affected by cold and warm thermoreceptor discharge rates, which, in turn, are correlated with human body temperature. Therefore, the most accurate thermal comfort prediction techniques attempt to fully characterize the heat transferred between the human and its environment, requiring the use of a human thermophysiology model (2, 3, 4). Most thermophysiological-based comfort models, however, do not consider the influence of air movement on passenger discomfort, which is correlated with the response of subcutaneous mechanoreceptors rather than thermoreceptors. Discomfort due to draught is a significant concern in the design of automotive environments since climate control systems typically rely on the movement of air to modulate the ambient temperature within the passenger cabin.

2. METHODS

In this study, a thermal comfort model (5) and a draught discomfort model (6) were used to assess a vehicle passenger environment by relying on measurements obtained from a test manikin. Draught discomfort was predicted as a function of the air velocity, turbulence intensity and temperature, all of which were measured across a representative human form using a passive sensor (HVAC) manikin (7). The HVAC manikin also had sensors for measuring thermal radiation and relative humidity, which were input to the human thermal model, along with the measured air temperature and velocity, to predict thermal comfort. The comfort predictions from both the thermal comfort and the draught discomfort models were used to evaluate a vehicle cabin environment undergoing a transient cooldown. Model predictions were compared to the thermal comfort and draught discomfort reported by a human subject who was exposed to the same thermal environment.

2.1 Test Protocol

A test protocol was devised to assess the thermal environment within a vehicle experiencing a transient cool down scenario at Volvo Car Corporation (Gothenburg, Sweden). This was accomplished by exposing a midsize Volvo vehicle to an overnight 40°C ambient environment (i.e., a hot soak) within a climate controlled wind tunnel. The test began when the human subject entered the vehicle, started the engine, and activated its climate control system. The wind tunnel was not equipped with solar lamps and the ambient conditions remained at the soak conditions during the cooldown period. In order to perform multiple repetitions of the test within a single day, subsequent soaking of the vehicle was limited to a period of 2 to 3 hours. The transient cool down portion of the test was limited to 60 minutes.

Two repetitions were performed, one involving an experienced human test subject and a second involving a test manikin. Both the human and the manikin were positioned in the driver's seat during their respective exposures. Overall sensation and comfort were reported at regular intervals, which were determined by prompting from the test operator. Additionally, the subject was directed to report draught discomfort when experienced, as well as the location on the body where draught was occurring. Prior to the first repetition, the subject examined the vehicle and adjusted the seat and vent positions, after which these were to remain fixed for the human subject test and that of the second repetition involving the manikin.

Fifteen minutes prior to the start of the test, the subject entered the chamber and sat in position with the door open while the test operator connected the thermocouples on his skin to the data acquisition system. At test start, the door was closed, the vehicle was engaged, and the climate control system was set to full automatic control with a setpoint of 22 °C. Once a steady-state environment within the cabin was expected to be attained (after 42 minutes), the climate control set point was decreased to 18 °C. This cooler exposure was to last 10 minutes, after which the set point was increased to 24 °C and remain at this setting until the completion of the test (60 minutes from start).

The test subject, who was familiar with vehicle thermal comfort assessment, was similar in size to the HVAC Manikin, which was designed to correspond to that of a 50th percentile male. Anthropometric measurements were not obtained as part of this study. The human subject was outfitted with light summer clothing (0.34 CLO), and was in a seated, though slightly reclined position, for the duration of the test. This was expected to correspond to an activity level of ~0.9 MET.

2.2 Test Manikin and Thermal Comfort Models

The test manikin used in this study was an HVAC Manikin (Thermetrics, LLC). The TAITherm® Human Thermal Model (8) was used to simulate thermo-physiological response and predict thermal comfort based on the measurements obtained from the HVAC Manikin. The physical manikin and the virtual manikin, as well as the coupling process methodology, have been previously described in detail (9).

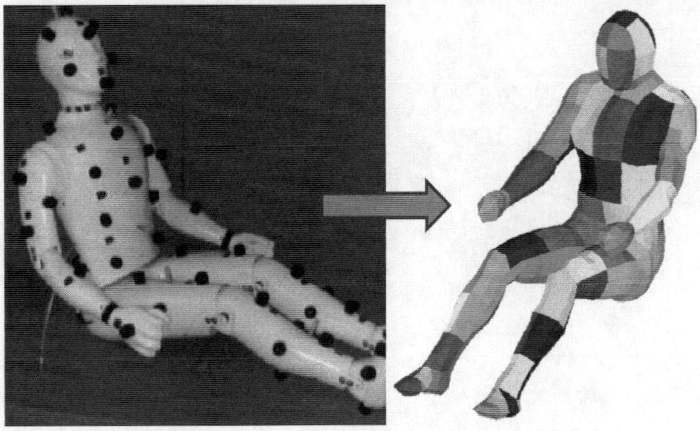

Figure 1. The human thermal model's parts map to zones on the manikin.

2.2.1 Draught Discomfort

Draught is an undesired local cooling of the human body caused by air movement, often experienced by passengers in automobiles, trains, and aircraft. Fanger et al. developed a relationship between the percentage of the population experiencing draught discomfort and environmental conditions, defined in terms of mean air velocity, air temperature, and turbulence intensity (10). The output of the model is expressed in terms of predicted percentage dissatisfied (PD):

$$PD = (34 - t_a)(V - 0.05)^{0.62}(0.37T_u + 3.14)$$

In this formulation, t_a, is the air temperature (°C), V is the air velocity (m/s), and turbulence intensity, T_u, is defined as

$$T_u = 100\frac{V_{sd}}{V}$$

where V_{sd} is the standard deviation of the velocity.

The intended application of the draught discomfort model was for evaluating near neutral environments, given that the ranges of air temperatures and air velocities considered were quite narrow, i.e., $20 < \boldsymbol{t_a} < 26$ °C, $0.05 < \boldsymbol{V} < 0.5$ m/s. The valid range for turbulence intensity is $0 < T_u < 70\%$.

Although Fanger et al. derived their correlation based on experiments that focused the air flow over the back of the neck, in the vehicle study, the measurements from individual sensors on the HVAC Manikin were combined so that draught discomfort could be evaluated over other regions of interest. For example, Figure 2 illustrates how the sub-regions associated with individual sensors can be combined into larger regions, in this example, the face, chest and lower body.

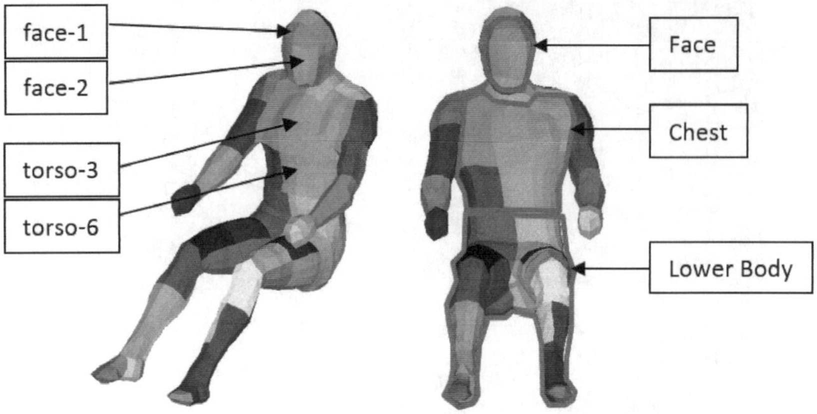

Figure 2. Grouping of manikin body segments for draught discomfort evaluation.

Since the environment within the vehicle was being actively controlled by a climate control system, the air velocity was not constant as it was during the experiments performed during development of the PD model. Further, the air velocity measurements were quite noisy, mainly due to turbulence. A numerical filter was consequently needed to obtain an approximation of the instantaneous mean, which was used as a direct input to the PD and for calculating the turbulence intensity. The instantaneous mean or filtered velocity was calculated from

$$V_{mean_i} = \frac{1}{N_{window}} \sum_{k=-N_{window}/2}^{N_{window}/2} V_{meas_{i+k}}$$

N_{window} is the size of the sliding window used for the numerical filter. $V_{meas_{i+k}}$ is the instantaneous measured velocity.

T_u can vary depending on how the mean velocity and standard deviation of the velocity are defined. In this study, three methods for approximating T_u were considered, the derivation of each involving increasing complexity.

Method 1:
A simple, single standard deviation of the air velocity is calculated over the entire dataset, which is then used to calculate a single T_u that can be used to calculate PD for every instantaneous air temperature and air velocity pair. Rather than calculating the standard deviation from the instantaneous velocity and the mean velocity over the entire data set, Method 1 relies on the instantaneous measured raw velocity and the filtered instantaneous velocity, which is used as a proxy for the mean, as follows:

$$V_{sd} = \sqrt{\frac{1}{N_{sample}} \sum_{i=1}^{N_{sample}} \left(V_{meas_i} - V_{mean_i}\right)^2}$$

In this case, V_{meas_i} is the instantaneous, raw, velocity and V_{mean_i} is the instantaneous, filtered velocity. This method uses N_{sample} equal to the total number of data samples (i.e., 720) over the 60 minute period. The turbulence intensity $\overline{Tu}$ is calculated as a mean over the entire sample size:

$$\overline{Tu} = \frac{1}{N_{sample}} \sum_i \frac{V_{sd}}{V_{mean_i}}$$

Then

$$PD_i = (34 - T_{a_i})(V_{mean_i} - 0.05)^{0.62}(0.37V_{mean_i}\overline{Tu} + 3.14)$$

Method 2:
This method uses the same, constant V_{sd} (calculated with N equal to 720) as Method 1, but uses an instantaneous Tu instead of an average value, calculated from

$$T_u = \frac{V_{sd}}{V_{mean_i}}$$

Now

$$PD_i = (34 - T_{a_i})(V_{mean_i} - 0.05)^{0.62}(0.37V_{mean_i}Tu + 3.14)$$

Neither Method 1 nor Method 2 take into account that the raw input velocity may fluctuate at different intensities during different phases of the test. In other words, the differences between the instantaneous and the filtered velocities, which are only considered in the calculation of standard deviation, are effectively spread out over the entire data set.

Method 3:
To allow for changes in V_{sd} to impact PD during the test, the data set must be subdivided, or segmented, so that the standard deviation of air velocity can be recomputed over each period of interest. For example, a period of 5 minutes at a data acquisition frequency of 5 seconds, results in a window size of 60 data points.

Thus

$$V_{sd_i} = \sqrt{\frac{1}{N_{window}} \sum_{k=-N_{window}/2}^{N_{window}/2} (V_{mean_{i+k}} - V_{meas_{i+k}})^2}$$

The calculation for T_u now uses the instantaneous standard deviation instead of the averaged value:

$$Tu_i = \frac{V_{sd_i}}{V_{mean_i}}$$

The expression for PD is now

$$PD_i = (34 - T_{a_i})(V_{mean_i} - 0.05)^{0.62}(0.37V_{mean_i}Tu_i + 3.14)$$

For this study, Method 3 was used with a segmentation period of 65 seconds (N=13). It is important to note that one of the disadvantages of this approach is that although it is intended to improve the accuracy of the instantaneous PD calculation, the accuracy of the standard deviation that is computed within each segment decreases (due to decreasing numbers of samples) as the segmentation increases.

3. RESULTS AND DISCUSSION

3.1 Preliminary Thermal Comfort Results

The thermal comfort ratings obtained from the human subject were compared to the manikin-model predictions. Thermal sensation and comfort were reported on 9-point scales consistent with those used by the Berkeley Comfort Model. The thermal sensation scale ranges from +4 (very hot) to -4 (very cold) with 0 corresponding to a neutral thermal sensation. The thermal comfort scale defines positive values to indicate comfort and negative values to indicate discomfort. A value of +4 corresponds to "very comfortable," +0 corresponds to "just comfortable," -0 corresponds to "just uncomfortable," and a value of -4 corresponds to "very uncomfortable."

Overall thermal sensation results are provided in Figure 3 and overall thermal comfort results are provided in Figure 4. The results show that both the human and the manikin-model report strong discomfort at the start of the test due to hot sensation, however, the sensation rapidly decreases, and the comfort rapidly increases as soon as the climate control system is activated.

The sensation predictions are often within 1 unit of the reported sensations for the remainder of the test, including when the climate control setpoint is suddenly decreased at the 42 minute mark, when a slightly cool to cool sensation is reported by the human subject. Both the human and the manikin report a large increase in sensation when the climate control setpoint is increased for the final stage of the test.

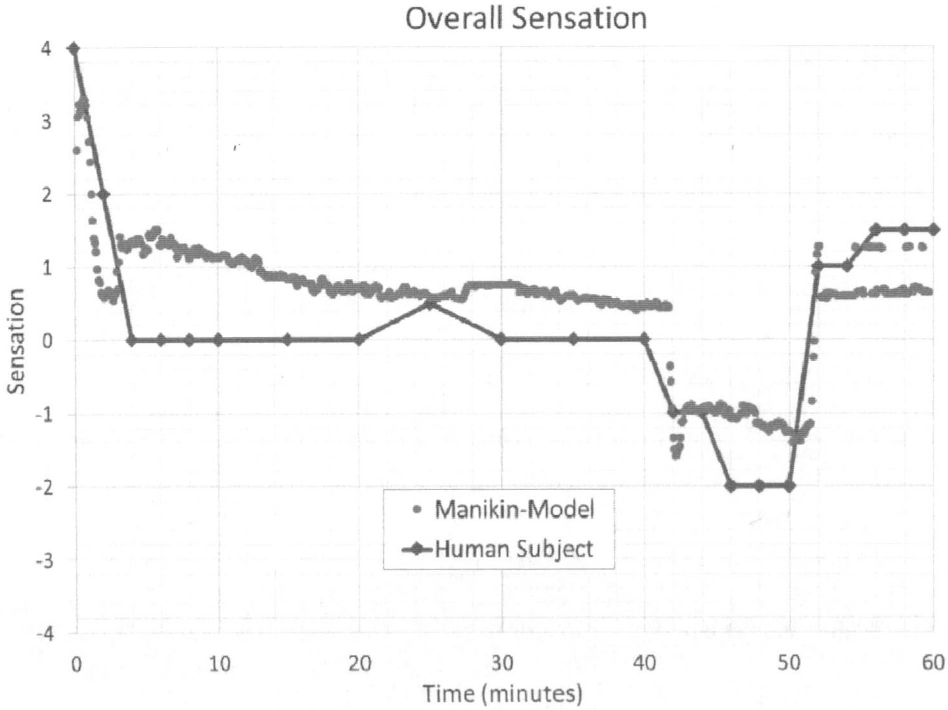

Figure 3. Overall thermal sensation as predicted by the manikin-model combination compared to human subject reported values.

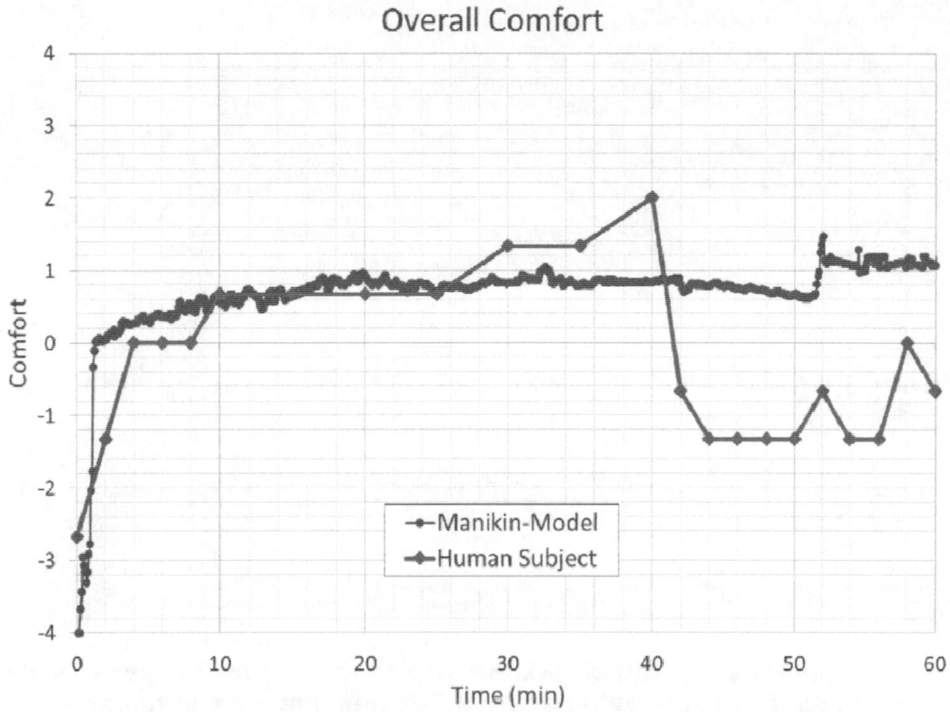

Figure 4. Overall thermal comfort as predicted by the manikin-model combination compared to human subject reported values.

The comfort predictions are also within one unit for the first half of the test. It is interesting to note that a sharp decrease in thermal sensation at the 42 minute mark does not impact the predicted comfort, despite the human subject reporting a significant drop in overall comfort. This is the same point in time at which the human subject reports that his most uncomfortable body segment (chest/left arm) is experiencing draught discomfort. The subject continues to report draught discomfort in the chest/left arm until the set point is increased again for the final phase. The relative increase in comfort during this period is also predicted by the manikin-model system.

3.2 Draught Discomfort Results

The HVAC Manikin measurements from the upper body were supplied to the PD draught discomfort model. The results are provided in Figure 5. The plot of PD is colored according to when the model inputs are within their expected ranges. Specifically, the plot is colored in red when the air temperature is far outside the expected range (t_a>30 °C), yellow when the air temperature is just outside the range (26 °C < t_a < 30 °C), and green when the air temperature is within its expected range.

The graph shows that the period over which maximum draught discomfort occurs is during the lowest air temperature set point and the highest air velocity. This is also the phase in which the human subject reported maximum draught discomfort. The input temperature is within the valid range for the model during this entire period.

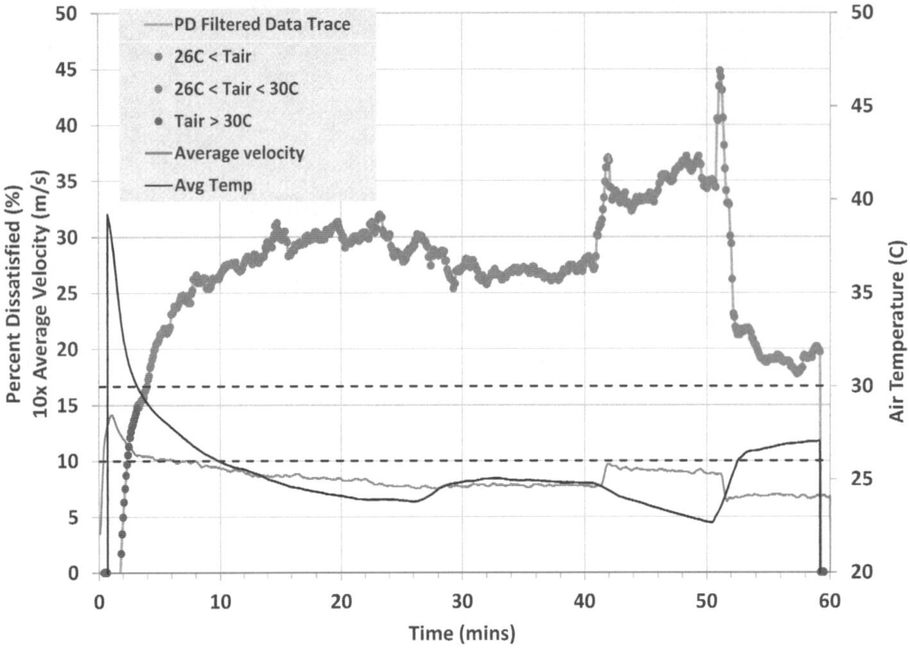

Figure 5. Predicted draught discomfort (PD Method 3) for the upper body evaluated from data obtained from test manikin measurements.

The effect of grouping the measurements from different sensors so that local draught discomfort can be evaluated on a regional basis is illustrated in Figure 6. When comparing the draught discomfort for the head, left arm and upper body, it is evident that the most uncomfortable segment due to draught is the left arm, while the least amount of draught is experienced on the head. This regional variation in draft discomfort was likely a result of the positioning of the vents, which the human subject was not permitted to adjust during transient cool down period.

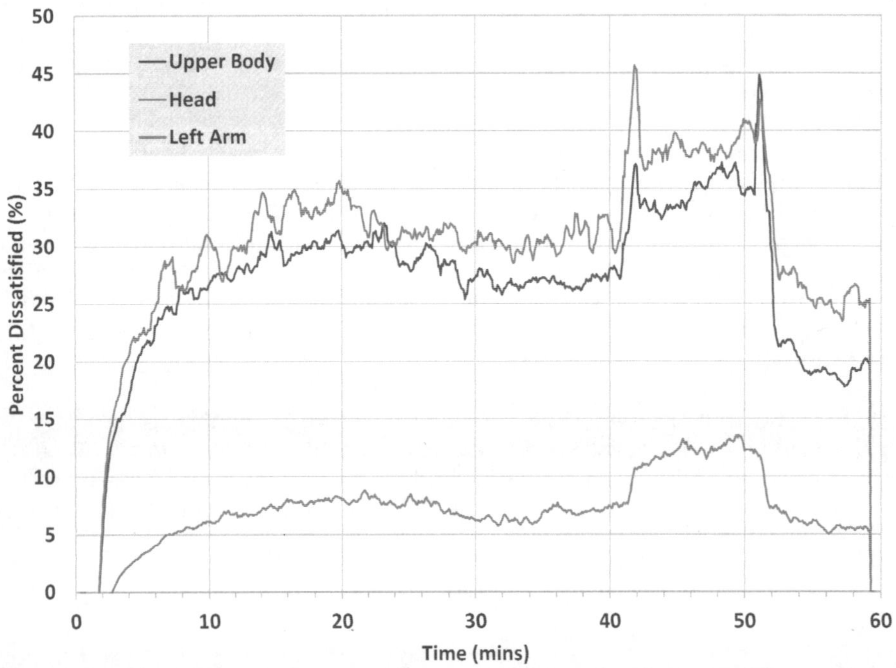

Figure 6. Predicted draught discomfort (PD Method 3) for different regions on the body.

The PD exhibits some spikiness throughout the test. While it was expected that the segmentation technique employed by Method 3 would be capable of bringing out the effect of the temporal variation in turbulence intensity, it was speculated that the large spikes at both the start and end of the lowest air temperature setpoint period may be simply an artifact of the data processing technique that was used. Specifically, that a discontinuity in air velocity may not be represented properly in its running average and may have biased the calculation V_{sd}, which would have consequently led to errors in the calculation of turbulence intensity.

A comparison of draught discomfort when evaluated using the different methods for calculating turbulence intensity is provided in Figure 7. As expected, the spikes are not present when PD is evaluated using Methods 1 and 2. Given that the difference between the PD when evaluated using Methods 1 and 2 are negligible, and the use of Method 3 only improves the prediction by a few percent, it may not be worth the extra effort to extensively process the measurement data, especially in light of the potential for errors that can be introduced when a discontinuity in air velocity is present.

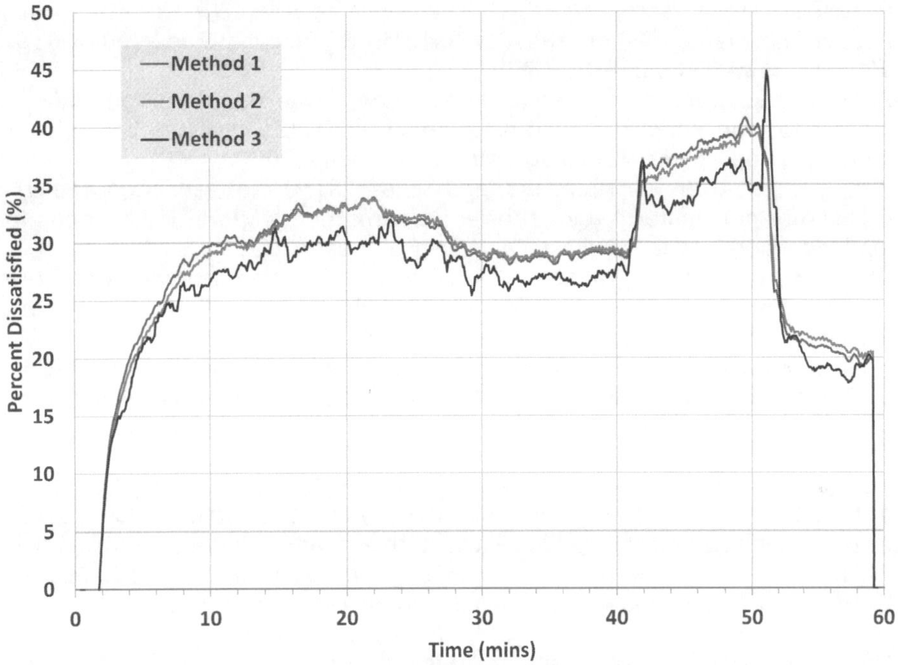

Figure 7. Predicted draught discomfort for the upper body evaluated using different methods for calculating turbulence intensity (Tu).

4. CONCLUSION

This paper demonstrates how a passive sensor (HVAC) manikin can be used in tandem with a thermal comfort model and a draught discomfort model to accurately predict human subject perception within a vehicle cabin environment. The HVAC Manikin was well suited for draught discomfort analysis given the large number of air velocity and air temperature sensors positioned across its surface, which facilitated the evaluation of draught discomfort over various regions of the body.

While the thermal comfort model used in this study was able to accurately predict the sensation of a human subject experiencing an equivalent environment, its comfort predictions were only accurate in so far as draught was not occurring.

The PD model has been successfully shown to predict changes in draught discomfort in a vehicle passenger compartment, however, it is limited in its applicability to automotive scenarios given its relatively narrow range of valid inputs, which may be an indication that the PD model may be better suited for architectural settings.

This study emphasizes the need to consider draught discomfort, and not just thermal comfort, when evaluating a vehicle cabin environment in terms of passenger comfort.

REFERENCE LIST

1. Hepokoski, M., Curran, A., Burke, R., Rugh, J. et al., "Simulating Physiological Response with a Passive Sensor Manikin and an Adaptive Thermal Manikin to Predict Thermal Sensation and Comfort," SAE Technical Paper 2015-01-0329, 2015, doi:10.4271/2015-01-0329.
2. Zhang H., E. Arens, C. Huizenga, and T. Han, "Thermal sensation and comfort models for non-uniform and transient environments: Part I: Local sensation of individual body parts." Building and Environment, 2009.
3. Zhang H., E. Arens, C. Huizenga, and T. Han. "Thermal sensation and comfort models for non-uniform and transient environments: Part II: Local comfort of individual body parts." Building and Environment, 2009.
4. Zhang H., E. Arens, C. Huizenga, and T. Han. "Thermal sensation and comfort models for non-uniform and transient environments: Part III: Whole-body sensation and comfort." Building and Environment, 2009.
5. Hepokoski, M., Gibbs, S. and Curran, A., "A new anatomical and thermophysiological description of a 50th percentile adult western male," Digital Human Modeling (DHM2016) Symposium, Montreal (2016).
6. ASHRAE (2005) Fundamentals handbook. American Society of Heating, Refrigeration and Air-Conditioning Engineers, Atlanta, pp 8.1-8.29.
7. Automotive HVAC Manikin System (2016). Retrieved October 12, 2016, from http://thermetrics.com/products/automotive-hvac-manikin-system
8. TAITherm Human Thermal Model (2016). Retrieved October 12, 2016, from http://www.thermoanalytics.com/pdf/human-thermal-flyer
9. Hepokoski, M., Curran, A., Burke, R., Rugh, J., Chaney, L., Maranville, C. 2014 Use of a Passive Sensor Manikin Coupled with a Thermoregulation Model to Predict Human Thermal Response to Environmental Conditions. Proceedings of the 10th International Meeting for Manikins and Modeling, Tampere, Finland.
10. Fanger, P.O., Melikov, A.K., Hanzawa, H., and Ring, J. 1989. Turbulence and draft. ASHRAE Journal 31(4):18-25.

Active charge air cooling

C D Whelan
Air Cycle Technology Ltd, UK
J M Allport
University of Huddersfield, UK

ABSTRACT

Boosted IC engine development, both diesel and gasoline, is driving the need for better charge air temperature control. Charge air temperature directly and indirectly influences engine performance, fuel economy & emissions. To date charge air temperature control has, in the main, been passive; aiming only to reduce temperatures towards, but still above, ambient temperature. Driven by increasing specific output, the opportunities for sub-ambient charge cooling need to be investigated. For some real world driving conditions it is also desirable to raise the charge temperature.

Air Cycle Technology ltd (ACT), working with the University of Huddersfield (UoH) are developing a controllable, or active, charge air cooling system (ACC), which will deliver charge air at sub-ambient temperature over a wide engine operating range. The system uses a turbo-expander to deliver air-cycle cooling in a simple and practical way. The system can be enhanced by combination with a charge air heating process. Accurate charge temperature control over as wide an engine operating range as possible is also addressed. The system is fully compatible with the operational, packaging and cost needs of vehicle OEMs.

The paper presents the principle, process and application of a controllable charge air cooling/heating system. It covers the design and operation of the turbo-expander which delivers sub-ambient charge temperatures, the option of charge air heating and possible control methods. Examples of specific engine applications are presented to clearly indicate the opportunities for improvements in engine performance and fuel economy.

INTRODUCTION

The use of charge air cooling using turbo expansion techniques was first developed in the 1950s (1) and one of its first applications was during the 1960s in prevention of detonation in natural gas engines caused by the variability in fuel properties (2). Over the 50 years since then, the principle has been revisited several times, for racing applications (3, 4) as well as road vehicle powertrains where a study (5) showed that the application of charge cooling for emissions improvement on medium and heavy-duty diesel as well as gasoline engines was feasible.

Current legislation demanding improved fuel economy and emissions has reinvigorated interest in the technology, and more recent theoretical studies demonstrated that there were significant benefits to be gained when applying charge cooling to downsized boosted gasoline engines. It was shown that a charge air temperature reduction of over 30 degrees C could be achieved without fuel penalty (6).

Further work (7) delivered a working prototype turbo-expander, based on production turbocharger design and manufacturing technology. The design was validated by rig testing and the target performance and efficiency was achieved. The prototype has been installed on a demonstrator vehicle and is continuing to amass durability data.

Although cooling charge air as a principle has been demonstrated to have advantages, there are also associated disadvantages. Overcooling of the air can result in ice formation in the intake, with potential for catastrophic effects if this is ingested into the engine. A more optimal solution would be to provide air at a constant temperature to the engine intake, and that is the objective of this work.

PRINCIPLES OF CONTROLLABLE CHARGE AIR COOLING AND HEATING

The operating principle of turbo-expansion is well known; the flow path and main components are shown in Figure 1. The sequence of 'events' for the charge air is:

- Compression by the compressor of the main turbocharger
- Cooling by the first charge air cooler (CAC1)
- Re-compression in the compressor of the turbo-expander
- Cooling by the second charge air cooler (CAC2)
- Expansion through the turbine of the turbo-expander

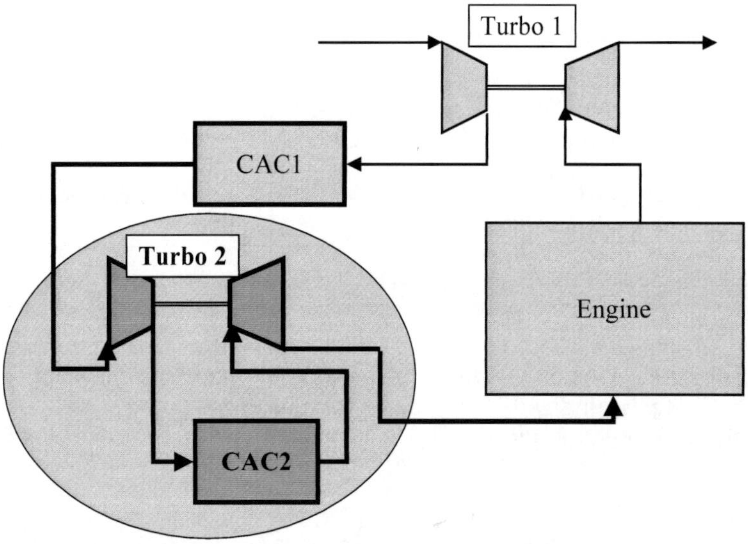

Figure 1: Turbo-Cooling concept Layout and Gas Flowpath

Air expansion through the turbo-expander turbine provides the power to drive its compressor. The overall loss of enthalpy through this second stage of compression and expansion is compensated for by increased work in the first compressor. The net effect on the engine is increased exhaust pressure into the main turbine.

Therefore, for a given charge air flow and density, which generally defines the indicated engine power, the intake pressure and temperature can be lowered at the expense of high primary boost pressure and increased engine pressure gradient (and pumping loss). The benefits of turbo-expansion result from the various ways of exploiting reduced intake pressure and temperature.

Any turbocharged engine can benefit from turbo cooling, although the range of potential benefits varies with both the combustion type and the application. The key principle is to combine increased intake charge density coupled with reduced charge air temperature. Using conventional boosting systems, these tend to be contradictory, with air temperature rising with increased boost pressure. It is also not possible to reduce the charge temperature to below that of the ambient air using either air to air or air to water heat exchangers. Reducing the charge air temperature offers benefits relating to output, detonation, thermal loading and emissions, whilst offering additional control over combustion rates and cylinder pressures.

It has been shown previously (7) that it is possible to supply charge air to the engine at over 30 deg C below the temperature that the system without the turbo expander would achieve. However, from a combustion control perspective, a more consistent inlet temperature would be preferable.

PROCESS

Given that the turbo expander unit can provide a supply of cool air, it is then possible to control the temperature of the intake air by a variety of means. The cool air can be mixed with warmer air from the turbocharger compressor, or could be warmed directly using heat from elsewhere as shown in figure 2.

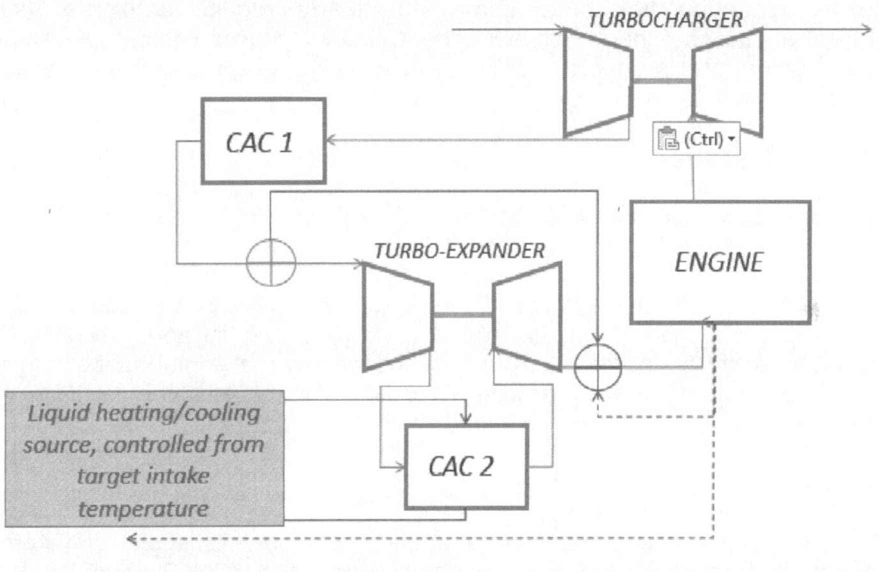

Figure 2: Temperature controlled concept Layout and Gas Flowpath

In this way, the temperature of the intake air can be managed to provide optimum conditions for combustion across a range of operating conditions. For example, the air could be heated under cold startup conditions to decrease engine warmup time, hence reducing emissions on startup. Conversely, when running under high load conditions, cooled air can be used to reduce combustion temperatures and prevent knock as an alternative to overfuelling, therefore reducing fuel consumption as well as emissions.

Requirement and Application

The opportunity for air heating is available in conventional charge cooling systems, either by charge air cooler (CAC) by-pass at high load or engine coolant heating transfer within the CAC to raise temperatures. These techniques are coming into use, driven by the need to minimise condensation in intake systems resulting from high EGR levels, particularly with low pressure loop systems.

The significant advance presented here is the application of charge air 'refrigeration' using air-cycle cooling, and the study concentrates on this aspect of ACC. The dominant technical driver is the combustion limitation on the increase of specific output in increasingly highly boosted gasoline engines. As specific outputs rise, detonation becomes the major limitation. The predominant mitigation techniques are ignition retardation combined with excess fuel input, and ultimately compression ratio reduction. Ignition retardation and excess fuelling are inseparable; the retardation reduces detonation but reduces combustion efficiency and increases exhaust temperature and thermal loads on critical components. Fuel is added, above the combustion requirement, to limit the exhaust temperature increase. This technique is acceptable in full load operation outside normal driving, but as engine capacities reduce, vehicles are spending greater proportions of real-world drive cycles at high load, with consequent significant increase in total fuel consumption. Similarly compression ratio reduction to limit detonation impacts on fuel economy at all load-speed conditions.

Charge air reduction has been shown to clearly reduce detonation so the opportunity to cool the charge air below the 'passive' charge cooling limits offers a significant potential to enable increases in specific outputs and minimise fuel consumption penalties (8).

System and Operation

The system is based on the schematic shown in Figure 2. The core element is the turbo-expander which delivers the charge air temperature drop by expansion. The efficiency of the turbine and, to a lesser extent, the compressor is critical to the overall performance of the system. There is an energy loss through the cooling process, which reduces charge density at exit from the turbo-expander; this is compensated for by an increase in primary boost from the engine turbocharger, so component efficiency becomes critical. The ACC turbine efficiency is 85% at the design point and above 80% over a wide operating range.

Modelling

The AMESim 1D modelling platform is used extensively to support system and components performance prediction. At the component level ACT have developed a 1D turbine design and performance prediction 'module' running within AMESim. This has been validated in previous studies and supports integration of 'dynamic' turbine performance within the ACC system. The ACC system has been modelled both to correlate with rig testing (9) and then to predict the potential for charge temperature control for specific engine applications.

Figure 3 shows a 1D engine model incorporating the charge air refrigeration components of ACC. The turbo-expander and its associated charge cooler are inserted between the main CAC and the engine intake manifold. The ACC application is very simple; the entire engine air charge passes through the turbo-expander and its cooler, resulting in uncontrolled charge air temperature reduction.

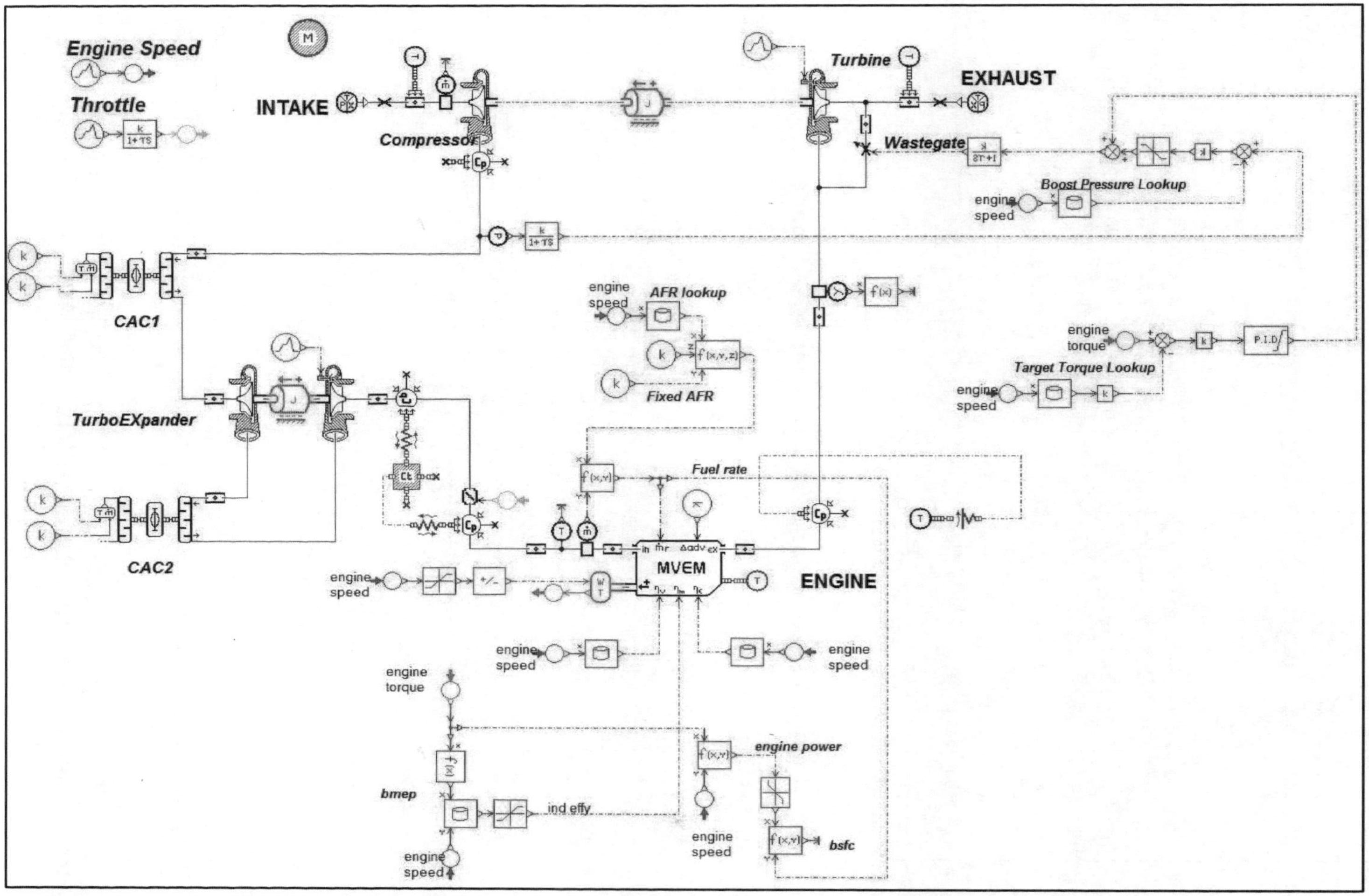

Figure 3: 1D system model with uncontrolled ACC

The engine chosen for the modelling example is a 1.6 litre 4 cylinder gasoline, normally rated at 240 Nm and 140 kW. To emphasise the potential of ACC the BMEP has been increased over the speed range to achieve 330 Nm and 158 kW maximum torque and power. This is the practical limit using the standard turbocharger. The enhanced torque curve is shown in Figure 4.

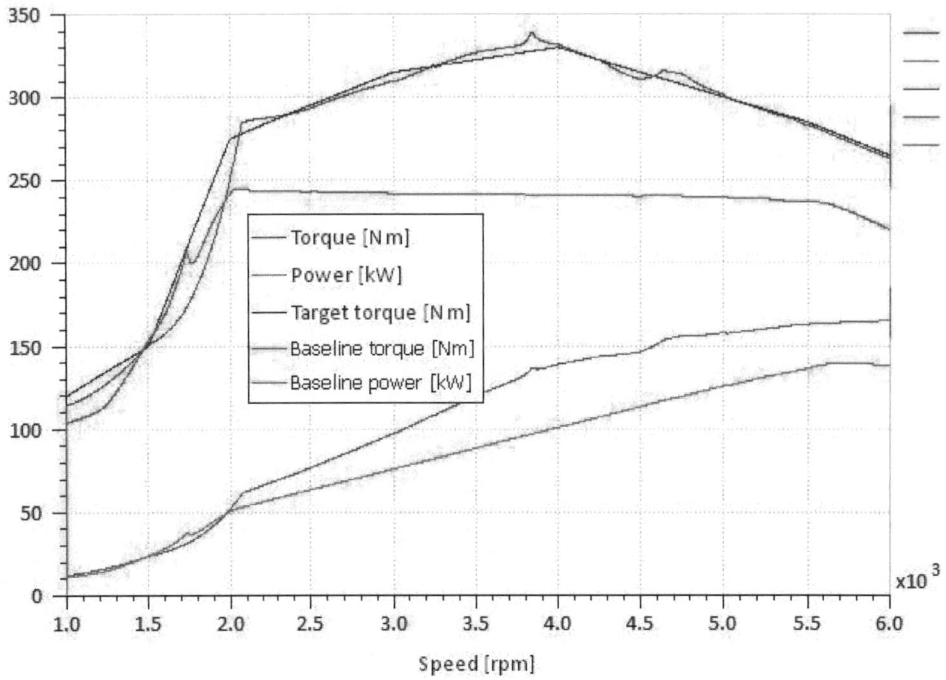

Figure 4: Performance at WOT over speed range

Figure 5 shows the key charge air temperature over the speed range; namely the temperature out of the primary charge air cooler (CAC1) and in the intake manifold, at exit from the turbo-expander turbine.

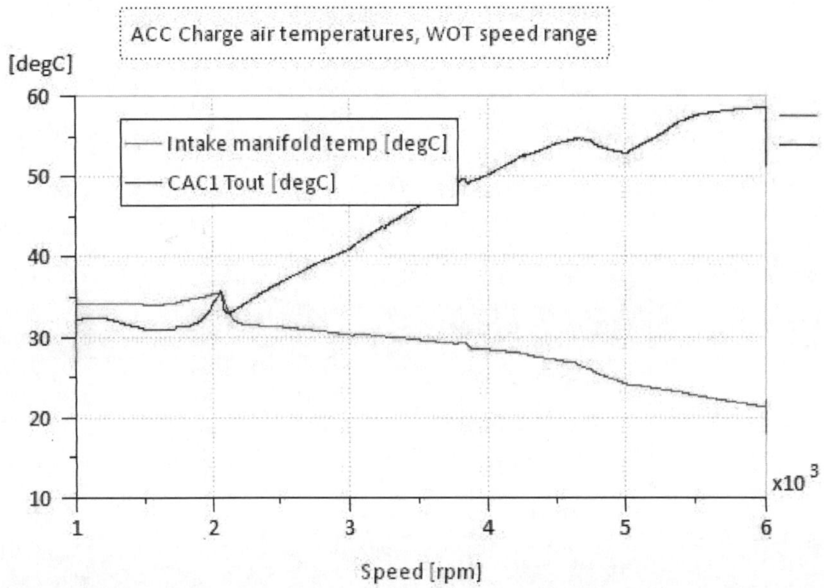

Figure 5: Charge air temperatures with baseline ACC, WOT speed range

Significant air temperature reductions are achievable; the manifold temperature at maximum engine speeds being approximately 35 degC. lower than at the main charge cooler exit. As mentioned above, the temperature reduction is uncontrolled, and increases with engine speed; at low engine (and turbo-expander) speed the mass flowrate though the turbo-expander is insufficient to generate useful pressure ratios, so no temperature reduction takes place.

A practical ACC system clearly requires a degree of air temperature control in order to optimise the engine performance in response to varying ambient and boost-related air temperatures. In the simplest form this can be achieved by diverting air from the main CAC around the turbo-expander and mixing it with the turbo-expander exit air. This can use a control system, discussed below, which aims to deliver air a target temperature at the intake manifold.

Figure 6 shows a local section of the 1D system model with a simple by-pass loop added.

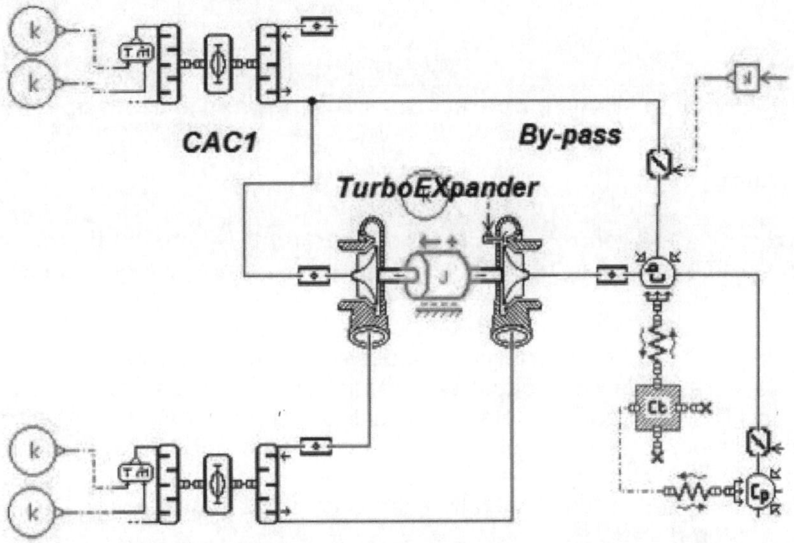

Figure 6: ACC by-pass flow path

The impact on air temperature of the by-pass system is shown in fig. 7.

In the model the target intake manifold temperature (Target Tman) has been set at 30 degC. Over the engine speed range above 2000 rpm, the by-pass control system modulates the air flow to maintain the target temperature, producing up to a 33 degC temperature reduction relative to the exit temperature the main charge air cooler (CAC1). The exit air temperature from the turbo-expander falls to approximately 15 degC. Below 2000 rpm, there is insufficient air flow for the turbo-expander to provide effective cooling. This is not considered to be a major issue in the context of real-world driving. ACT are working on more sophisticated controls systems which will extend the envelope of effective temperature control.

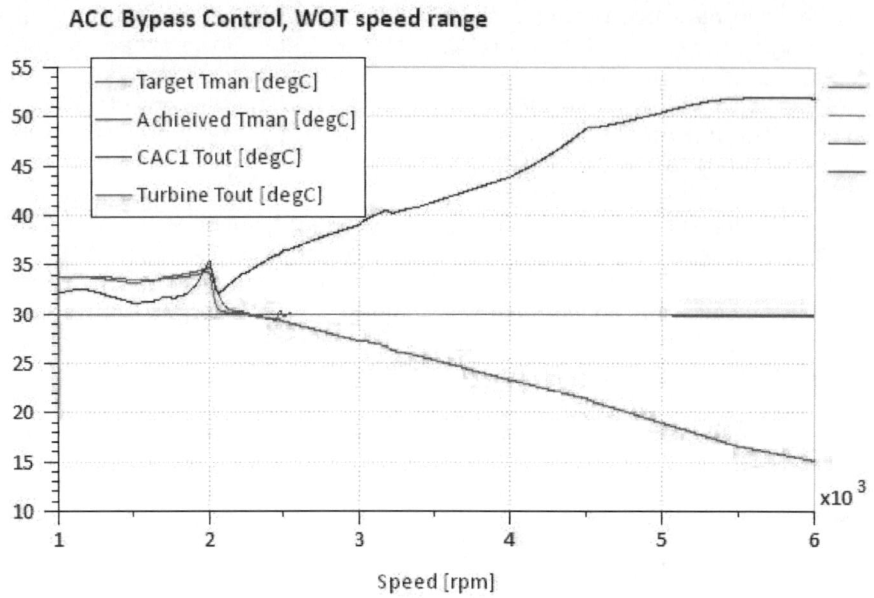

Figure 7: Key system temperatures with by-pass system operating

Control Strategy

The objective is to control manifold temperature when both the ambient temperature and the boost related temperature rise are varying under driving conditions.

At a given driving condition there will be a desired manifold temperature; examples being to reduce warm-up time, optimise combustion or minimise mechanical friction or emissions. The desired temperature could be higher than the 'available' un-cooled boost temperature or lower than the 'achievable' cooled boost temperature. With current charge cooling systems these limits are either 'hot' by-passing the main charge air cooler or 'cold' by fully charge cooling, (which in turn is a function of the ambient temperature).

The control strategy for active charge cooling is therefore based on achieving a target intake manifold temperature over the widest possible speed-load envelope of the engine, which may either be hotter or colder than that achieved by the passive cooling typical of current systems.

The approach is to utilise engine coolant temperatures to raise the charge air temperatures and air cooling turbo-expansion to extend the limits of a passive system.

The strategy can simply be a control loop based on the error between a measured temperature in the intake manifold and a target temperature, stored in the ECU, which could typically be a function of engine/vehicle speed and load. This has been presented for WOT operation in the modelling study above. It may also be desirable to provide effective temperature control under part-load conditions in relation to, for example, EGR temperature management, combustion stabilisation or emissions control.

Vehicle demonstration

A prototype ACC system has been fitted to a gasoline engine passenger car, for the purposes of demonstrating the basic principles and typical results achievable. The C/D class car has a 1.8 litre 4 cylinder turbocharged engine, generating a maximum of 132 kW. The prototype ACC system was installed with minimum modification to the production engine package. In this particular application the turbo-expander charge cooler (CAC2) is liquid cooled, with an independent cooling circuit and front mounted low temperature radiator (LTR). The turbo-expander and charge cooler sub-assembly is shown in fig. 8. This prototype is based on a production turbocharger, with the exhaust gas turbine and housing replaced by the cold air turbine and housing designed by ACT.

Figure 8: Turbo-expander and cooler sub-assembly

The vehicle installation is shown in fig. 9, where the turbo-expander and its charge cooler can clearly be seen. The by-pass was inoperative, and no changes were made to the engine calibration.

Figure 9: Vehicle installation of turbo-expander

The system was fitted with sensors and a logging system and driven on the road and a test track. High load operation best demonstrates the performance potential of the system and sample results are shown in Fig. 10. The test sample was WOT acceleration through the gears, from 76 to 121 mph vehicle speed. The key temperatures; at turbo-expander (TEX) intake and exit are presented, also the engine air box temperature and the engine and turbo-expander speeds.

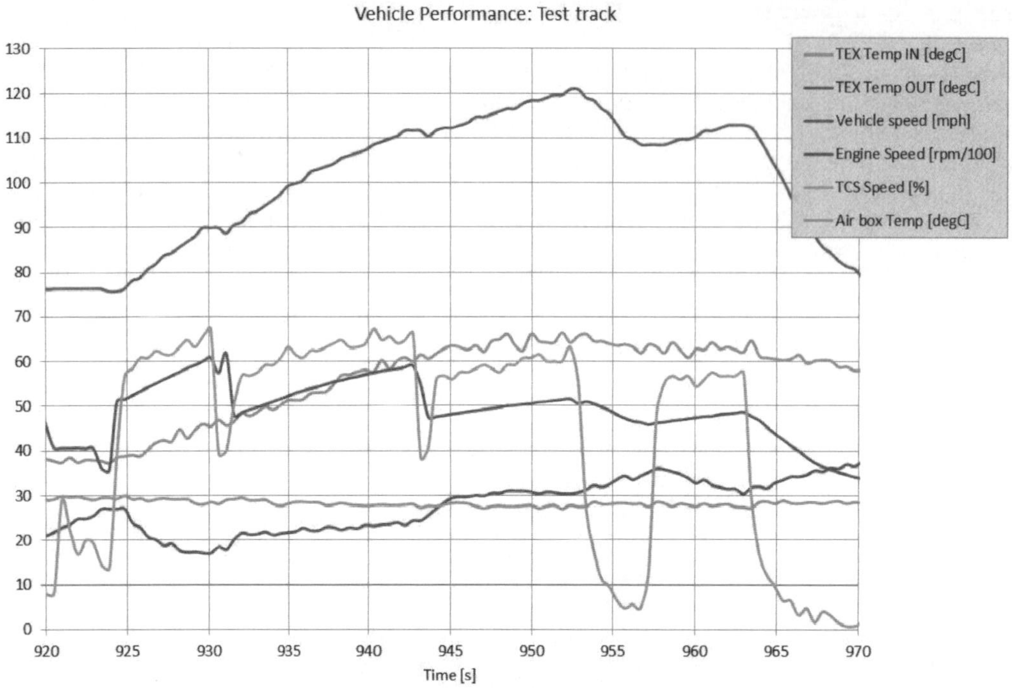

Figure 10: Vehicle performance

The temperature drop through the turbo-expander can clearly be seen, reaching a maximum of 35 degC, which is consistent with the results of the modelling study. At t = 950 s a TEX intake temperature (same location as CAC1 exit) of 65 degC resulted in an engine intake temperature of 30 degC. Also noteworthy is that during the early part of the acceleration (t= 925 > 945 s), the TEX exit temperature is lower than the engine air-box temperature, confirming that sub-ambient charge cooling is feasible. The TEX shaft speed indicates the excellent transient response of the turbo-expander, which has a low rotational inertia.

SUMMARY

The work presented here has covered the thermodynamic principles of an 'active' charge air heating and cooling system, followed by a more detailed study of charge air cooling to levels below those achievable with current 'passive' systems. The operating principle is the air-cycle refrigeration process. A system comprising a turbo-expander with dedicate cooler and a by-pass loop for temperature modulation, has been developed. 1D engine and charge air system thermo-fluid modelling confirmed the technical feasibility. This led to a prototype turbo-expander being developed, derived from automotive turbocharger technology. This was rig tested then installed on a passenger car demonstrator vehicle. The vehicle testing confirmed the predicted performance, demonstrating significant charge air temperature reductions. Temperature control using a by-pass process has been modelled, and will be tested in the near future.

CONCLUSION

The potential benefits, in engine performance fuel economy and emissions, of charge air temperature modulation are being evaluated by OEMs, driven by both customer and legislative demands. The potential for moving from 'passive' to 'active' charge air temperature control has been demonstrated by the work presented here. The ability to sub-cool charge air is the most innovative feature of the work, to which the addition of air heating would be relatively simple. Sub-cooling has been achieved with 1D modelling, prototype hardware manufacture and vehicle testing. The potential of the system has been clearly demonstrated. Air-cycle cooling principles have been applied, using a turbo-expander and related charge cooler which are capable of cost-effective high volume production using current automotive processes and even existing components. The development of a high volume product from the prototype work presented here can be carried out quickly with low technical and commercial risk, resulting in a system capable of manufacture by existing automotive Tier 1 suppliers.

REFERENCES

(1) Crooks, W. R., "Combustion Air Conditioning Boosts Output 50 Percent" CIMAC 1959 a15, p475

(2) Helmich, M.J., "Development of Combustion Air Refrigeration System Enabling Reliable Operation at 220 psi BMEP for a Large Four-Cycle Spark-Ignited Gas Engine" ASME 66-DGEP-7

(3) C Whelan and N Hayes – 'Formula 1 intake refrigeration system, in combination with a turbo-compound system'. Cosworth Engineering ltd, internal report, 1984

(4) R W G Turner, R J Pearson, M D Bassett and J Oscarsson, "Lotus: Performance and Fuel Economy Enhancement of Pressure Charged SI Engines through Turbo-expansion – An Initial Study'" SAE 2003-01-0411, 2003

(5) R C Meyer and S M Shahed, "An Intake Charge Cooling System for Application to Diesel, Gasoline and Natural Gas Engines" SAE910420, 1991

(6) C D Whelan and R A Richards, "Turbo-Cooling applied to Light Duty Vehicle Engines" IMechE 8th Conference on Turbochargers and Turbocharging, May 2006

(7) C D Whelan, W C Tang, R A Richards and M J Saville, "Charge Air Refrigeration for Engine Performance and Fuel Economy Enhancement", I Mech E Internal Combustion Engines conference 2015

(8) Turner, J., Popplewell, A., Patel, R., Johnson, T. et al., "Ultra Boost for Economy: Extending the Limits of Extreme Engine Downsizing," *SAE Int. J. Engines* 7(1):387-417, 2014, DOI: 10.4271/2014-01-1185

(9) C D Whelan, R A Richards, S W Spencer and A Young, "Design and development of a turbo-expander for charge air cooling" IMechE 9th Conference on Turbochargers and Turbocharging, May 2010

AUTHOR INDEX

Agurto-Goya, A	171
Allport, J M	243
Bogarra, M	97
Chalet, D	47
Cheng, L	29
Cipollone, R	221
Cormerais, M	47
Curran, A	233
Di Battista, D	221
Dickes, R	109
Dunne, J F	17
Dutta, N	195
Fennell, D	97
Fukumoto, A	79
Garland, M	185
Gilchrist, R	3, 29
Gingras, C	233
Guillaume, L	109
Gullman, S	233
Hepokoski, M	233
Hergueta, C	97
Herreros, M	97
Hétet, J-F	47
Huang, K	141
Hughes, I	195
Itoh, Y	79

Jacobsson, D	233
Jafari, S	17
Jansen, W	195
Khalane, H	127
Kitte, J	171
Knox, A	185
Kotrba, A	127
Langari, M	17
Lauzon, D	87
Legros, A	109
Lemort, V	109
Li, B	141
Li, J	141
Lion, S	205
Long, C A	17
Longo, S	155
Marimbordes, T	47
Momesso, M G	205
Montecucco, A	185
Morganti, M V	155
Mullen, P	185
Naiknaware, A	127
Orman, L	87
Owen, R	57
Peck, S	233
Pirault, J-P	17
Price, A	57
Rouaud, C	205
Sahoo, D	127
Sara, H	47
Seseke-Koyro, U	87
Shah, R M	155
Shayler, P J	3, 29

Siviter, J 185
Staelens, V 57
Stegmann, B 195
Swidersky, H W 87

Taccani, R 205
Tate, E 57
Thalackottore Jose, J 17
Tietze, T 171
Trivic, I 171
Tsolakis, A 97

Uppuluri, S 127

Vlaskos, I 205

Whelan, C D 243
Wyszynski, M 97

Yamayoshi, T 79
Yan, Y 141
Yang, Z 17
Yang, Z 57

Zhao, C 3, 29